MECHANICS AND PROPERTIES OF MATTER

Mechanics and Properties of Matter

Third Edition

REGINALD J. STEPHENSON

HARN PROFESSOR OF PHYSICS
THE COLLEGE OF WOOSTER

JOHN WILEY & SONS, INC.,

NEW YORK · LONDON · SYDNEY · TORONTO

10 9 8 7 6 5 4 3 2 1
Library of Congress Catalog Card Number: 69-16130
SBN 471 82270 1
Printed in the United States of America

To the memory of my wife
HELEN

PREFACE

This book is designed to be used after a year of general physics and with a knowledge of calculus generally represented by a two-year course in the mathematics department. It is intended to be used in a one-semester course, although this may necessitate leaving some of the more elementary material for the student to read by himself. Actually, there is some repetition of the physics covered in the general course, but this is done deliberately as an introduction to the more mathematical formulations of the topics.

Physics is essentially an experimental and mathematical science. Mechanics deals with only three variables, mass, length, and time, so that the depth of analysis can be considerable. This, however, presents a double difficulty to students, first with the physical concepts themselves and second with their mathematical formulation. It is only after the sophistication that comes with increased knowledge that students can appreciate the beauty of mathematical physics. The underlying philosophy of the book is that students need help in overcoming the mathematical hurdles in physics. Although mathematics is indispensable in physics, it must be the servant and not the master. It is the physical concepts that are of primary importance.

A number of changes have been made in this third edition that are largely a result of the reactions to the earlier editions over the past sixteen years. In the selection of topics careful consideration has been given to which concepts and mathematical techniques will be made use of in later physics courses. While a great deal of the earlier material has been unaltered, there has been the addition of several topics: relativity, alpha particle scattering, the Lagrange and Hamilton equations, Hamilton's variational principle, and a number of minor changes. There is no longer a consideration of phenomena concerned with statics.

In presenting the Lagrange and Hamilton equations in generalized co-ordinates, the question arose as to how these should be presented so as to be most easily understood. The method adopted has been to present the material in two parallel columns, one in which the generalized coordinates are used and the other with similar equations for the simple example of central force motion. For easy transference from one column to the other, similar equations are given the same numbers, so that if the generalized equations prove too difficult at first the theory and method can be appreciated with the aid of the example. Thus, for a student whose knowledge of calculus is limited, it should be possible for him to follow the example and later come to appreciate the more general formulation.

Hamilton's variational principle has been introduced in a similar manner with the two contrasting columns, the parallel example being concerned with the motion of an object projected vertically upwards in a uniform gravitational field. This variational principle involves the calculus of variations and may be new to the student. Examples are given showing the application of the Lagrange and Hamilton equations as well as two examples in the calculus of variations. The alpha particle scattering, involving a repulsive inverse square force, has made use of much of the gravitational theory developed for the motion of a planet about the sum.

In eliminating the chapter on statics I have looked at it from the point of view of a physicist and what a student needs for going on to more advanced courses in physics. It is not wise or possible to include every topic that a professor desires. Rather, I feel the author has to make a choice and rise or fall with that choice. As for statics versus elasticity and the Lagrange and Hamilton equations, there is no question for me. I am not trying to compete with Synge and Griffith or Goldstein. Instead, I am trying to give an understandable introduction to these difficult topics so that when the student comes to the more advanced texts he will be better prepared to master them.

Vectors and some simple forms of vector analysis have been used throughout the book, these being introduced where they first arise. Many examples have been given in the text to aid in the understanding of the underlying theory. A number of new problems are presented at the end of each chapter and the answers to the odd-numbered problems are given at the end of the book. The more problems that a student solves himself, the better is his understanding of the subject.

A rigorous selection of topics has been made in properties of matter. Elasticity has been mathematically limited to a discussion of homogeneous isotropic materials. Since there has been an enormous advance in the understanding of the solid state, a brief qualitative section has been added on some of the findings in this field. Historical references have been made throughout the book to add that touch of human interest which often helps a student to feel more familiar with the laws and principles he is studying.

It is impossible to thank the many people who, either as teachers, authors, students, colleagues, have contributed to the writing of this book. In particular, I express my deep appreciation to my colleagues Professor B. R. Russell and Mr. C. B. Wenger who have read much of the manuscript and given valuable criticism and help.

Reginald J. Stephenson

Wooster, Ohio, 1968

CONTENTS

Translational and rotational motion of rigid bodies about fixed axis   181

Chapter 6

Wave motion   **326**

Chapter 9

MECHANICS AND PROPERTIES OF MATTER

1

KINEMATICS

1.1 Introduction

Physics, like all the sciences, has evolved from the simple to the more complex. As new observations have been made, theories have generally been modified rather than discarded. Each new theory must account for all the old phenomena as well as the new. Thus one finds the theoretical structure of physics becoming more and more complicated and correspondingly more capable of interpreting the results of a highly developed experimental structure. To understand modern science one must begin in a humble manner to master those basic ideas on which it is founded. To a large degree they are the concepts of mechanics which Galileo and Newton gave us and which were later developed in breadth and depth by other scientists. Essentially the science of mechanics gives a description of how objects change their positions with respect to time.

The science of mechanics and our modern scientific era may be considered to have started in 1638 with the publication of Galileo's book, *Two New Sciences*.* This book contains some of the fundamental concepts of mechanics and even today is worthy of study. Galileo's ideas were extended by Newton, who in 1686 published *The Mathematical Principles of Natural Philosophy*, generally called *The Principia*.† In this book the laws of motion, the law of gravitation, the theory of tides, and the theory of the solar system are presented. Newton shows how these laws can be used to describe the motion of the moon about the earth and the planets about the sun. At that time these laws appeared to be universal in application, and provided a stimulus for further investigation which has continued to the present. That fundamental

---

* *Two New Sciences*. Translated by Crew and de Salva, Northwestern University Press, 1932.
† *The Principia*. A revision of Motte's translation by F. Cajori, University of California Press, 1934.

laws underly the behavior of nature has provided the guiding principle of natural science ever since the time of Newton.

After Newton many advances were made by the French and German mathematical physicists. In 1743 D'Alembert published an important book, *Traité de Dynamique*, in which momentum and kinetic energy are clearly distinguished. Lagrange (1736–1813) published *Mécanique Analytique*, and Laplace (1749–1827) *Mécanique Céleste*. In these, the concept of potential and many modern methods of analysis were first introduced. During the first half of the nineteenth century Joule performed many experiments, the interpretation of which led to the important principle of conservation of energy. The culmination of many of the advances in mechanics came with the publication in 1905 of the theory of relativity by Einstein. Although the Newtonian theory has been superseded by the Einstein theory, nevertheless for objects moving with velocities much less than the velocity of light, namely 186,000 mi/sec or 3×10^8 m/sec, the Newtonian theory is sufficiently accurate.

The discoveries in mechanics did much to further our knowledge of the sciences of electricity, sound, heat, and light. As will be shown later, there is a close analogy between the equations representing the motion of the forced oscillation of a spring, the motion of an electron in an atom acted on by electromagnetic radiation, the motion of a galvanometer coil, and the alternating current in a circuit consisting of an inductance and capacitance in series. In each of these equations both force and motion are represented, though in each the nature of the force and of the object moved is different.

The study of wave motion has likewise proceeded from the simple to the more complex. It now appears that all matter in motion exhibits wave properties. This combination of the particle and wave theories is dealt with in more advanced texts under the name of quantum mechanics. Fourier's studies in periodic heat flow gave rise to a form of mathematical analysis called the Fourier series. This analysis now has applications in all periodic phenomena, ranging from vibrating strings to radar. With this brief review of physics one can perhaps glimpse something of the unity of this science as well as the importance of mechanics in it.

At this point it might be well to comment on the role of mathematics in physics. Without mathematics the physical sciences would have been almost stillborn; they could only have been qualitative and incapable of much development. Mathematics is the tool employed in the precise analysis of physical concepts. However, without a clear understanding of the physical concepts involved, mathematics is of limited value in solving a physical problem. It is a good rule to try to express in words the physical concepts represented in any equation expressing a relationship between physical quantities.

1.2   Space and Time

A discussion of kinematics or the "geometry" of motion involves an understanding of the concepts of space and time. Many of us feel that space and time are in no way connected with each other. That this was Newton's idea is shown by his definitions of absolute space and absolute time given at the beginning of *The Principia*. These definitions were not seriously challenged for over 200 years. However, difficulties arose in the interpretation of certain optical phenomena which led Einstein to examine the concepts of space and time very carefully. He assumed in the theory of relativity that the laws of physics have the same mathematical form when formulated relative to any coordinate systems which are in uniform translational motion with respect to each other, and that the velocity of light in a vacuum is constant and independent of the motion of the source and the observer. The restricted theory applies only to those coordinate systems that move relative to each other with constant velocity. As we shall see later in this chapter the Einstein transformation reduces to the Newtonian one for all velocities which are small compared to the velocity of light. These small velocities include all moving macroscopic objects on our earth, even the modern jet plane. It is with such relatively small velocities that the Newtonian or classical physics is concerned.

As a simple example of the kind of relativity dealt with in classical physics consider the motion of a stone dropped from a car moving along a straight road with constant speed. If air resistance is neglected, the stone appears to the driver to fall vertically down in a straight line. To a person standing beside the road the stone appears to follow a parabolic path. Which of these descriptions is correct? Both are correct if the observer's coordinate system is taken into account. In a coordinate system moving with the car the path is a straight line, whereas in a coordinate system attached to the earth the path is parabolic.

For most of our work we shall use a coordinate system attached to the earth, though, as we shall see, the rotation of the earth must be taken into account in some instances.

1.3   Measurement of Space and Time

We are all familiar with the simple direct measurement of length in which a rod of some arbitrary length is taken as a unit and is compared with the object to be measured. In contrast to this comparatively simple method of measurement are the much more complicated and less direct methods of measuring very short and very large distances. Unfortunately there are two systems in common use which have different unit lengths as their standard. These are the British or engineering system, based on the foot, and the metric, based on the meter. To these unit lengths are added their multiples and submultiples, giving derived lengths. The relationship between the British and the metric units of length was originally

$$1 \text{ m} = 39.37 \text{ in.}$$

but this has been slightly changed so that the relationship between the inch and the centimeter is exactly

$$1 \text{ in.} = 2.54 \text{ cm}$$

In the metric system the following prefixes are used to denote decimal multiples and submultiples:

kilo—(k), thousand, 10^3

mega—(M), million, 10^6

giga—(G), billion, 10^9

tera—(T), trillion, 10^{12}

millimicro—(mμ) or nano—(n), 10^{-9}

deci—(d), 0.1 or 10^{-1}

centi—(c), 0.01 or 10^{-2}

milli—(m), 0.001 or 10^{-3}

micro—(μ), 0.000001 or 10^{-6}

micromicro—($\mu\mu$) or pico—(p), 10^{-12}

The measurement of time is based on the rotation of the earth. One complete rotation of the earth relative to the fixed stars is an interval of time called a sidereal day and is assumed to be constant. Relative to the sun one rotation is called a true solar day. Since the sun moves eastward at a slightly varying rate with respect to the fixed stars, the true solar days are of different lengths. An average of all true solar days in a year is called a mean solar day. Our watches and clocks indicate mean solar hours, minutes, and seconds.

Since these standards of length and time may not be so constant nor so easily duplicated as is desirable for today's science, it has been recommended that atomic standards be adopted. About the beginning of the present century Michelson compared the standard meter with the wavelength of the red cadmium line in air. This work has been extended. The International Conference on Weights and Measures in 1960 endorsed the definition of the meter as equal to 1,650,763.73 times the wavelength of the orange light emitted by the isotope of kryton of mass number 86. It was also suggested by the International Conference in 1964 that time be measured by an 'atomic' clock, which is based on the frequency of vibration of the cesium atom. Measurements show that a time interval of one second is equal to the time of 9,192,631,770 vibrations of the cesium atom. The accuracy of time measurements* are such that at the Radio Standards Laboratory of the National

* *The Physics Teacher*, September 1966. "Atomic Standards of Frequency and Time" by John M. Richardson and James F. Brockman.

Bureau of Standards, time intervals are measured to an accuracy of about 5 parts in 10^{12}, that is, about one second in 6000 years. This is a much greater accuracy than can be attained for measurements of length and mass which can only be determined to a few parts in 10^9.

1.4   Linear Velocity and Acceleration

In the following discussion we shall be concerned with the motion of a geometrical point or of an object whose dimensions are negligible compared to the distance necessary to fix its position. Such an object is called a particle. Thus the earth may be regarded as a particle when its motion relative to the sun is considered, but it is very far from being a particle to us moving about on its surface. It should be noted that motion is always relative to some reference system that is usually considered stationary.

Let us consider a particle moving along a straight line, which for convenience we shall take as the X axis. As the time increases, the position of the particle along the X axis changes. The origin O of the X axis, (Fig. 1.1) is the reference point for locating the particle on the X axis. In agreement with the usual convention, positions to the right of O are taken as positive and to the left of O as negative. At time t_1 the particle has the position x_1, and at time t_2 the position x_2. The distance $x_2 - x_1$ is called the displacement of the particle in the time interval $t_2 - t_1$. If, as shown in the figure, x_2 is larger than x_1, then $x_2 - x_1$ is a positive displacement, whereas $x_1 - x_2$ would represent a negative displacement. The average velocity $\bar{v}$ of the particle during the time interval is defined as

$$\bar{v} = \frac{x_2 - x_1}{t_2 - t_1}$$

Suppose now that we attempt to find the actual or instantaneous velocity of the particle at the instant that it is moving past position x. If the velocity of the particle is not constant, then its instantaneous velocity varies with x. To obtain the instantaneous velocity at x, we find the average velocities over smaller and smaller intervals of time and then proceed to the limit where the interval approaches zero. In order to indicate small finite changes in any variable, the symbol delta Δ is used in front of the variable. Thus, if the

Fig. 1.1   Motion along the X axis.

particle is at position x at time t, and after a short interval of time Δt the particle is at position $x + \Delta x$, then the average velocity $\bar{v}$ during the interval Δt is

$$\bar{v} = \frac{\Delta x}{\Delta t}$$

The instantaneous velocity v is the limit of this expression as Δt approaches zero; that is, it is given by the derivative of x with respect to t. Thus

$$v = \lim_{\Delta t \to 0} \frac{\Delta x}{\Delta t} = \frac{dx}{dt} = \dot{x}$$

where $\dot{x}$ (x dot) is used to indicate the first derivative of x with respect to t. This is the instantaneous rate of change of the displacement x with respect to time t. If the displacement is increasing with increase in time, then $\dot{x}$ is positive; if the displacement is decreasing with increase in time, then $\dot{x}$ is negative. Notice that some convention must always be adopted for the positive direction of x.

As an example of a particle moving with non-uniform motion, consider one projected vertically upward under gravity. Its displacement x from some fixed point at any time t is

$$x = ut - \tfrac{1}{2}gt^2$$

where g is a constant, x is positive in an upward direction, and at $t = 0$ and $x = 0$ the object is moving with an upward velocity u. The instantaneous velocity of the particle at any time t is

$$\dot{x} = \frac{dx}{dt} = u - gt$$

This instantaneous velocity $\dot{x}$ decreases as the time t increases, becoming zero at a time u/g, and negative thereafter. Notice that the velocity $\dot{x}$ can be negative even though the displacement is positive.

When the velocity of a particle changes, the particle is said to be accelerated. By definition, *acceleration is the rate of change of velocity with time.* In the present example we are considering motion along a straight line where the acceleration consists in a change in the magnitude of the velocity and not in the direction. Later we shall discuss accelerations where the direction changes but the magnitude of the velocity, or the speed, remains constant. Finally we shall consider accelerations where both the magnitude and the direction of the velocity of a particle change.

Suppose that a particle is moving along the X axis in such a manner that its velocity at some point x and time t is v, and at $x + \Delta x$ and time $t + \Delta t$

is $v + \Delta v$. The average acceleration $\bar{a}$ in the interval of time Δt is

$$\bar{a} = \frac{\Delta v}{\Delta t}$$

and the instantaneous acceleration a at time t is the limit of this as Δt approaches zero, or

$$a = \lim_{\Delta t \to 0} \frac{\Delta v}{\Delta t} = \frac{dv}{dt} = \dot{v}$$

where $\dot{v}$ (v dot) is the time derivative of v. Since the instantaneous velocity of this particle at position x and time t is $v = dx/dt$, the instantaneous acceleration along the X axis may be expressed as

$$a = \frac{d}{dt}\left(\frac{dx}{dt}\right) = \frac{d^2 x}{dt^2} = \ddot{x}$$

where $\ddot{x}$ (x double dot) is the second derivative of x with respect to t. From the previous example in which $\dot{x} = u - gt$, we see that $\ddot{x}$ is a constant negative quantity, $-g$. In the upward motion of the particle the displacement and velocity are positive, or upward, whereas the acceleration is negative, or downward. For the downward motion of the particle the displacement is positive and the velocity and acceleration are both negative.

Suppose that the acceleration $\ddot{x}$ along the X axis is a constant quantity, a. By integrating a simple differential equation, the velocity at any time or the distance traveled in any time may be easily obtained.

$$\ddot{x} = \frac{d^2 x}{dt^2} = \frac{d}{dt}\left(\frac{dx}{dt}\right) = a$$

Integration gives

$$\int d\left(\frac{dx}{dt}\right) = \int a\, dt$$

The evaluation of this expression depends on the initial conditions. Let us assume that at time zero the particle is at the origin and is moving along the positive X axis with an instantaneous velocity u. At $t = 0$, $dx/dt = u$, and at $t = t$, $dx/dt = v$. Thus

$$\left[\frac{dx}{dt}\right]_u^v = [at]_0^t \qquad \text{or} \qquad v - u = at \qquad (1.1)$$

This is the familiar equation for a particle moving with a constant acceleration, and follows directly from the definition of acceleration. The velocity

v at any time t may be given as dx/dt so that

$$\frac{dx}{dt} = u + at$$

where u and a are constants in this example. Integrating this equation and from the conditions that the displacement is zero at time $t = 0$ and that the displacement is x at time t,

$$\int_0^x dx = \int_0^t (u + at)\, dt \qquad \text{or} \qquad x = ut + \frac{at^2}{2} \qquad (1.2)$$

The two equations, Eqs. 1.1 and 1.2, are two independent relationships between four of the five quantities, x, t, a, u, v. Thus three other equations may be obtained from these two by eliminating one quantity at a time. These equations are the familiar ones from elementary physics, namely:

$$x = vt - \tfrac{1}{2}at^2 \qquad v^2 - u^2 = 2ax \qquad x = \frac{u + v}{2}\, t$$

1.5 Vectors

Quantities, such as velocity and force, that require both magnitude and direction to be completely described, and that are added by the parallelogram law of addition, are called *vector* quantities. They may be contrasted with such quantities as time, volume, and speed that have magnitude only and are called *scalar* quantities. Scalar quantities expressed in the same units may be added or subtracted by the simple rules of arithmetic, but this does not hold for vector quantities. (In order to distinguish between scalar and vector quantities it is usual to print scalars in ordinary type and vectors in boldface type. For the blackboard or notebook it is convenient to denote a vector by placing a line, having an arrow on one end, above the appropriate symbol such as $\vec{r}$.)

A vector may be represented graphically by an appropriately directed line segment whose length, drawn to a suitable scale, represents the magnitude of the vector and whose direction is indicated by an arrow drawn on the line. Thus the addition of two vector displacements **A** and **B**, to form the vector resultant **C**, is accomplished by placing the tail of vector **B** at the head of vector **A** as shown in Fig. 1.2a. Completion of the parallelogram of vectors as in Fig. 1.2b shows that

$$\mathbf{C} = \mathbf{A} + \mathbf{B} = \mathbf{B} + \mathbf{A}$$

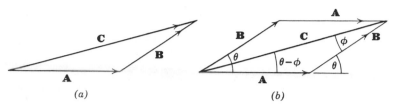

Fig. 1.2 (a) Addition of two vectors. (b) Completion of the parallelogram showing $\mathbf{C} = \mathbf{A} + \mathbf{B} = \mathbf{B} + \mathbf{A}$.

This commutative law of addition is an important condition imposed on the definition of a vector as having both magnitude and direction. A finite angular rotation has both magnitude and direction and yet, as shown in Fig. 1.3, the result of adding rotations A and B does not give the same result as adding rotation B first and then A. In this figure A represents a rotation through 90° about the y axis and B a rotation of 90° about the z axis. Thus finite rotations cannot qualify as vectors. However, an infinitesimal rotation and an angular velocity both qualify as vectors, as shown in Section 6.11.

The magnitude and direction of the resultant vector $\mathbf{C}$ produced by adding vectors $\mathbf{A}$ and $\mathbf{B}$ (Fig. 1.2b) can be obtained from the trigonometric relation of cosines and sines. These are:

$$C^2 = A^2 + B^2 + 2AB \cos \theta$$

$$\frac{A}{\sin \phi} = \frac{B}{\sin (\theta - \phi)} = \frac{C}{\sin \theta}$$

To subtract one vector from another, the negative of the one to be subtracted is added to the first vector. The negative of a vector is a vector of the same magnitude but of opposite direction.

1.6   Components of a Vector. Unit Vectors

A vector in the x, y plane such as $\mathbf{B}$ in Fig. 1.4 can be considered to be made up of the components B_x along the x axis and B_y along the y axis. In many problems it is useful to introduce unit vectors that have a constant magnitude of unity and a definite direction. In the x, y Cartesian coordinate plane it is usual to denote the unit vector along the x axis as $\mathbf{i}$ and along the y axis as $\mathbf{j}$. Thus the vector $\mathbf{B}$ can be written as

$$\mathbf{B} = \mathbf{i}B_x + \mathbf{j}B_y$$

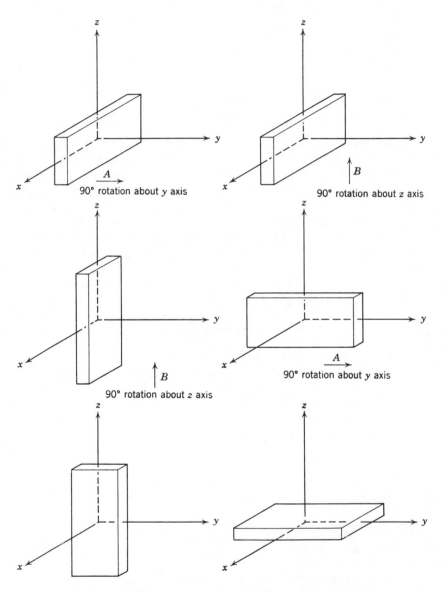

Fig. 1.3 Showing that a finite rotation cannot be represented as a vector, since **A** + **B** ≠ **B** + **A**, or the commutative law of addition is not valid for finite rotations.

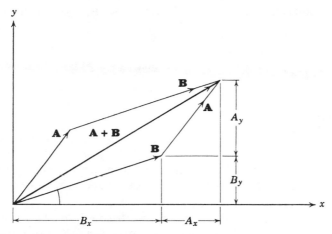

Fig. 1.4 Components of vectors in a plane. $A = iA_x + jA_y$; $B = iB_x + jB_y$ and $A + B = i(A_x + B_x) + j(A_y + B_y)$.

where B_x is the component of the vector **B** along the x axis and B_y that along the y axis. It should be noted that B_x and B_y are numbers. For example, the vector **B** could be

$$B = 4i + 3j$$

and would have a length of 4 units along the direction of **i** or the x axis and 3 units along the **j** direction or the y axis. The magnitude of this vector **B**, written as $|B|$ is

$$|B| = \sqrt{(B_x^2 + B_y^2)} = \sqrt{4^2 + 3^2} = 5$$

The component of a vector in a given direction is the product of the magnitude of the vector and the cosine of the angle between the direction of the vector and the chosen direction. If the angle between the vector **B** and x axis is denoted by (B, x) then in Fig. 1.4,

$$\cos (B, x) = \frac{B_x}{\sqrt{(B_x^2 + B_y^2)}} = B_x/|B|$$

and the component B_x is given by

$$B_x = \sqrt{(B_x^2 + B_y^2)} \cos (B, x) = |B| \cos (B, x)$$

In vector analysis there are two entirely independent definitions for the product of two vectors, the one a scalar product and the other a vector or cross product. Following the notation introduced by Professor Willard Gibbs of Yale about 1880, the scalar product of two vectors **A** and **B** is written as **A · B** (**A** dot **B**) which is given by the product of the magnitude of the two vectors **A** and **B** and the cosine of the angle $(\mathbf{A} \cdot \mathbf{B})$ between them, that is

$$\mathbf{A} \cdot \mathbf{B} = |\mathbf{A}| \times |\mathbf{B}| \cos (\mathbf{A} \cdot \mathbf{B})$$

If $\mathbf{A} = \mathbf{i}A_x + \mathbf{j}A_y$ and $\mathbf{B} = \mathbf{i}B_x + \mathbf{j}B_y$, then

$$\mathbf{A} \cdot \mathbf{B} = (\mathbf{i}A_x + \mathbf{j}A_y) \cdot (\mathbf{i}B_x + \mathbf{j}B_y) = A_x B_x + A_y B_y \qquad (1.3)$$

since by definition

$$\mathbf{i} \cdot \mathbf{i} = \mathbf{j} \cdot \mathbf{j} = 1 \qquad \text{and} \qquad \mathbf{i} \cdot \mathbf{j} = \mathbf{j} \cdot \mathbf{i} = 0$$

As is readily seen from Fig. 1.4, the sum of the components of two vectors in any direction is equal to the component of the resultant in the chosen direction.

A space vector needs three coordinates to specify it. These may be the x, y, z coordinates in a Cartesian coordinate system, or a distance and two angles in a spherical coordinate system. In the Cartesian coordinate system **k** is used for the unit vector along the Z axis. Thus the space vector $\mathbf{OP} = \mathbf{r}$, in Fig. 1.5, having coordinates x, y, z, may be written

$$\mathbf{OP} = \mathbf{OA} + \mathbf{AC} + \mathbf{CP}$$

or

$$\mathbf{r} = \mathbf{i}x + \mathbf{j}y + \mathbf{k}z \qquad (1.4)$$

The magnitude of the vector **r** is

$$|\mathbf{r}| = \sqrt{(x^2 + y^2 + z^2)}$$

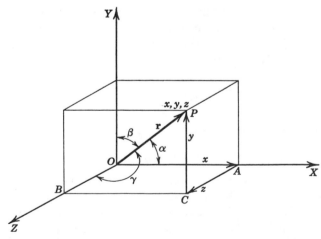

Fig. 1.5   Figure showing the space vector **r** with its components.

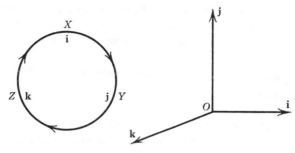

Fig. 1.6 The cyclical order for the positive set of axes.

In general we shall employ a right-handed system of coordinate axes X, Y, Z. By this we mean that, if the X axis is rotated through 90° into the Y axis, this rotation will cause a right-handed screw to progress along the Z axis. Similarly, rotation of the Y axis into the Z axis causes movement of a right-handed screw along the positive X axis. This rule may be easily remembered by means of Fig. 1.6 in which the X, Y, Z axes are placed in clockwise order.

If the angles that the vector **r** makes with the X, Y, Z axes are α, β, γ respectively, then from Fig. 1.5

$$\cos \alpha = \frac{x}{r} \qquad \cos \beta = \frac{y}{r} \qquad \cos \gamma = \frac{z}{r}$$

and

$$r^2 = x^2 + y^2 + z^2$$

Hence

$$\frac{x^2}{r^2} + \frac{y^2}{r^2} + \frac{z^2}{r^2} = 1$$

or

$$\cos^2 \alpha + \cos^2 \beta + \cos^2 \gamma = 1 \tag{1.5}$$

The cosines of these angles are called the direction cosines of the vector **r**. Thus the sum of the squares of the direction cosines is unity.

From the definition of the dot or scalar product of two vectors it follows for the unit vectors **i**, **j**, **k** that:

$$\mathbf{i} \cdot \mathbf{i} = \mathbf{j} \cdot \mathbf{j} = \mathbf{k} \cdot \mathbf{k} = 1 \qquad \text{and} \qquad \mathbf{i} \cdot \mathbf{j} = 0 = \mathbf{j} \cdot \mathbf{k} = \mathbf{k} \cdot \mathbf{i}$$

The other special product of two vectors, called the vector of cross product, is given in Chapter 6. It is of historical interest to note that prior to the introduction of vector analysis the scalar and vector products had been used in the algebra of quaternions. Clerk Maxwell made limited use of the vector part of quaternions in his *Treatise on Electricity*, and it was largely due to this that Gibbs came to invent his form of vector analysis. Quaternions are a form of quadruple algebra while vector analysis uses a triple algebra.

1.7 Relative Motion

As we have already seen, all velocities and accelerations are relative to some coordinate system which we have so far considered to be at rest. There are situations in which one has to consider the motion of one body relative to another moving body. If we are sitting in a train A, which is moving along tracks parallel to a second moving train B, then the motion of A relative to B depends on the velocities of both trains. Let A and B be two trains moving along the same straight line. At time t let the position of A be x_1 and of B be x_2 (Fig. 1.7a). Then $x_2 - x_1$ is the displacement of B relative to A, and $dx_2/dt - dx_1/dt$ is the velocity of B relative to A. Thus the velocity of B relative to A is equal to the velocity of B minus the velocity of A. If we were on train A, then we would consider ourselves at rest and train B moving relative to us with a velocity equal to the difference of the velocities, $dx_2/dt - dx_1/dt$. Notice that we are adopting the Newtonian point of view here, which is sufficiently accurate for all velocities that are small compared to the velocity of light.

In a more general manner let us consider two points A and B (Fig. 1.7b), whose position vectors at some instant of time are $\mathbf{r}_1$ and $\mathbf{r}_2$; then the displacement of B relative to A is $\mathbf{r}_2 - \mathbf{r}_1$, and the velocity of B relative to A is $d\mathbf{r}_2/dt - d\mathbf{r}_1/dt$ or $\dot{\mathbf{r}}_2 - \dot{\mathbf{r}}_1$.

As an example of relative motion, consider two swimmers A and B such that at a given instant of time A is 0.25 mi south of B and B is swimming northwest at 3 mph while A is swimming at 4 mph (Fig. 1.8). Assuming the swimmers maintain constant speeds and directions, find the direction that A must take in order to reach B, and also find the time taken.

Consider a set of axes with the origin through the initial position of A, and let the velocity of A be $\mathbf{v}$ making an angle α with the y axis (Fig. 1.8a). If $\mathbf{i}$ and $\mathbf{j}$ are

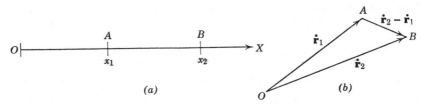

(a) $\qquad\qquad\qquad\qquad$ (b)

Fig. 1.7 (a) Relative motion of B with respect to A when both are moving along the X axis. (b) Velocity of B relative to A is $\dot{\mathbf{r}}_2 - \dot{\mathbf{r}}_1$.

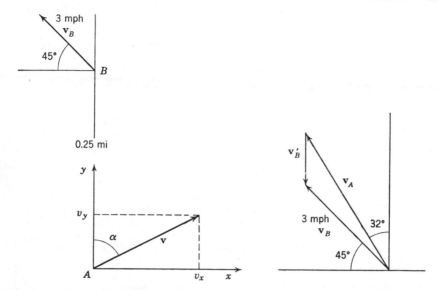

Fig. 1.8 (a) Initial positions and velocities of swimmers. (b) Velocity of B relative to A is $v_B{}'$.

unit vectors along the x and y axes respectively, then the velocity of A is given by $\mathbf{i} v_x + \mathbf{j} v_y$. The velocity of B relative to A (Fig. 1.8b) is

$$\mathbf{v}_B{}' = \mathbf{v}_B - \mathbf{v}_A$$
$$= -\mathbf{i}\, 3 \cos 45° + \mathbf{j}\, 3 \sin 45° - \mathbf{i}\, v_x - \mathbf{j}\, v_y$$
$$= -\mathbf{i}(2.12 + v_x) + \mathbf{j}(2.12 - v_y)\,\text{mph}$$

In order for A to reach B the relative velocity $\mathbf{v}_B{}'$ must be along the $-\mathbf{j}$ direction since A is directly south of B. Thus the $\mathbf{i}$ component of $\mathbf{v}_B{}'$ must be zero, that is, $2.12 + v_x = 0$ or $v_x = -2.12$ mph. The component v_y is obtained from v^2 which is equal to 16 (mph)². Thus $v_x{}^2 + v_y{}^2 = 16$, so that $v_y{}^2 = 16 - 4.5 = 11.5$ and $v_y = 3.39$ mph. The direction of A's velocity is given by

$$\tan \alpha = v_x/v_y = -2.12/3.39 = -0.626$$

or α is 32°W of N. The time for A to reach B is 0.25 mi/(3.39 − 2.12) mph = 0.197 hr or 11.8 min. As an exercise, show that the position where the swimmers meet has the coordinates $x = -0.42$, $y = 0.67$ mi.

1.8   Radial and Transverse Accelerations in a Plane

Another useful expression for the acceleration of a point moving along a curve can be given in terms of a vector $\mathbf{r}$ to the point. Consider a point

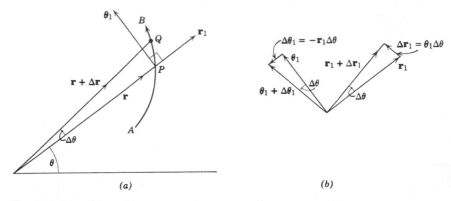

Fig. 1.9 (a) Point P, located at r, θ moving on curve AB; and showing unit vectors $\mathbf{r}_1$ and $\boldsymbol{\theta}_1$ at P. (b) Showing changes in the unit vectors $\Delta\mathbf{r}_1$ and $\Delta\boldsymbol{\theta}_1$. Since the unit vectors have a constant magnitude, any change such as $\Delta\mathbf{r}_1$ must be at right angles to $\mathbf{r}_1$, that is, $\Delta\mathbf{r}_1$ is along the direction of $\boldsymbol{\theta}_1$. Similarly $\Delta\boldsymbol{\theta}_1$ is along the negative direction of $\mathbf{r}_1$.

moving along a curve AB (Fig. 1.9a) such that for some point P on the curve

$$\mathbf{OP} = \mathbf{r} = r\mathbf{r}_1$$

where $\mathbf{r}_1$ is unit vector along the vector $\mathbf{r}$ and r is the magnitude $|\mathbf{r}|$ of the vector. Another unit vector $\boldsymbol{\theta}_1$ is drawn perpendicular to $\mathbf{r}_1$ as shown in the figure. The velocity $\mathbf{v}$ of the point P is given by

$$\mathbf{v} = \frac{d\mathbf{r}}{dt} = \dot{\mathbf{r}} = r\frac{d\mathbf{r}_1}{dt} + \mathbf{r}_1\frac{dr}{dt}$$

By drawing the unit vectors $\mathbf{r}_1$ and $\boldsymbol{\theta}_1$ at points P and Q, it is seen when they are placed together as in Fig. 1.9b that, approximately,

$$\Delta\mathbf{r}_1 = \boldsymbol{\theta}_1\,\Delta\theta \qquad \text{and} \qquad \Delta\boldsymbol{\theta}_1 = -\mathbf{r}_1\,\Delta\theta$$

hence

$$\frac{d\mathbf{r}_1}{dt} = \boldsymbol{\theta}_1\lim_{\Delta t\to 0}\frac{\Delta\theta}{\Delta t} = \boldsymbol{\theta}_1\frac{d\theta}{dt} : \qquad \frac{d\boldsymbol{\theta}_1}{dt} = -\mathbf{r}_1\lim_{\Delta t\to 0}\frac{\Delta\theta}{\Delta t} = -\mathbf{r}_1\frac{d\theta}{dt}$$

In the limit the vector velocity is given by

$$\mathbf{v} = \dot{\mathbf{r}} = r\dot\theta\boldsymbol{\theta}_1 + \dot{r}\mathbf{r}_1 \tag{1.6}$$

The acceleration a is given by

$$\mathbf{a} = \dot{\mathbf{v}} = \ddot{\mathbf{r}} = \dot{r}\dot\theta\boldsymbol{\theta}_1 + r\ddot\theta\boldsymbol{\theta}_1 + r\dot\theta\,d\boldsymbol{\theta}_1/dt + \ddot{r}\mathbf{r}_1 + \dot{r}\,d\mathbf{r}_1/dt$$

$$= \dot{r}\dot\theta\boldsymbol{\theta}_1 + r\ddot\theta\boldsymbol{\theta}_1 - r\dot\theta^2\mathbf{r}_1 + \ddot{r}\mathbf{r}_1 + \dot{r}\dot\theta\boldsymbol{\theta}_1$$

$$= \mathbf{r}_1(\ddot{r} - r\dot\theta^2) + \boldsymbol{\theta}_1(2\dot{r}\dot\theta + r\ddot\theta) \tag{1.7}$$

The acceleration **a** has the components $\ddot{r} - r\dot{\theta}^2$ along the radius vector **r** or **OP** (Fig. 1.9a) and $r\ddot{\theta} + 2\dot{r}\dot{\theta}$ along the direction at right angles to OP in a counterclockwise or positive rotation of OP. We can give a physical interpretation of some of the terms in this expression for the acceleration. The component $\ddot{r}$ is the outward acceleration of the particle along the radius vector, and $r\dot{\theta}^2$ is the centripetal or inward acceleration. For the transverse components along $\boldsymbol{\theta}_1$, the term $r\ddot{\theta}$ represents the linear acceleration of the particle at P due to the angular acceleration $\ddot{\theta}$. The last transverse component $2\dot{r}\dot{\theta}$, called the Coriolis acceleration, can be experienced if you walk radially out on a moving merry-go-round.

As a summary, the components of acceleration in a plane are as follows:

The radial component along $\mathbf{r}_1$ is $\ddot{r} - r\dot{\theta}^2$ \hfill (1.8)

The transverse component along $\boldsymbol{\theta}_1$ is $r\ddot{\theta} + 2\dot{r}\dot{\theta}$ \hfill (1.9)

Notice that when a point is moving with constant speed in a circle, ($\ddot{\theta} = 0$, $\dot{r} = 0$) the only acceleration is that towards the center of the circle; the centripetal acceleration $-r\dot{\theta}^2$.

1.9   Motion along the Tangent and Normal to a Curve

Another method of expressing the accelerations of a point moving along a curve is that of giving the components of the acceleration along and perpendicular to the curve. If **r** is a vector from the fixed origin O (Fig. 1.10a) tracing out the curve C, then the velocity along the curve is given by

$$\mathbf{v} = \lim_{t \to 0} \frac{\Delta \mathbf{r}}{\Delta t} = \frac{d\mathbf{r}}{dt} = \dot{\mathbf{r}}$$

Suppose s is the distance measured along the curve from some arbitrary point, then

$$\frac{d\mathbf{r}}{ds} = \lim_{\Delta s \to 0} \frac{\Delta \mathbf{r}}{\Delta s} = \mathbf{T}_1 \tag{1.10}$$

where $\mathbf{T}_1$ is the unit vector tangent to the curve at the point s.

The velocity **v** along the curve at the point s is instantaneously along the tangent $\mathbf{T}_1$; hence we may write $\mathbf{v} = v\mathbf{T}_1$, where v is the scalar speed. The acceleration **a** along the curve is given by

$$\mathbf{a} = \frac{d\mathbf{v}}{dt} = \frac{d}{dt}(v\mathbf{T}_1) = \mathbf{T}_1 \frac{dv}{dt} + v\frac{d\mathbf{T}_1}{dt} \tag{1.11}$$

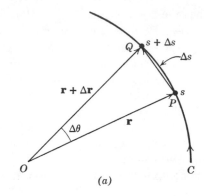

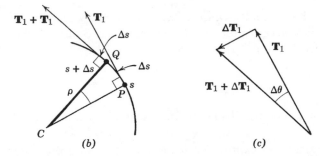

Fig. 1.10   (*a*) Particle moving along curve *C* from *s* to *s* + Δ*s* or *P* to *Q* and also specified by the vectors **r** and *r* + Δ**r** drawn from fixed origin *O*. (*b*) Unit tangent vectors $\mathbf{T}_1$ and $\mathbf{T}_1 + \Delta\mathbf{T}_1$ are drawn at *s* and *s* + Δ*s*, respectively. Normals are drawn to these tangents which intersect at the center curvature *C*. If Δ*θ* is very small then approximately Δ*s* = *ρ* Δ*θ* where *ρ* is the radius of the curve. (*c*) In the limit as Δ*θ* approaches zero, the direction of $\Delta\mathbf{T}_1$ is along the normal $\mathbf{N}_1$ to the curve, and since $\mathbf{T}_1$ is a unit vector $|\Delta\mathbf{T}_1| = |\mathbf{T}_1| \, \Delta\theta$ or $\Delta\mathbf{T}_1 = \mathbf{N}_1 \, \Delta\theta$ where $\mathbf{N}_1$ is unit vector along the normal.

Since $\mathbf{T}_1$ has the constant magnitude of unity, $d\mathbf{T}_1$ must be at right angles to $\mathbf{T}_1$.

Consider the two points *P* and *Q* on the curve such that their unit tangent vectors $\mathbf{T}_1$ and $\mathbf{T}_1 + \Delta\mathbf{T}_1$ make an angle Δ*θ* with each other. The direction of $\Delta\mathbf{T}_1$, in the limit as Δ*θ* approaches zero, is along the normal to the curve. Hence

$$\frac{d\mathbf{T}_1}{d\theta} = \lim_{\Delta\theta \to 0} \frac{\Delta\mathbf{T}_1}{\Delta\theta} = \mathbf{N}_1 \qquad (1.12)$$

where $\mathbf{N}_1$ is unit vector since its magnitude, the limit of $\Delta\mathbf{T}_1/\Delta\theta$ as Δ*θ* approaches zero, is unity.

By the chain rule for differentiation, the term $v(dT_1/dt)$ in the expression for the acceleration may be written as

$$v\frac{dT_1}{dt} = v\frac{dT_1}{ds}\frac{ds}{dt} = v^2\frac{dT_1}{ds} = v^2\frac{dT_1}{d\theta}\frac{d\theta}{ds} = N_1\frac{v^2}{\rho}$$

where $d\theta/ds = 1/\rho$ and ρ is the radius of curvature of the curve at the point s. From Fig. 1.10 b and c it is seen that the unit normal vector N_1 is directed from the concave side of the curve toward the center of curvature. Thus $N_1(v^2/\rho)$ is the centripetal acceleration. The expression for the acceleration along a curve is

$$\frac{d\mathbf{v}}{dt} = T_1\frac{dv}{dt} + N_1\frac{v^2}{\rho} = a_T T_1 + a_N N_1 \qquad (1.13)$$

in which the first term is the tangential acceleration along the curve and the second term is the centripetal acceleration along the inward normal to the curve, both calculated for the same point.

EXAMPLE.   The vector position $\mathbf{r}$ in meters, of a moving point at any time t, is given by

$$\mathbf{r} = 2t^2\mathbf{i} + 6t\mathbf{j} + 5\mathbf{k}$$

The problem is to find, at time of $t = 2$ sec, the velocity $\mathbf{v}$, acceleration $\mathbf{a}$, the tangential and normal accelerations a_T and a_N, the unit normals T_1 and N_1, and the radius of curvature ρ of the path of the moving point. Note that the point is moving in the x, y plane at a distance of 5 units along the z axis. The velocity $\mathbf{v}$ at any time t is

$$\mathbf{v} = 4t\mathbf{i} + 6\mathbf{j}$$

which at $t = 2$ sec gives $\mathbf{v} = 8\mathbf{i} + 6\mathbf{j}$, the magnitude of which is $|\mathbf{v}| = \sqrt{8^2 + 6^2} = 10$ m/sec. Since the vector velocity $\mathbf{v} = vT_1$, the unit tangent to the curve at $t = 2$ sec is $T_1 = 0.8\mathbf{i} + 0.6\mathbf{j}$, which has a magnitude of unity. The acceleration $\mathbf{a}$ of the moving point, at $t = 2$ sec, is

$$\mathbf{a} = \ddot{\mathbf{r}} = \dot{\mathbf{v}} = 4\mathbf{i} \text{ m/sec}^2$$

This acceleration can be expressed in terms of an acceleration a_T along the tangent to the curve and a_N along the normal so that $\mathbf{a} = a_T T_1 + a_N N_1$. Now a_T is the component of the acceleration $\mathbf{a}$ along the curve, or

$$a_T = \mathbf{a} \cdot T_1 = (4\mathbf{i}) \cdot (0.8\mathbf{i} + 0.6\mathbf{j}) = 3.2 \text{ m/sec}^2$$

From the relationship $\mathbf{a} = a_T T_1 + a_N N_1$, it follows that

$$4\mathbf{i} = (3.2)(0.8\mathbf{i} + 0.6\mathbf{j}) + a_N N_1 = 2.56\mathbf{i} + 1.92\mathbf{j} + a_N N_1$$

and $a_N N_1 = 1.44\mathbf{i} - 1.92\mathbf{j}$. The magnitude of the normal acceleration a_N is v^2/ρ and this is equal to $\sqrt{1.44^2 + 1.92^2} = \sqrt{2.07 + 3.69} = \sqrt{5.76} = 2.4$ m/sec^2. Thus $v^2/\rho = 2.4$ m/sec^2 and $\rho = 100/2.4 = 41.7$ m. The unit normal N_1 is obtained from the relationship $a_N N_1 = 1.44i - 1.92j$ so that $N_1 = (1.44i - 1.92j)/2.4 = 0.6i - 0.8j$, which has a magnitude of unity. As an exercise, show that the unit vectors T_1 and N_1 are at right angles to each other, that is, show $T_1 \cdot N_1 = 0$. Note also that $a_N = a \cdot N_1$, as may be readily verified numerically.

1.10   The Relativity of Galileo and Newton

It is well to recognize that wherever there is motion there is some form of relativity associated with it. The form which this took with Galileo and Newton is illustrated in Fig. 1.11 in which there are two coordinate systems, S and S' or x, y and x', y'. For simplicity consider the system S' moving with constant speed v relative to S along the x, x' direction. If the two coordinate systems coincide at time zero, then after a time t system S' has moved a distance vt. A point P has the coordinates x, y in the S system and x', y' in the S' system. The relativity of Galileo and Newton gives

$$x' = x - vt; \qquad y = y'; \qquad t = t' \tag{1.14}$$

Consider a beam of light originating at the point P (Fig. 1.11). This beam travels with a speed c of about 3×10^8 m/sec in the S system. From Eq. 1.14 it follows that in the S' system the speed of light is

$$\dot{x}' = \dot{x} - v = c - v$$

that is, the speed of light is less in the S' system than in the S system.

During the latter part of the last century, there were many experiments performed to determine whether the speed of light depended on the coordinate system. The most famous of these experiments was that made by Michelson

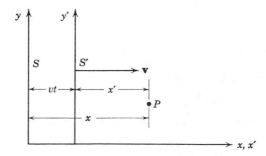

Fig. 1.11   Motion of coordinate system S' relative to S along the x, x' axes. Relativity of Galileo and Newton gives $x' = x - vt$, $y = y'$, $z = z'$, $t = t'$. Relativity of Einstein gives

$$x' = (x - vt)/\sqrt{1 - v^2/c^2}, \qquad y' = y, \qquad z' = z, \qquad t' = (x - vx/c^2)/\sqrt{1 - v^2/c^2}$$

and Morley* in 1887 in Cleveland, Ohio. Actually Michelson had first tried the experiment in Berlin in 1881, but the apparatus was not sufficiently rigid to give reliable results. The Michelson-Morley experiment, often called the ether drift experiment, showed that the hypothesis of the ether was untenable and from this and other experiments it was shown that the speed of light in free space was constant, independent of the motion of source or observer. The essential theory of the ether drift experiment is contained in Problem 14, in which the water is the equivalent of the ether and the boat of the light signal. In this there is a difference of times for the boat, making a round trip first across the river and second down the river. In the corresponding optical experiment no difference of times was observed.

Another development, of the greatest importance, was that of the electromagnetic theory of light, given by J. C. Maxwell in 1864. Several eminent physicists, Larmor, Fitzgerald, Lorentz, and Poincaré, were working, about 1900, on the theory of relativity in which both Michelson's experiment and Maxwell's theory were being used. From these investigations came new transformation equations called the Lorentz-Fitzgerald equations. However, these transformation equations were obtained from somewhat limited considerations. It was Einstein who obtained the transformation equations and other important results in 1905 from general principles of a philosophical character.†

1.11   The Special Relativity of Einstein

It was in 1905 that Einstein presented the special theory of relativity, which was concerned with systems moving with constant velocity relative to one another. This was followed some ten years later by the general theory dealing with accelerated systems. In the special theory, Einstein first analysed the concepts of simultaneity and time, showing that both of these are different in moving coordinate systems. His two fundamental assumptions were the following

a. The speed of light is constant in free space and is independent of the motion of the light source or the receiver.

b. The laws of physics are the same in coordinate systems moving with constant velocity relative to one another.

* "The Michelson-Morley Experiment." R. S. Shankland. *Sci. Am.*, **211**, 107, November 1964.
† *The Principle of Relativity.* A collection of original papers by Lorentz, Einstein, Minkowski, and Weyl (in translation). Dover Publications, New York.

From the latter it follows that there is no such physical concept as a "stationary system," and that all one can meanfully use is the concept of systems in uniform relative motion. There is no physical experiment that one can perform in a uniformly moving car which will demonstrate this motion.

The necessary space and time transformation equations, called the Lorentz transformation equations, can be derived in a number of ways. One of the less mathematical, as given by Einstein, is that of considering a light signal sent out in the S and S' systems, which are moving in uniform relative motion. Consider the propagation of the light along the x axis, given by the equation

$$x = ct \qquad \text{or} \qquad x - ct = 0$$

and in the S' system along the x' axis by the equation

$$x' = ct' \qquad \text{or} \qquad x' - ct' = 0$$

Notice that an event in the S system at x and t is also given in the S' system at x' and t', that is both the space and time coordinates are different in the two systems, whereas for the Newtonian system $t = t'$. If system S' is moving with velocity v in the x direction relative to system S we shall assume a linear relationship such that Eq. 1.14 becomes

$$x' = \gamma(x - vt) \tag{1.15}$$

where γ is a constant depending on the relative velocity v. Similarly for the S system moving with velocity $-v$ relative to the S' system, Eq. 1.14 becomes

$$x = \gamma(x' + vt') \tag{1.16}$$

Now consider the light signal being propagated along the x, x' axes such that $x = ct$ and $x' = ct'$, then

$$ct' = \gamma(ct - vt) \qquad \text{and} \qquad ct = \gamma(ct' + vt') \tag{1.17}$$

Combining Eqs. 1.17 so as to eliminate t and t' it follows that

$$\frac{ct'}{\gamma} = (ct - vt) = \gamma(ct' + vt') - \frac{v\gamma}{c}(ct' + vt')$$

$$\frac{c}{\gamma} = \gamma\left(c + v - v - \frac{v^2}{c}\right) = \gamma c(1 - v^2/c^2)$$

Hence

$$\gamma = 1/\sqrt{1 - v^2/c^2} \tag{1.18}$$

so that

$$x' = \gamma(x - vt) = (x - vt)/\sqrt{1 - v^2/c^2} \tag{1.19}$$

The relationship between t and t' can be obtained by eliminating x' between Eqs. 1.15 and 1.16, giving

$$\gamma v t' = x - \gamma x' = x - \gamma^2(x - vt)$$

$$= x - \frac{c^2}{c^2 - v^2}(x - vt) = \frac{x(c^2 - v^2 - c^2) + vtc^2}{c^2 - v^2}$$

or

$$\gamma t' = (c^2 t - xv)/(c^2 - v^2) = (c^2 t - vx)/c^2(1 - v^2/c^2)$$

$$= \gamma^2(t - vx/c^2)$$

Hence

$$t' = \gamma(t - vx/c^2) = (t - vx/c^2)/\sqrt{1 - v^2/c^2} \tag{1.20}$$

Equations 1.19 and 1.20 together with $y = y'$ and $z = z'$ constitute the Lorentz transformation equations for uniform motion of system S' relative to S along the x, x' axes. An equivalent system is that of S moving along the $-x$ axis with uniform velocity $-v$ such that the Lorentz transformation equations become

$$x = \gamma(x' + vt') \quad \text{and} \quad t = \gamma(t' + vx'/c^2) \tag{1.21}$$

Equations 1.21 may be derived algebraically from Eqs. 1.19 and 1.20 as given in Problem 26 at the end of the chapter.

EXAMPLE 1. Apparent Contraction in the Length of a Rod. Consider a rigid rod placed along the x' axis in the S' system, which is moving with a constant velocity v in the x direction relative to system S. The rigid rod of length L_0 has its end points at x_1' and x_2' so that $L_0 = x_2' - x_1'$. The problem is to find the length of the rod when observed in the S system. In order to do this it is necessary to find the x values of the ends of the rod at the *same instant of time* t *in the* S *system*, where the observer is located. For convenience, consider the time t to be zero. From Eq. 1.19, with $t = 0$,

$$x_2' = \gamma x_2; \quad x_1' = \gamma x_1 \tag{1.22}$$

Now $x_2' - x_1' = L_0$ is the length of the rod in the S' system and $L = x_2 - x_1$ is the apparent length of the rod measured in the S system. Using Eq. 1.22 it follows that

$$L_0 = \gamma L \quad \text{or} \quad L = L_0/\gamma = L_0\sqrt{1 - v^2/c^2} \tag{1.23}$$

Thus the rod appears to be shortened.

This result must be symmetrical, namely, if the rod of length L_0 is placed in the S system along the x axis, parallel to the direction of motion, then the apparent length L of this rod in the S' system is $L = L_0/\gamma$. It was this change in length which

Fitzgerald gave in order to satisfy the null result in the Michelson-Morley experiment.

EXAMPLE 2.   Time Dilation.   A further surprising result arising from the uniform motion of S' relative to S is that a clock beating seconds in S' appears to be going slowly relative to a similar clock in S. This is often called time dilation of a moving clock, where the verb "to dilate" means "to enlarge beyond normal size." The clock in S' is at rest in system S' and must remain at some particular position x', which for convenience may be taken as $x' = 0$. From Eq. 1.21, and setting $x' = 0$, it follows that $t = \gamma t'$. If the clock in S' is beating seconds such that $t_2' - t_1' = 1$ sec, then since $t_2 = \gamma t_2'$ and $t_1 = \gamma t'$, it follows that

$$t_2 - t_1 = \gamma(t_2' - t_1') = \gamma \text{ sec} = 1 \text{ sec}/\sqrt{1 - v^2/c^2} \qquad (1.24)$$

Thus $t_2 - t_1$ is larger than 1 second, that is, the interval of time between the ticks of the seconds clock in S' appear longer than 1 second to an observer in system S. This result is again symmetrical, that is, a clock beating seconds in the S system appears, to an observer in S', to take a longer period of time than seconds from his own similar clock. It should be noticed that placing the clock in S' at $x' = 0$ is not necessary; what is necessary is that x' remain constant or the clock remain at rest in the S' system. If this time dilation appears somewhat mysterious we must bear in mind that the clock in S' is moving with respect to an observer in system S. Any mystery must be a result of the original assumption that the speed of light is constant and that this is the maximum speed at which any signal can be transmitted.

An experimental check on the phenomena of time dilation has been given by the decay of positive pions, π^+ mesons. The clock in S' is now the π^+ meson, which decays with a mean life of about 2.5×10^{-8} sec. Beams of π^+ mesons have been produced with speeds of close to the speed of light such that

$$v/c = (1 - 5 \times 10^{-5})$$

Thus $(1 - v^2/c^2)$ is very approximately equal to 10^{-4}, and

$$\gamma = 1/\sqrt{1 - v^2/c^2} \approx 10^2$$

To an observer in the laboratory system S, the mean life of these high speed mesons is about a hundred times that observed for a π^+ meson at rest. The laboratory observer finds the mean life of these high speed π^+ mesons to be about 2.5×10^{-6} sec. If no time dilation occurred, the high speed π^+ mesons would, on the average, travel a distance of about $(2.5 \times 10^{-8}) c$ or about 750 cm before decaying. What is observed is that the π^+ mesons travel a distance of about 100×750 cm or 750 m before decaying.*

---

* Educational Services Incorporated, E. S. I. film, "Time Dilation. An Experiment with π^+ Mesons." F. Friedman, D. Frisch, and J. Smith. Report given by the authors in Am. J. Phys. **31**, 342 (1963).

Addition of Velocities

A third result of the Lorentz transformation equations is that occurring when an object moves with a speed u_x' in the x direction in the S' system. The S' system is moving with constant speed v in the x direction relative to the S system (Fig. 1.12a). From the transformation equations of Galileo and Newton, the speed u_x observed in the S system is given by

$$u_x = u_x' + v \qquad (1.25)$$

This is not the result given by the Lorentz transformation equations for no signal can travel with a speed greater than that of light c, as would occur with Eq. 1.25 when $u_x' = c$. The Einstein equation for the addition of velocities is obtained from Eq. 1.21:

$$x = \gamma(x' + vt') \qquad \text{and} \qquad t = \gamma(t' + vx'/c^2)$$

so that

$$dx = \gamma(dx' + v\,dt') \qquad \text{and} \qquad dt = \gamma(dt' + v\,dx'/c^2)$$

Setting $dx/dt = u_x$ and $dx'/dt' = u_x'$ it follows that

$$u_x = \frac{dx}{dt} = \frac{\gamma(dx' + v\,dt')}{\gamma(dt' + v\,dx'/c^2)} = \frac{(u_x' + v)}{(1 + u_x'v/c^2)} \qquad (1.26)$$

EXAMPLE 3.   Consider a light signal traveling along the x' direction with a speed c. If the S' system is moving in the x direction with a speed $c/2 = v$ relative to the S system and $u_x' = c$, then according to the Galilean transformation equation the speed of the light signal as measured by an observer in the S system would be $3c/2$. This is not the result given by the Einstein equation for the addition of

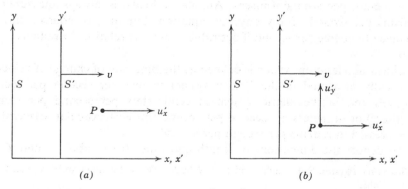

(a)   (b)

Fig. 1.12   (a) Object P moving with speed u_x' in S' system. (b) Object P moving with speeds u_x' and u_y' in S' system.

velocities, Eq. 1.26. This equation gives for the speed of light in the S system

$$u_x = \frac{c + c/2}{1 + c^2/2c^2} = \frac{1.5c}{1.5} = c$$

That is, the speed of light in both the S and S' systems is c.

As a further example of the addition relationship, consider the object P to be moving in the y' direction with a speed $u_y' = dy'/dt'$, and also moving in the x' direction with a speed u_x' (Fig. 1.12b) with system S' moving in the x, x' direction relative to S. The speed of object P as measured in the S system in the y direction is $u_y = dy/dt$. Now $y = y'$ and $dy = dy'$, hence

$$u_y = \frac{dy}{dt} = \frac{dy'}{\gamma(dt' + v\,dx'/c^2)} = \frac{u_y'}{\gamma(1 + vu_x'/c^2}$$

$$= \frac{u_y'\sqrt{1 - v^2/c^2}}{(1 + vu_x'/c^2)} \tag{1.27}$$

If $u_x' = 0$, that is, the object P is moving only in the y' direction in system S' with speed u_y', then the speed of P measured in the S system is

$$u_y = u_y'\sqrt{1 - v^2/c^2} \tag{1.28}$$

There are a great many elementary books on relativity, among which are the two mentioned below.*

1.12   Units and Dimensions of Velocity and Acceleration

Linear velocity is measured in terms of length and time. The common units of the British system are feet per second (ft/sec) and miles per hour (mph). In the metric system the units are centimeters per second (cm/sec) and meters per second (m/sec). Angular velocity is always measured in radians per second in any physical equation, but this, of course, may be changed to degree per second if desirable, using the relation 2π radians equal $360°$.

Linear acceleration, which is defined as the time rate of change of velocity, is usually measured in the British system in feet per second per second (ft/sec²) and in the metric system in centimeters per second per second (cm/sec²) or in meters per second per second (m/sec²). Angular acceleration is expressed in radians per second per second.

To denote the dimensions of length and time, the symbols L and T are

* *Spacetime Physics*. E. F. Taylor and J. A. Wheeler. W. H. Freeman and Co., San Francisco, Calif.

Relativity: The Special and General Theory. A. Einstein. An early (1916) general discussion with the mathematical derivations in an appendix. Great Books Foundation, 307 N Mich. Ave. Chicago, Ill.

adopted. In general, symbols in square brackets represent dimensions of quantities. Thus velocity has the dimensions $[L/T]$ or $[LT^{-1}]$, and linear acceleration has the dimensions $[L/T^2]$ or $[LT^{-2}]$. Since an angle in radians is defined as an arc of a circle divided by its radius, it follows that an angle has no dimensions. Thus angular velocity has the dimensions $[1/T]$ or $[T^{-1}]$ and angular acceleration of $[1/T^2]$ or $[T^{-2}]$.

Every term in an equation expressing a relationship between physical quantities, that is, a physical equation, must have the same units and dimensions. This basic principle of dimension theory, called the principle of dimensional homogeneity, was stated by Fourier in 1822 and by Maxwell in 1863. A discussion of the dimensions of physical concepts is given by Moon and Spencer.* For those interested in this subject many other references will be found in this article. The restriction on physical equations does not in general apply to mathematical equations. For instance, consider the linear distance s covered by a body starting from rest and moving with a linear acceleration a for a time t. The relation between these quantities is $s = at^2/2$, and the velocity attained by the body in the time t is $v = at$. Both equations are dimensionally correct. Since, if equals are added to equals, the results are equal, it follows mathematically that

$$s + v = \frac{at^2}{2} + at$$

On the other hand, this equation is not dimensionally homogeneous and is of no value as a physical equation.

Obviously the units employed in the various terms of a physical equation must be consistent; that is, in one term the acceleration must not be expressed in centimeters per second per second and in another in feet per second per second. It is also quite easy to make rather foolish errors in changing the units in which a given quantity is expressed. As a simple method for avoiding such errors consider changing an acceleration of 50 mph/min to feet per second per second. To do this we must change the miles to feet, the minutes to seconds, and the hours to seconds, thus,

$$\frac{50 \text{ mi}}{\text{hr} \times \text{min}} = \frac{50 \text{ mi}}{\text{hr} \times \text{min}} \times \frac{1 \text{ min}}{60 \text{ sec}} \times \frac{1 \text{ hr}}{3600 \text{ sec}} \times \frac{5280 \text{ ft}}{1 \text{ mi}}$$

$$= \frac{50 \times 5280 \text{ ft}}{60 \times 3600 \text{ sec} \times \text{sec}} = 1.22 \text{ ft/sec}^2$$

Each term after the first 50 mph/min is a conversion factor whose value is unity, containing suitable units so placed that after all units are canceled

---

* P. Moon and D. E. Spencer, "Dimensions of Physical Concepts," *Am. J. Physics*, **17**, 171 (1949).

out the answer is obtained in the desired units. This method of explicitly writing out all units may appear cumbersome, but it will help to avoid many careless errors.

1.13 Setting Up and Checking of Physical Equations

Although no rules can be given for solving physical problems which will replace a sound knowledge of physical principles, the following suggestions should be of real aid.

a. After reading the problem and understanding its implications, draw a neat diagram if the situation calls for it.

b. Clearly indicate on the diagram which direction or directions you are choosing for positive, such as the positive direction for the coordinate axes, forces, velocities, accelerations, etc.

c. Set up a physical equation and explicitly state what assumptions you are making in doing this.

d. Check this equation and any derived from it for consistency of dimensions and units. It should be noted that this does not check any multiplying factors such as the 1/2 in the equations $s = at^2/2$.

e. After solving the equation to obtain the final answer, examine it to see if it is qualitatively correct, that is, of the right order of magnitude. This can usually be done by simplifying the original equation by omitting one or more of the less important terms and then solving the simplified equation. From the nature of the problem it should be possible to say whether the answer to the simplified equation should be greater or less than the correct answer. For example, if the time of fall of a body is to be found, taking into account the resistance of the air, this may be compared with the time of fall with no air resistance. Another method applied especially in research problems is to compare the theoretical answer with an experimental one. Though this method may not be possible with many textbook problems, it should be followed whenever practical.

PROBLEMS

1. An object falls from rest from a height of 6400 ft with an acceleration of $g = 32$ ft/sec². (*a*) Assuming g constant and no air resistance, find the time of fall and the speed with which the object strikes the ground. (*b*) Owing to air resistance,

the falling object is subjected to an opposing acceleration of $4v$ ft/sec², where v is the speed of the object at any instant of time. Briefly state how the speed and acceleration of the object change during the motion. (c) Find the constant terminal speed with which the object ultimately falls.

2. The position y of a moving point is given for various times t in the table below:

t (sec)	0	1	2	3	4	5
y (cm)	2	5	6	5	2	−3

(a) Plot these values with t as the abscissa or x coordinate. (b) Find the average velocity of the moving point during each of the intervals 0–1, 1–2, 2–3, 3–4, 4–5 sec. (c) Find the rate of change of the average velocity during each succeeding second. (d) From the graph, determine the average velocity during each of the intervals 1–2, 1–1.5, 1–1.25 sec and the instantaneous velocity at $t = 1$ sec. (e) Find the instantaneous velocity at $t = 2, 3, 4$ sec. (f) Find the rate of change of these instantaneous velocities and show that this is an acceleration of -2 cm/sec². (g) Assuming the acceleration to be -2 cm/sec² and the moving point to have an initial displacement of 2 cm and a velocity of 4 cm/sec, find the equation for the displacement at any time. (h) From this equation find the displacement at $t = 0, 1, 2, 3, 4, 5$ sec.

3. A particle has a linear acceleration of $\ddot{x} = 4 - t^3$ m/sec². Initially, the particle has a displacement of 10 m and a speed of 6 m/sec. Find the speed and displacement of the moving particle after 3 seconds.

4. For a time $t \geqslant 0$, the acceleration of a particle is given by $\ddot{\mathbf{r}} = \mathbf{i}3te^{-t} + \mathbf{j}6(t^2 - 2) - \mathbf{k}4t \sin 2t$. If the displacement $\mathbf{r}$ and the velocity $\dot{\mathbf{r}}$ are zero at $t = 0$, find the expressions for $\mathbf{r}$ and $\dot{\mathbf{r}}$ at any time t. (This is an exercise in integration and $\ddot{\mathbf{r}} = \mathbf{i}\ddot{x} + \mathbf{j}\ddot{y} + \mathbf{k}\ddot{z}$—the integration makes use of $\int u\, dv = uv - \int v\, du$.)

5. A moving point has a displacement given by $x = A[\cos(\beta t + \phi) + \sin(\beta t + \phi)]$. (a) Show that the acceleration is $\ddot{x} = -\beta^2 x$. (b) Show that the velocity and displacement differ in phase by $\pi/2$ radians at every instant of time. (c) Show that the acceleration and displacement differ in phase by π radians at every instant of time.

6. A moving point has a displacement given by $x = Ae^{i\beta t}$, where $i = \sqrt{-1}$, and $e^{i\beta t} = (\cos \beta t + i \sin \beta t)$. See Appendix. Now x is a complex number and may be set equal to $x_1 + ix_2$, where x_1 and x_2 are real. Show the velocity and acceleration differ in phase from displacement by $\pi/2$ and π radians respectively at every instant of time. Show velocity $\dot{x} = i\beta Ae^{i\beta t}$ is 90° ahead of displacement given by $x = Ae^{i\beta t}$.

7. Find the magnitude and direction cosines of the vector $3\mathbf{i} + 6\mathbf{j} - 7\mathbf{k}$.

8. A vector $\mathbf{A}$ makes angles with the X, Y, Z coordinate axes whose cosines, called direction cosines, are l_1, m_1, n_1 respectively, and vector $\mathbf{B}$ makes direction

cosines of l_2, m_2, n_2. If θ is the angle between the two vectors **A** and **B**, show that $\cos \theta = l_1 l_2 + m_1 m_2 + n_1 n_2$. (*Hint:* Take the dot product of vectors **A** and **B** and expand in terms of the components noting that $A_x/|A| = l_1$, etc., where $|A|$ is the magnitude of vector **A**.)

9. Find the cosine of the angle between the two vectors $2\mathbf{i} - 4\mathbf{j} + 4\mathbf{k}$ and $4\mathbf{i} + 4\mathbf{j} - 2\mathbf{k}$.

10. Show that

$$\frac{d}{dt}(\mathbf{P} \cdot \mathbf{Q}) = \mathbf{P} \cdot \frac{d\mathbf{Q}}{dt} + \mathbf{Q} \cdot \frac{d\mathbf{P}}{dt}$$

(*Hint:* Expand $\mathbf{P} \cdot \mathbf{Q}$ in terms of components and carry out the differentiation on the scalar quantities.)

11. Show that the sum of the components of vectors **A** and **B** along **C** is equal to the component of the resultant $(\mathbf{A} + \mathbf{B})$ along **C**, and also show, by expanding the vectors along the three axes x, y, z, that $(\mathbf{A} + \mathbf{B}) \cdot \mathbf{C} = \mathbf{A} \cdot \mathbf{C} + \mathbf{B} \cdot \mathbf{C}$.

12. A motorboat having a speed of 16 mph is headed directly across a river 200 ft wide with a current of 4 mph. (*a*) How far down the opposite bank from the point directly opposite to the starting point does the boat land? (*b*) At what angle upstream must the boat be headed if it is to land at a point directly opposite its starting point, and how long does it take to cross?

13. A river, of width L, is flowing with a speed v; a boat travels at a speed c relative to the river. (*a*) The boat moves across the river so as to reach a point directly opposite to the starting point and then immediately returns to its starting point. Show the total time for the round trip is $t_1 = 2L/\sqrt{c^2 - v^2}$. (*b*) The boat now goes distance L downstream and returns to the starting point. Show the time for this round trip is $t_2 = 2Lc/(c^2 - v^2)$. (If $t_1 = t_2$, show the lengths L_1 and L_2 are given by $L_1\sqrt{1 - v^2/c^2} = L_2$. This is analogous to the Fitzgerald contraction.)

14. A river is 0.5 miles wide and is flowing at 6 mph. A person in a boat, which has a speed relative to the river of 4 mph, wishes to cross the river to a point directly opposite from the starting point. To do this he crosses the river to some place on the opposite bank and walks at 3 mph to the point directly opposite from the starting point. Find the angle at which the boat must head out relative to the bank, in order to reach the opposite point in a minimum amount of time, and also find the minimum time.

15. A person in a boat is moving westward at 4 mph and finds that the wind appears to blow directly from the north. If the speed of the boat is doubled, the wind appears to come from the northwest. Show the velocity of the wind is $4\sqrt{2}$ mph northeast.

16. A molecule is moving upwards with a velocity v_y in a container having an airtight piston that is moving vertically downwards with a velocity V, both

velocities being measured with respect to the earth. The molecule rebounds from the piston with the same velocity, relative to the piston, as it strikes the piston (elastic impact). Show that after impact, the molecule moves downward with a velocity $v_y + 2V$, relative to the earth.

17. After a free fall of 16 ft, a ball strikes a platform moving vertically upward with a velocity of 4 ft/sec, both velocities being measured with respect to the earth. If the impact is elastic, that is, the ball rebounds with the same velocity as it strikes, relative to the platform, show the height to which the ball rises is 25 ft above the initial place of impact.

18. The radius of the earth is approximately 4000 mi. (a) Find the angular velocity of the earth about its axis in radians per hour. (b) Find the linear velocity in miles per hour at the equator. (c) Find its linear velocity at latitude 60° N. Find the centrifugal acceleration for a point at (d) the equator; (e) a latitude of 60° N.

19. The apparent acceleration due to gravity g is the vector sum of the earth's gravitational attraction g_0 which is directed toward the center of the earth and the centrifugal acceleration away from the axis of the earth. If R is the radius and ω is the angular velocity of the earth, then show that the resultant acceleration g at any latitude θ is approximately

$$ g = g_0 \left(1 - \frac{\omega^2 R \cos^2 \theta}{g_0} \right) $$

(Note that ω is a relatively small quantity so that terms such as $\omega^4 R^2/g_0^2$ may be neglected as compared to 1.)

20. The equations for the radial and transverse accelerations, Eqs. 1.8 and 1.9, can be obtained by starting with the relationships $x = r \cos \theta$ and $y = r \sin \theta$. Find the values of $\dot{x}$ and $\dot{y}$, and show that the velocity in the direction of $\mathbf{r_1 r}$ is $(\dot{x} \cos \theta + \dot{y} \sin \theta) = \dot{r}$, and in the direction of $\mathbf{\theta_1}$ is $(\dot{y} \cos \theta - \dot{x} \sin \theta) = r\dot{\theta}$. In Fig. P20 note the components of $\dot{x}$ and $\dot{y}$ along $\mathbf{r_1}$ are $(PD + PC)$ and along $\mathbf{r_1}$ are $(PE - PF)$. Find $\ddot{x}$ and $\ddot{y}$ and show the acceleration in the direction of $\mathbf{r_1}$ is $(\ddot{r} - r\dot{\theta}^2)$ and in the direction $\mathbf{\theta_1}$ is $(r\ddot{\theta} + 2\dot{r}\dot{\theta})$. Show this latter expression can be written as $(1/r)\, d(r^2\dot{\theta})/dt = (2/r)\, dA/dt$ where dA is an element of area swept out by an angular change $d\theta$ in θ.

21. A wheel whose radius is 0.15 m has a constant angular acceleration of 60 rpm/sec and an initial angular velocity of 15 rpm. (a) Find the angular acceleration and the velocity in radians per second per second and radians per second respectively. (b) Find the angular velocity of a point on the periphery at the end of 0.2 sec. (c) Find the resultant linear acceleration of a point on the periphery at the end of 0.2 sec. Give both the magnitude and the direction of the resultant acceleration. (d) Find the angle through which the wheel turns in the 0.2 sec.

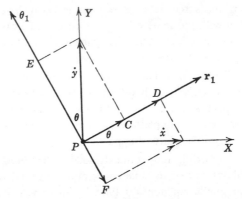

Fig. P20   Components of velocities $\dot{x}$ and $\dot{y}$ along directions $\mathbf{r}_1$ and $\mathbf{\theta}_1$.

22. A point moving in a plane has the coordinates $x = 3$, $y = 4$ and has the components of speed $\dot{x} = 5$ m/sec, $\dot{y} = 8$ m/sec at some instant of time. Find the components of speed in polar coordinates r, θ, along directions $\mathbf{r}_1$ and $\mathbf{\theta}_1$. The magnitude of the velocity is given by $\sqrt{\dot{x}^2 + \dot{y}^2}$ or by $\sqrt{(r\dot{\theta})^2 + \dot{r}^2}$. Show that these two expressions give the same numerical result.

23. Consider a point $P(r, \theta)$, having $r = 13$ m and $\theta = $ arc sin 5/13, moving in a plane with a speed whose components along $\mathbf{r}$ are $\dot{r} = 6$ m/sec, and at right angles to r, along $\mathbf{\theta}_1$ are $r\dot{\theta} = 5$ m/sec. Find the components of speed along the X and Y coordinates, that is, $\dot{x}$ and $\dot{y}$, and find the magnitude of the speed from the two expressions given in Problem 22.

24. The vector position $\mathbf{r}$ in meters of a point moving along a curve is such that at time t seconds, $\mathbf{r} = 6(4t - \cos 4t)\mathbf{i} + 6(2 - \sin 4t)\mathbf{j}$. Show that the value of the time for which the unit tangent vector is along the $\mathbf{i}$ or x axis is $\pi/8$ sec and the velocity $\mathbf{v}$ at this time is 48 m/sec.

25. The vector position $\mathbf{r}$ (in feet) of a point moving along a curve is such that at time t seconds, $\mathbf{r} = 2t^2\mathbf{i} + 1.5\mathbf{j} - 6t\mathbf{k}$. At time $t = 2$ sec find the values of: $\mathbf{v}$, v, $\mathbf{a}$, a_T, a_N, ρ, $\mathbf{T}$, and $\mathbf{N}$.

26. From the Lorentz transformation equations, show that for an event (x', t') in the S' system, its corresponding coordinates (x, t) in the S system are

$$x = \frac{x' + vt'}{\sqrt{1 - v^2/c^2}} \quad \text{and} \quad t = \frac{t' + vx'/c^2}{\sqrt{1 - v^2/c^2}}$$

27. A rod in the S coordinate system has a length L_0. Show that the length of this rod when measured from the S' system is $L = L_0\sqrt{1 - v^2/c^2}$, or the rod appears to be shortened.

28. A clock in the S coordinate system is beating seconds. Show that this time interval measured in the S' system is $1/\sqrt{1 - v^2/c^2}$ seconds.

29. A rod of length 1 m is moving parallel to its length with a speed of 10^8 m/sec. Find the apparent length of this rod in the laboratory system of coordinates.

30. A star, being eclipsed once every 3 min by a dark companion, is known to be moving toward the earth with a speed of 3×10^7 m/sec. Find how often the eclipses appear when viewed from the earth.

31. A star appears to be eclipsed once every 5 min when viewed from the earth. The star is moving away from the earth with a speed of 3×10^7 m/sec. Find how often the star is being eclipsed as viewed from a system of coordinates attached to the star.

32. A meter stick is placed along the x' axis in the S' coordinate system which is moving along the x' axis with a velocity v relative to the S system. Find the length of this meter stick in system S for $v = 0.1c$, $0.01c$, and $0.001c$. What is the length of this meter stick when measured in the S' system at these three velocities? If the meter stick is placed along the y' axis, find its apparent length when viewed from the S coordinate system. (Use $(1 + x)^n = 1 + nx \ldots$).

33. Show that two simultaneous events, $t_1 = t_2$, taking place at different places, x_1 not equal to x_2, in the S coordinate system, are not simultaneous in system S'.

34. A π^+ meson has a mean life of about 2.5×10^{-8} sec in a system in which it is at rest. Find the mean life of these mesons when traveling with velocities of $0.1c$, $0.01c$, $0.001c$, and find the distances traveled in one mean life in coordinate systems attached to the mesons and also in coordinate systems that are at rest relative to the mesons.

35. If a particle moves along the circumference of a circle by being attracted to a point on the circle, show that the force of attraction varies as the inverse fifth power of the distance between the particle and the fixed point. [*Note.* $r = 2R \cos \theta$ where θ is the angle between the diameter $2R$ and r, the distance from the fixed point to any point on the circumference, and $f(r) = \ddot{r} - r\dot{\theta}^2$ while $r^2\dot{\theta}$ is a constant.]

2

FORCE AND MOTION
OF PARTICLES

2.1 Introduction

In the last chapter we discussed motion and changes in motion without any regard to the manner in which these are produced. Now we turn our attention to a study of dynamics, which is concerned with forces and their effects on the motion of bodies.

Intuitively we think of changes in the motion of a body as caused by a push or a pull. A football is set in motion by a kick, a bullet by the force of expanding gases, a falling body by the pull of the earth. In each of these examples we can identify some force with the change in motion. The change in motion, or acceleration, depends not only the size of the force but also on a property of the body moved, which Newton first called the "quantity of matter" and which later became known as the inertia or mass of the body. The difference between setting a brick and a football in motion by means of a kick will readily convince one of this property of inertia.

Furthermore, the forward motion of the ball does not cease when the kick is over. The ball continues to move forward and, owing to the action of the gravitational pull of the earth, downward. The forward motion diminishes somewhat, owing to the resistance force of the air. Suppose now that the air resistance force and the gravitational pull on the earth could be eliminated. What would happen to the motion of the ball? Though it is impossible to reduce the forces on a moving body to zero, experiments show that, as these forces are reduced, the velocity of the body becomes more and more nearly constant. It is assumed that if the resultant force on a body were zero its velocity would remain unchanged. A force is required to change the velocity of a body, that is, to start or stop a body, to speed it up or slow it down, or to change the direction of its motion. The acceleration depends both on the force applied and on the inertia of the body. For the present we are concerned with the motion of a body as a whole, and not with any rotation about an axis through the body. For such purposes it is convenient to consider the body as a particle. The quantitative measure of inertia is mass, and a simple experiment will illustrate how this may be measured.

2.2   Inertia and Mass

Suppose that two cars, A and B in Fig. 2.1, have frictionless bearings and are placed on a horizontal track. The cars are held together by a taut thread which prevents them from separating when a compressed spring is placed between them. This compressed spring produces an internal force on the system which exerts an equal and opposite force on the two cars. When the thread is burned, the cars move apart, car A moving to the right with a velocity v_A and car B to the left with a velocity v_B. The cars would then continue to move indefinitely with these velocities if the resultant force on them were zero.

If the changes in velocity from zero to v_A and v_B respectively take place in some short interval of time Δt, then the acceleration of car A is

$$a_A = \frac{v_A}{\Delta t}, \quad \text{and of car } B \text{ is} \quad a_B = \frac{v_B}{\Delta t}$$

Although Δt may not be known, it is the same for both cars. Hence

$$\frac{a_A}{a_B} = \frac{v_A}{v_B}$$

Since the velocities can be measured, it follows that the ratio of the accelerations can be determined. Suppose that the experiment is repeated with a different tension in the compressed spring. It will be found that the ratio of the accelerations is the same as that previously obtained. Further experiments would show that, no matter what the tension on the spring, the ratio of the accelerations would be constant. These accelerations depend on the inertias

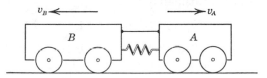

Fig. 2.1   Comparison of masses of cars A and B.

or masses of the cars; the larger the inertia or mass of a car, the smaller its acceleration for a given force. If m_A is the mass of car A and m_B the mass of car B, the ratio of these masses is defined as

$$\frac{m_A}{m_B} = \frac{a_B}{a_A}$$

Suppose now that car B is replaced by a car C and the experiment is repeated. Then

$$\frac{m_A}{m_C} = \frac{a_C}{a_A}$$

If car B is placed on top of car C, then

$$\frac{m_A}{m_{B+C}} = \frac{a_{B+C}}{a_A}$$

From these experiments each of the masses m_B, m_C, m_{B+C} can be obtained in terms of m_A. The results show that the total mass m_{B+C} is equal to the sum of the masses m_B and m_C. Thus mass is a scalar quantity. The quantity called mass is here the inertial mass in contrast to the gravitational mass involved in weighing. The word mass has a long and interesting history as its uses in widely different situations testify.*

Suppose that we arbitrarily assume the mass of car A to be unity. Then the inertial mass of any object can be determined by this method. Actually this dynamic method is not followed in practice, but, instead, some static method depending on the extension of a uniform spring or on an equal arm balance. In all these methods some arbitrary standard of mass is adopted, and then an unknown mass is determined in terms of the standard. Consistent results for the ratio of masses are obtained by any of these methods.

The unit of mass in the meter-kilogram-second (mks) system is the kilogram, which is the mass of a lump of platinum-iridium carefully preserved in France. A submultiple of this, a gram or one-thousandth of a kilogram, is the unit of mass in the centimeter-gram-second (cgs) system. It was originally intended that the gram should be the mass of 1 cm³ of pure water at its temperature of maximum density, 4°C. However, more precise experiments have shown that this is not exactly correct. In the British system the unit of mass, the pound, is the mass of a lump of platinum-iridium kept in London. It turns out approximately that

$$1 \text{ lb} = 453.6 \text{ gm} \qquad \text{or} \qquad 1 \text{ kg} = 2.2 \text{ lb}$$

Another unit of mass, called the slug, has been adopted in the British engineering system of units and will be discussed later.

Concepts of Mass; Max Jammer, Harv. Univ. Press, Camb. Mass. (1961).

2.3   Force and Mass

In the discussion of inertia we stated that the velocity of a body is constant when the resultant force acting on the body is zero. This is the basis of Newton's first law of motion, which in more formal language may be stated as follows.

Every body continues in a state of rest or in uniform motion in a straight line unless acted on by some force.

This law implies that the motion of a body is changed when an unbalanced force acts on it.

Newton's second law deals with the relationship of the force involved and the change in motion. In order to state this law Newton coined a new term, "quantity of motion" of a body. This is now replaced by the term "momentum" of a body. *Momentum* $\mathbf{p} = m\mathbf{v}$ is a vector quantity having the direction of the velocity and a magnitude equal to the product of the mass and the magnitude of the velocity of the body. Just as in the discussion of velocity and acceleration we had to choose carefully the coordinate system in which the motion took place, so in using Newton's second law we must also be careful. The coordinate system to which Newton's laws are applicable is called an *inertial system* and any system moving with constant velocity relative to this is an equivalent system. Although coordinates attached to the earth do not constitute an inertial system, yet for most of our problems it is sufficiently accurate to assume that they do. In a later section we shall see what effects the rotation of the earth produces. For an inertial coordinate system. Newton's second law of motion may be stated as follows.

The time rate of change of momentum of a body is proportional to the resultant force acting on the body:

$$\mathbf{F} \propto \frac{d}{dt}(m\mathbf{v}) = k\frac{d(m\mathbf{v})}{dt} = k\frac{d\mathbf{p}}{dt}$$

where $\mathbf{F}$ is the resultant force acting on a body of mass m and instantaneous velocity $\mathbf{v}$. The proportionality factor k is a constant whose value can be made unity by a suitable choice of units. With such units Newton's second law may be written

$$\mathbf{F} = \frac{d(m\mathbf{v})}{dt} = m\frac{d\mathbf{v}}{dt} + \mathbf{v}\frac{dm}{dt} \tag{2.1}$$

From the time of Newton to early in the present century, the mass of a body was considered a constant, independent of its velocity. With the advent of relativity in 1905 Einstein showed that the mass of a body relative to an observer varies with its velocity. However, for bodies moving with velocities small compared with the velocity of light, the change in mass with velocity is negligibly small, as discussed in Section 2.14.

Since the velocities dealt with in classical or Newtonian physics are all negligibly small compared to the velocity of light, we shall assume that the mass of any body is a constant quantity, independent of its velocity. Thus we may write Newton's second law for a constant mass as

$$\mathbf{F} = m\frac{d\mathbf{v}}{dt} = m\mathbf{a} \qquad (2.2)$$

where $\mathbf{a}$ is the acceleration of a particle of mass m acted on by a resultant force $\mathbf{F}$.

Equation 2.2 deals with the vector quantities $\mathbf{F}$ and $\mathbf{a}$. Suppose that a force $\mathbf{F}$ has components F_x and F_y along the X and Y axes respectively. Then, using unit vectors as in Section 1.6,

$$\mathbf{F} = \mathbf{i}F_x + \mathbf{j}F_y$$

If this force is acting on a particle of mass m, its accelerations $\ddot{x}$ and $\ddot{y}$ are given by

$$\mathbf{i}F_x = \mathbf{i}m\ddot{x} \qquad \text{and} \qquad \mathbf{j}F_y = \mathbf{j}m\ddot{y}$$

Also, if the vector $\mathbf{r}$ has the components $\mathbf{i}x$ and $\mathbf{j}y$, then

$$\mathbf{r} = \mathbf{i}x + \mathbf{j}y \qquad \text{and} \qquad \ddot{\mathbf{r}} = \mathbf{i}\ddot{x} + \mathbf{j}\ddot{y}$$

so that

$$\mathbf{F} = m\ddot{\mathbf{r}} = \mathbf{i}m\ddot{x} + \mathbf{j}m\ddot{y}$$

Forces and accelerations may be compounded by the vector method of addition, as was done above. When a particle is acted on by several forces simultaneously, the product of its mass and the resultant acceleration is equal to the resultant of the forces compounded vectorially. When the force and acceleration are resolved in any direction, the product of mass and acceleration in that direction is equal to the component of force in the same direction.

2.4   Fictitious Forces. Centrifugal Force

Newton's second law may also be written $\mathbf{F} - m\mathbf{a} = 0$. The vector quantity $-m\mathbf{a}$ is called a *fictitious force* or sometimes the *kinetic reaction* or the *inertial reaction*.

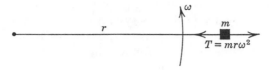

Fig. 2.2 Centripetal and centrifugal forces, T and $mr\omega^2$ respectively.

In his *Traité de Dynamique*, published in 1743, D'Alembert enunciated a principle now known as D'Alembert's principle, which may be stated as follows: When any particle is acted on by an external force, the resultant of this external force and the kinetic reaction of the particle is zero. Familiar examples of kinetic reaction are observed when we are thrown backward by a streetcar's starting suddenly, or forward by a car's stopping suddenly, or outward by a car's rounding a corner at high speed. In the last example it is usual to call this kinetic reaction the centrifugal force.

Suppose that a particle of mass m is whirled around on a cord of length r with an angular velocity ω (Fig. 2.2). From Eq. 1.8 the centripetal acceleration of the particle is $r\omega^2$. Hence the tension in the string, that is, the centripetal force, is $mr\omega^2$, which is the inward pull of the string on the mass m. The force that the moving mass m exerts on the string or on the hand holding it is the kinetic reaction $mr\omega^2$ acting outward. This force, which does not act on mass m but on the means of constraint, is called the centrifugal force. Thus we may consider the moving mass m to be in equilibrium under the action of the force T inward and the kinetic reaction outward. Essentially the application of D'Alembert's principle changes a problem of dynamics into one of statics.

A vivid illustration of the centrifugal force is that provided by a circular horizontal platform with vertical sides attached on the periphery and rotating about a vertical axis through the center of the platform. These are frequently seen at an amusement park. When the speed of rotation is sufficient the persons on the platform are moved to the sides and then if the platform, but not the sides, is lowered the people continue the rotation and appear fixed to the sides.

2.5   Units of Force

The manner in which the force and mass units are defined differs in the metric and the British engineering systems. In the metric system it is usual to take mass as the fundamental quantity and force as a derived quantity. The unit of force in the mks system is called the *newton*. It is that force which when acting on a mass of 1 kg gives it an acceleration of 1 m/sec^2. Similarly, the unit of force in the cgs system is called the *dyne*, and is that force which when acting on a mass of 1 gm gives it an acceleration of 1 cm/sec^2. From these definitions it follows that

$$1 \text{ newton} = 10^5 \text{ dynes}$$

If the unit of mass in the British system is taken as 1 lb, then the unit of force in this system is called the poundal, and is that force which when acting on a mass of 1 lb gives it an acceleration of 1 ft/sec^2. In practice the poundal is not employed as a unit of force.

The above units are sometimes called the absolute system of units, although there is really nothing absolute about them. They refer to a system in which mass is taken as fundamental and force is a derived quantity. For such a system the dimensions of force are given in terms of the dimensions of mass $[M]$, length $[L]$, and time $[T]$ from Newton's second law as

$$[F] = [MLT^{-2}]$$

There is also the gravitational system of units which we shall discuss in the next section.

2.6   Mass and Weight

We have previously stated that, at any particular location on the earth, all bodies fall vertically downward with the same acceleration if air resistance is absent. Newton demonstrated this by the so-called feather and guinea experiment in which a feather and a small coin were allowed to fall in an evacuated tube. It was found that the light and the heavy body fell with the same acceleration. This acceleration, called the acceleration due to gravity and denoted by g, is not exactly the same at different elevations and places on the earth. For sea-level elevation, the value of g is about 983.21 cm/sec^2 at the poles and decreases to about 977.99 cm/sec^2 at the equator of the earth. In the middle latitudes the value of g may be taken as approximately 980 cm/sec^2 or 32 ft/sec^2.

Since a body of mass* m falls with an acceleration $\mathbf{g}$, it follows that there is a vertically downward force of $m\mathbf{g}$ on this body. This force due to the earth's gravity is called the weight $\mathbf{w}$ of the body. Thus

$$\mathbf{w} = m\mathbf{g}$$

* It is the gravitational mass that is involved here, in contrast to the inertial mass, as given in Section 4.1.

A body whose mass is 1 gm has a weight of 1 gram weight (gmw) and exerts a force of approximately 980 dynes:

$$1 \text{ gmw} = 980 \text{ dynes} \qquad 1 \text{ kgw} = 9.8 \text{ newtons}$$

In the British absolute system it similarly follows that 1 pound weight (lbw), the gravitational force on 1 pound mass, is approximately equal to 32 poundals.

It is the practice in engineering to use the British engineering or gravitational system of units. In this system, force is taken as a fundamental quantity and the unit of mass is derived. The unit of force in the engineering system is taken as the gravitational pull on a pound mass at sea level and 45° N latitude; that is, the unit of force is 1 lbw. For this unit of force, what unit of mass has to be used in Newton's second law? This unit of mass is called a *slug*. It is that mass which, when acted on by a force of 1 lbw, acquires an acceleration of 1 ft/sec². Since a force of 1 lbw gives a mass of 1 lb an acceleration of 32 ft/sec², it follows that the slug is 32 times as large as the pound mass.

Writing $[F]$ for the dimension of force, it follows that, in the gravitational system, mass $[M]$ has the dimensions $[FT^2L^{-1}]$. In this system Newton's second law is written

$$\mathbf{F} = \frac{w}{g}\,\mathbf{a}$$

The forces $\mathbf{F}$ and w are in pounds weight, and the accelerations $\mathbf{a}$ and g are in feet per second per second. In general we shall use the metric system, but most of the equations may be written immediately in the engineering system by replacing the mass m by w/g. Since the earth is rotating, the value of g observed is that resulting from the gravitational attraction of the earth and the centrifugal force, as shown in Problem 19 of Chapter 1.

2.7   Newton's Third Law. Conservation of Momentum

You have probably noticed that forces always occur in pairs, although acting on different objects. For example, when a gun is fired, there is a forward force on the bullet and an equal but opposite force on the gun. This force is sometimes called the "back-kick" of the gun. In walking, one pushes back on the earth and the earth exerts a forward force on the person. If there is no friction, as on perfectly smooth ice, one cannot push backwards. There is then no forward force, and one cannot make any headway. A propeller pushes back on the air, and the air pushes the propeller and airplane forward. When we observe any one force, it is an interesting game to pick out the other force acting on some other body. The principle involved in these and other examples is summarized in *Newton's third law of motion*, which may be stated as follows.

> *To every force acting on one body there is an equal and opposite force acting on another body.*

If $\mathbf{F}_{12}$ is the force exerted by body 2 on body 1, and $\mathbf{F}_{21}$ is the force exerted by body 1 on body 2, then Newton's law may be mathematically stated:

$$\mathbf{F}_{12} = -\mathbf{F}_{21}$$

where the negative sign indicates that the forces are in opposite directions.

Let us now consider the interaction of two bodies. The forces of interaction may be due to spring connections, to electric charges, to collisions, etc. If the forces of interaction are the only forces acting on the bodies, the system is a closed or isolated one. From Newton's second and third laws, the important principle of conservation of momentum can be derived for an isolated system.

Consider an isolated system consisting of two particles m_1, m_2 moving with velocities $\dot{\mathbf{x}}_1$, $\dot{\mathbf{x}}_2$ respectively in which the forces of interaction are $\mathbf{F}_{12}$ and $\mathbf{F}_{21}$. Then by Newton's third law

$$\mathbf{F}_{12} = -\mathbf{F}_{21}$$

and by Newton's second law

$$\frac{d}{dt}(m_1\dot{\mathbf{x}}_1) = -\frac{d}{dt}(m_2\dot{\mathbf{x}}_2)$$

or

$$\frac{d}{dt}(m_1\dot{\mathbf{x}}_1 + m_2\dot{\mathbf{x}}_2) = 0$$

But $m_1\dot{\mathbf{x}}_1$ is the momentum of particle 1, and $m_2\dot{\mathbf{x}}_2$ that of particle 2, so that the rate of change of the total momentum is zero and independent of time. This is the principle known as the conservation of momentum. It may be stated as follows.

> *If no external forces act on a system, that is, if the system is isolated, the total momentum in any direction remains constant.*

Notice that in the discussion of mass in Section 2.2, Newton's third law was assumed; in fact, there is no completely logical argument that separates mass and force.

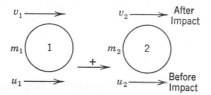

Fig. 2.3 Collision of two spheres and conservation of momentum.

As an example of the interaction between two bodies, consider the impact of two spheres moving along the same straight line. Let the spheres have masses m_1, m_2 and velocities u_1, u_2 before impact and v_1, v_2 after impact, as shown in Fig. 2.3. If the positive direction of motion is to the right, then all the velocities shown are positive. In order for there to be an impact, u_1 must be larger than u_2, and, if the spheres are to separate, v_2 must be larger than v_1. The forces at impact on the two masses are

$$\mathbf{F}_{12} = -\mathbf{F}_{21}$$

By Newton's second law

$$\frac{m_1(v_1 - u_1)}{\Delta t} = -\frac{m_2(v_2 - u_2)}{\Delta t}$$

where Δt is the short time interval of impact during which the velocity of m_1 changes from u_1 to v_1 and of m_2 changes from u_2 to v_2. Transposing and canceling, we have

$$m_1 u_1 + m_2 u_2 = m_1 v_1 + m_2 v_2 \tag{2.3}$$

The left-hand side of the equation represents the total momentum of the two balls before collision, and, similarly, the right-hand side the total momentum after the collision. Thus there is no momentum lost in the collision; that is, the momentum is conserved at the collision.

One cannot overemphasize the importance of the law of *conservation of momentum* in science. It is applied universally in physics whether such large objects as stars or such submicroscopic objects as atoms are dealt with. While Newton's third law is valid for all contact forces and objects at rest it is not valid in some problems in electromagnetism, although the law of conservation of momentum appears to hold in all situations. Momentum is a vector quantity and has the same direction as the velocity of the object. For a shell exploding in the air, the momentum of the shell immediately before the explosion is equal to the total momentum of the system just after the explosion. Fragments of the shell are presumably sent in every direction. In any direction the total momentum of the fragments after the explosion is equal to the momentum of the shell in the given direction before the explosion. If just before the explosion the shell is moving downward, then after the explosion the total momentum of the fragments added vectorially would be downward and the total momentum in the horizontal direction would be zero.

Before procceding with an example of the components of momentum, it is worth while noting that Eq. 2.3 does not give the velocities v_1, v_2 of the two masses, since a single equation can give the value of only one unknown. Further information concerning the impact is necessary if these velocities are to be found. This information is given by what Newton termed the *coefficient of restitution e*, which is defined as the ratio of the relative velocity of separation to the relative velocity of approach. For the two impacting balls shown in Fig. 2.3, the coefficient of restitution e is

$$e = \frac{v_2 - v_1}{u_1 - u_2} \tag{2.4}$$

There are two limiting values for e, namely $e = 0$ and $e = 1$. An impact for which $e = 0$ is called an *inelastic impact*, and $v_2 = v_1$ or the bodies stick together after impact. The other case, $e = 1$, is an ideal one never actually attained in the macroscopic world. For this case the bodies would move apart with the same relative velocity as that of approach, and the impact would be *perfectly elastic*.

EXAMPLE. As an example of impact consider two smooth balls A and B. Suppose that ball A has a mass of 100 gm and on impact an initial velocity of 20 cm/sec at an angle of 30° with the line of centers. Similarly, ball B has a mass of 50 gm and a velocity of 10 cm/sec at an angle of 60° with the line of centers. The velocities after impact are given in Fig. 2.4. With a coefficient of restitution of 0.9 between the balls, the problem is to find the velocities after the impact.

Since the spheres are assumed to be smooth, there is no force perpendicular to the line of centers. Thus the component of the momentum of either ball does not

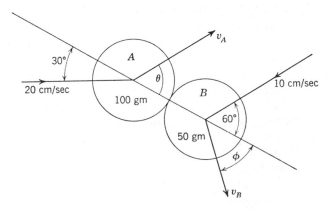

Fig. 2.4 Collision of two smooth spheres.

change along this direction. Hence, resolving the velocities of momenta along the line perpendicular to the line of centers of impact, we have for ball A

$$20 \sin 30° = v_A \sin \theta \quad \text{or} \quad v_A \sin \theta = 10 \text{ cm/sec}$$

and for ball B

$$10 \sin 60° = v_B \sin \phi \quad \text{or} \quad v_B \sin \phi = 5\sqrt{3} \text{ cm/sec}$$

By the principle of conservation of momentum, there is no change at impact in the total momentum of the two balls in any direction. The most convenient direction to choose is that along the line of centers, and from the law of conservation of momentum it follows that

$$2000 \cos 30° - 500 \cos 60° = 100v_A \cos \theta + 50v_B \cos \phi$$

or

$$2v_A \cos \theta + v_B \cos \phi = 20\sqrt{3} - 5 = 29.64$$

From the definition of the coefficient of restitution it follows that

$$0.9 = \frac{v_B \cos \phi - v_A \cos \theta}{20 \cos 30° + 10 \cos 60°}$$

or

$$v_B \cos \phi - v_A \cos \theta = 9\sqrt{3} + 4.5 = 20.09 \text{ cm/sec}$$

There are now four unknowns and four independent equations. Solving these equations gives

$$\tan \theta = 3.142 \qquad \theta = 72° 21' \qquad v_A = 10.48 \text{ cm/sec}$$
$$\tan \phi = 0.372 \qquad \phi = 20° 24' \qquad v_B = 24.80 \text{ cm/sec}$$

Hence, as a result of the collision, ball A and ball B move approximately in the directions shown in Fig. 2.4.

2.8 Momentum Change and Impulse

In practice there are many examples of a large force acting for a very short time to produce a momentum change in some object. A few familiar examples are a hammer striking a blow, a bullet propelled from a gun, and a charge of electricity setting the coil of a ballistic galvanometer in motion. Even though the variation of the force with time may not be known, the force-time integral is a definite quantity. If the variation of force with time is that shown in Fig. 2.5 the integral $\int \mathbf{F} \, dt$ is given by the area under the curve. This integral is called the *impulse* of the force. Suppose that at any time t a force $\mathbf{F}$ is acting on a particle mass m and velocity $\mathbf{v}$. By Newton's second law

$$\mathbf{F} = \frac{d(m\mathbf{v})}{dt} = \frac{d\mathbf{p}}{dt}$$

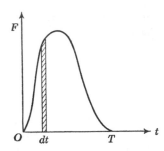

F

O dt T t

Fig. 2.5 Impulse represented as the area under a force-time curve.

By integration

$$\int_0^T \mathbf{F}\, dt = \int_{\mathbf{v}_0}^{\mathbf{v}} d(m\mathbf{v}) = \int_{p_0}^{p} d\mathbf{p}$$

where $\mathbf{v}_0$ is the velocity at the time $t = 0$ when the force starts to act, and $\mathbf{v}$ is the velocity at time $t = T$ when the force ceases to act. Thus if the mass m is constant, the impulse is

$$\int_0^T \mathbf{F}\, dt = m\mathbf{v} - m\mathbf{v}_0 = \mathbf{p} - \mathbf{p}_0 \qquad (2.5)$$

The impulse of a force acting on a particle is equal to the momentum change of the particle. Impulse is a vector quantity, as is momentum. It follows that the component of the impulse in any direction is equal to the momentum change in that direction. When applying the impulse-momentum relationship, one must be careful to maintain a consistent set of units, namely, those of Newton's second law.

As an example of the impulse-momentum relationship let us consider the motion of a rocket moving vertically upwards from the earth. In order to simplify the problem we shall assume that the acceleration due to gravity g is constant and that air resistance forces can be neglected. Suppose M_0 is the constant mass of the empty rocket and m_0 is the initial mass of the fuel. At some particular instant of time the mass of fuel remaining is m, and the total mass of the rocket at this instant of time $M = M_0 + m$. The fuel is lighted at the tail and the hot gases rush out with a velocity u_0, assumed constant, relative to the rocket. The rocket (Fig. 2.6) is moving relative to the earth with a velocity v in a positive upward direction so that the velocity of the hot gases relative to the earth is $(v - u_0)$ upward. Suppose that in a small time interval dt a mass of fuel dm is burned and the velocity of the rocket is increased by an amount dv. The mass of fuel $dm = dM$ burned is equal to the mass of hot gases ejected during the time interval dt. By the impulse-momentum relationship the momentum change of the rocket $d(Mv)$ minus the momentum change $(v - u_0)\, dM$ of the hot gases is equal to the impulse, $-Mg\, dt$, during the interval, or

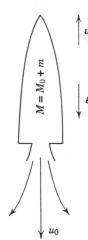

$M = M_0 + m$

v

g

u_0

Fig. 2.6 Rocket in vertical flight where m is the mass of the fuel and M_0 is the mass of the empty rocket.

$$-Mg\, dt = d(Mv) - (v - u_0)\, dM$$

$$= M\, dv + v\, dM - v\, dM + u_0\, dM$$

$$-g\, dt = dv + u_0 \frac{dM}{M}$$

The conditions at $t = 0$, when the rocket is fired, are $v = 0$, $M = M_0 + m_0$.

$$-g \int_0^t dt = \int_0^v dv + u_0 \int_{M_0+m_0}^M \frac{dM}{M}$$

or

$$-gt = v + u_0[\ln M - \ln (M_0 + m_0)] = v + u_0 \ln \left[\frac{(M_0 + m)}{(M_0 + m_0)}\right]$$

or the vertical velocity v at time t is

$$v = u_0 \left[\ln \frac{(M_0 + m_0)}{(M_0 + m)}\right] - gt$$

The maximum velocity the rocket can attain under the conditions of the problem occurs at time T, when all the fuel is burned, i.e., $m = 0$, thus:

$$v_{\max} = u_0 \left[\frac{\ln (M_0 + m_0)}{M_0}\right] - gT$$

2.9 Examples Involving Newton's Second Law

Our first problem is to determine the acceleration of two unequal masses hanging on the ends of a cord passing over a pulley (Fig. 2.7). Such a system is known as an Atwood machine after George Atwood, a professor at Cambridge University at about 1780. For simplicity, we shall make the following assumptions about this apparatus.

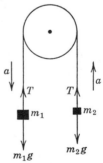

1. The cord is weightless.
2. The cord is perfectly flexible and inextensible.
3. The pulley is weightless and there is no friction at the center of the pulley.

Obviously these assumptions are impossible to attain in practice, but they may be so approximated that there is only a small error introduced.

Fig. 2.7 Atwood's machine.

Let us assume that the mass m_1 is larger than m_2. Then m_1 has a downward acceleration and m_2 an equal upward acceleration, since the string is inextensible. The forces T marked on the cords represent the forces exerted by the cords on the masses m_1, m_2. It requires all three assumptions to justify considering the tensions equal.

Consider first the mass m_1 on which there is a resultant force $(m_1g - T)$ acting downward. The downward acceleration a of mass m_1 is given by Newton's second law as

$$(m_1g - T) = m_1a$$

Similarly, for m_2 with its upward acceleration a and upward resultant force $(T - m_2 g)$, we have

$$T - m_2 g = m_2 a$$

Adding these two equations gives

$$a = \frac{(m_1 - m_2)g}{(m_1 + m_2)}$$

This equation could have been written down immediately for the system as a whole, since the resultant force is $(m_1 - m_2)g$ and the total mass moved is $(m_1 + m_2)$.

By eliminating the acceleration a, the tension T in the cord is

$$T = \frac{2m_1 m_2 g}{(m_1 + m_2)}$$

This tension must have a value between $m_1 g$ and $m_2 g$ and be expressed in units of either dynes or newtons if the metric system is employed.

In the British engineering system the above equations would be written

$$w_1 - T = \frac{w_1 a}{g} \qquad T - w_2 = \frac{w_2 a}{g}$$

$$a = \frac{(w_1 - w_2)g}{(w_1 + w_2)} \quad \text{and} \quad T = \frac{2w_1 w_2}{(w_1 + w_2)}$$

In this example T is in pounds weight since w is in pounds weight. The force on the axle at the center of the pulley is $2T$, which is less than $(m_1 + m_2)g$. Give the physical reason for this.

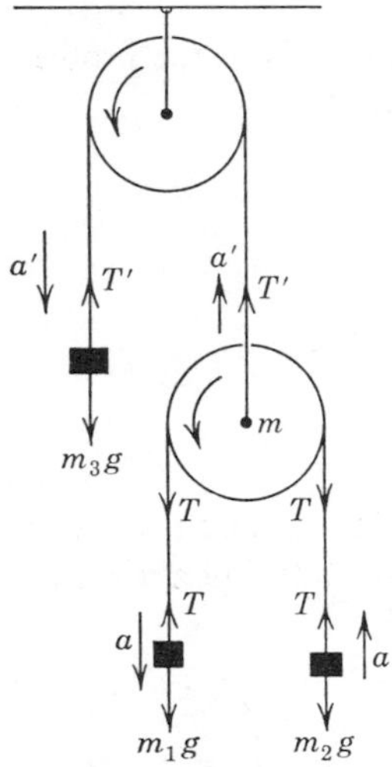

As a modification of Atwood's machine, let us consider the system shown in Fig. 2.8. All assumptions given above apply here except that we shall now consider the movable pulley to have a mass m but to have a negligible rotational inertia. Suppose the mass m_3 to be sufficiently large to move downward with an acceleration a'. Then the movable pulley of mass m moves upward with the acceleration a'. If m_1 is larger than m_2, then, relative to the center of the moving pulley, m_1 moves downward with an acceleration a and m_2 moves upward with the same acceleration.

Since the center of the moving pulley has an upward acceleration a', the acceleration of m_1 relative to a stationary object is $(a - a')$ downward and that of m_2 is similarly $(a + a')$ upward. By Newton's second law, the equations of motion are:

Fig. 2.8 Linear acceleration in a modified Atwood machine.

For m_3: $(m_3 g - T') = m_3 a'$

For m_1: $(m_1 g - T) = m_1(a - a')$

For m_2: $(T - m_2 g) = m_2(a + a')$

For m: $(T' - 2T - mg) = ma'$

If the movable pulley had zero mass, $m = 0$, T' would be equal to $2T$. Since this is not so, there must be a resultant force acting on m to give it the upward acceleration a'. It can be seen that in this problem a carefully drawn diagram with the forces and accelerations correctly placed is almost indispensable.

From the above equations involving m_1 and m_2, it follows that the acceleration a is:

$$a = \frac{m_1 - m_2}{m_1 + m_2}(g + a')$$

Compare this equation with the equation for a in the simple Atwood machine.

As another application of Newton's second law, let us determine the acceleration of a body on a frictionless inclined plane. This assumption means that there is no friction force parallel to the plane opposing the motion of the mass. Since the mass moves down the plane, its acceleration a is also in this direction. The resultant force along the plane is the component of the weight mg in this direction. By resolving the vertical force mg along and at right angles to the plane, we find that the component along the plane is $mg \sin \theta$ and the component at right angles to the plane is $mg \cos \theta$, as shown in Fig. 2.9. This latter component is balanced by the force N, called the normal reaction exerted by the plane on the mass m. Since the mass m has no motion at right angles to the plane, it follows from Newton's second law that

$$mg \cos \theta - N = 0$$

The equation of motion of the mass m down the plane is

$$mg \sin \theta = ma$$

or

$$a = g \sin \theta$$

It is impossible to have a frictionless plane. There is always a frictional force present, and this we shall now discuss.

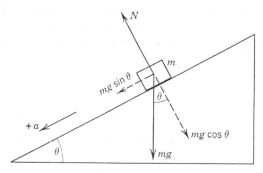

Fig. 2.9 Acceleration on a frictionless inclined plane.

2.10   The Force of Friction

In general, when any two objects are in contact and the one is moved with respect to the other, there is a force at the plane of contact opposing the motion. This force is called the force of friction. Though the force of friction has some annoying aspects, it also has many useful ones. Without friction we could not walk, automobile wheels would spin around without producing any forward motion, and power could not easily be transmitted from a motor to a machine. On the other hand, friction produces power losses in moving machinery. Mechanical energy is always transformed into heat by the work done against the force of friction.

It appears that Leonardo da Vinci (1452–1519), with amazing practical genius and insight, was the first to present the laws of friction. They may be stated in the following manner: The force of friction is (a) independent of the area of the sliding surfaces and (b) proportional to the load between them. Apparently these laws were not commonly known, for in 1699 a French engineer, Amontons, independently discovered them. About 1781 Coulomb verified Amonton's observations and also made a clear distinction between the force of static friction, that is, the force required to start sliding, and kinetic or sliding friction, that is, the force required to maintain sliding. It should be recognized that neither static nor sliding friction is constant. Static friction increases with the time of contact of the two objects while sliding friction decreases with the speed of travel.

To understand some of the characteristics of the force of friction, let us consider a block placed on a horizontal surface, as in Fig. 2.10. A horizontal cord is attached to the block, and the tension in it is increased until it reaches some value F_r at which sliding commences. If the force is smaller than F_r, no apparent motion results. When sliding starts without any appreciable acceleration, the frictional force F_r is equal to the applied force F. The normal reaction N exerted by the surface on the block is equal to the weight of the block in this example. The results of experiments show that, for sliding, there

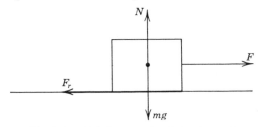

Fig. 2.10   Friction on a horizontal plane.

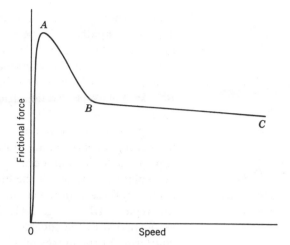

Fig. 2.11   Change in the frictional force with speed of sliding (Region OA greatly exaggerated along the speed axis.)

is an approximately constant ratio between the force of friction F and the normal reaction N. This ratio is called the coefficient of friction and is denoted by the symbol μ.

$$\mu = \frac{F}{N} \qquad \text{or} \qquad F = \mu N \qquad\qquad (2.6)$$

If a weight is placed on the upper surface of the block, both N and F are increased in the same ratio so that F/N or μ continues to have the same value.

It should be noted that the force $\mathbf{F}$ in $\mathbf{F} = \mu N$ is derived empirically and is not a fundamental law of physics in the sense in which Newton's second law, $\mathbf{F} = d\mathbf{p}/dt$, is.

During the present century there has been considerable experimental work on friction with a corresponding understanding of the phenomena underlying it. Experiments have shown that adhesion takes place between metal surfaces. This adhesion is considered to be the principal source of the frictional force, as opposed to a theory of roughness, since smooth surfaces can exhibit considerable frictional forces. Friction is a complex process; for instance it has been shown that there is really no such phenomenon as static friction. As the applied force is increased, the material at the surface of contact actually changes shape or creeps. Although this creep may be too small to measure for most metals, it has been exhibited in soft metals such as lead for speeds up to about 10^{-6} cm/sec. This is represented by the region OA in Fig. 2.11. It is found in some cases that the force represented by OA increases with the length of time the two surfaces are left in contact.

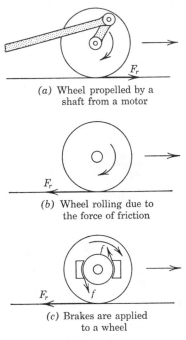

(a) Wheel propelled by a
shaft from a motor

(b) Wheel rolling due to
the force of friction

(c) Brakes are applied
to a wheel

Fig. 2.12 The frictional force F_r on a wheel under different conditions. In (c) if the braking forces f are large compared to the frictional forces F_r the wheel may lock and skidding take place.

Once sliding starts, the frictional force decreases rapidly with speed as shown in the region AB of Fig. 2.11. With further increase in speed the frictional force decreases very little, as shown in the region BC. In Fig. 2.11 "static" friction corresponds to point A while the dynamic or sliding friction corresponds to point B. An interesting phenomenon known as stick-slip or oscillations induced by friction can be explained in terms of the change in the frictional force with the change in speed in the region AB in Fig. 2.11. Stick-slip is associated with the squeals, squeaks, and chattering of the hinges of a slowly closed door, of a badly adjusted lathe tool, of automobile tires in a fast, sharp turn, etc. A piece of dry chalk held almost perpendicular to and pressed hard against a blackboard while being pushed along gives rise to a series of dots on the board and at the same time produces a high squeaking or squealing sound. For more information on this and other problems of friction consult the references given in the footnote.*

Figure 2.12 shows how the force of friction between the wheels of a locomotive and the ground changes during the operations of propulsion and braking.

2.11   Motion of Projectiles

Here the problem is to find the time of flight and range on a horizontal plane of a particle initially projected at some angle with the horizontal. Suppose that a particle of mass m is projected with a velocity v at an angle α with the horizontal (Fig. 2.13). We shall make the following assumptions in this problem.

* "Resource Letter F-1 on Friction." E. Rabinowicz, *Am. J. Phys.* **31**, 897, 1963, available from the Am. Inst. of Phys. New York. This gives a large number of references with some discussion of the problems of friction. Also by the same author an article on "Stick and Slip," *Sci. Am.*, May 1956.

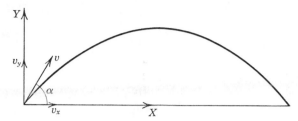

Fig. 2.13 Parabolic path of a projectile, assuming no air resistance.

1. The value of the gravitational acceleration g is constant throughout the motion.

2. The forces due to air resistance can be neglected.

3. Any effects due to the rotation of the earth may be neglected.

If F_x is the component of force acting on the particle along the X axis and F_y is the component along the Y axis, then, taking into consideration these three assumptions and Newton's second law, the equations of motion of the particle are

$$F_x = m\ddot{x} = 0$$

$$F_y = m\ddot{y} = -mg$$

Integrating the first of the equations gives

$$\dot{x} = C_1 \quad \text{and} \quad x = C_1 t + C_2$$

The initial velocity of the particle along the X axis is the resolved part of the velocity v along this direction. Thus at $t = 0$

$$\dot{x} = v \cos \alpha$$

Hence the constant C_1 is $v \cos \alpha$; that is, the horizontal velocity is constant. If the origin of the coordinates is at the point of projection, then, at $t = 0$, $x = 0$, and the constant $C_2 = 0$. The horizontal distance traveled in time t is

$$x = vt \cos \alpha \tag{2.7}$$

For motion in the positive Y direction we have

$$\ddot{y} = -g \quad \dot{y} = -gt + C_1$$

From the initial conditions of motion, $\dot{y} = v \sin \alpha$ when $t = 0$. Hence

$$C_1 = v \sin \alpha$$

or

$$\dot{y} = -gt + v \sin \alpha \tag{2.8}$$

Integrating again gives

$$y = -\frac{gt^2}{2} + vt \sin \alpha + C_2 \tag{2.9}$$

At $t = 0$, $y = 0$. Hence $C_2 = 0$.

The equation of the path traveled by the particle is obtained by eliminating the time t between Eqs. 2.7 and 2.9, giving

$$y = x \tan \alpha - \frac{gx^2}{2v^2 \cos^2 \alpha} \tag{2.10}$$

Since for any particular projection v, g, and α are constants, it follows that Eq. 2.10 represents a parabola.

The time t_f taken for the projectile to attain its maximum height is the time taken for the vertical velocity to become zero. From Eq. 2.8

$$t_f = \frac{v \sin \alpha}{g}$$

The total time the particle is in flight is $2t_f$, since the time of descent is equal to that of ascent. During this time the particle has been moving with a constant horizontal component of velocity $v \cos \alpha$ so that the total horizontal distance traveled by the particle, or its horizontal range R, is

$$R = 2t_f v \cos \alpha = \frac{2v^2 \sin \alpha \cos \alpha}{g} = \frac{v^2 \sin 2\alpha}{g} \tag{2.11}$$

The maximum range occurs when $\sin 2\alpha$ has its maximum value of unity, that is, when

$$2\alpha = 90° \quad \text{or} \quad \alpha = 45°$$

Thus for any angle of projection greater or less than 45° the range is smaller than the maximum value.

2.12   The Inclined Plane with Friction

Suppose that a body of mass m_1 is placed on an inclined plane whose angle of inclination is α. From the mass m_1 a cord is attached which passes over a frictionless peg to a hanging mass m_2 (Fig. 2.14). The coefficient of friction between the mass m_1 and the plane is μ. In any numerical problem one first has to determine the direction of motion of the masses. Depending on the relative values of m_1 and m_2 and the value of μ, there may be either no motion or an acceleration up or down the plane. To be specific let us

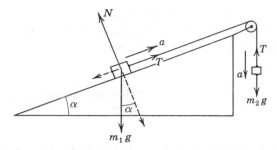

Fig. 2.14 Motion on an inclined plane.

assume that the mass m_1 is accelerated up the plane with an acceleration a. Then, assuming that the cord is inextensible, the mass m_2 has the same acceleration vertically downward. The force of friction which opposes the motion is

$$\mu N = \mu m_1 g \cos \alpha$$

By Newton's second law the equation of motion of m_1 is

$$T - m_1 g \sin \alpha - \mu m_1 g \cos \alpha = m_1 a$$

and of m_2 is

$$m_2 g - T = m_2 a$$

Hence

$$a = \frac{(m_2 - m_1 \sin \alpha - m_1 \mu \cos \alpha) g}{(m_1 + m_2)}$$

2.13 Motion in a Resisting Medium

When an object moves through a medium such as air, it is subjected to a resistance force. This resistance force is a function of the velocity and the size and shape of the moving object. There is no simple relationship between the velocity and the force, but at very low velocities it appears that the force varies approximately as the velocity, whereas at higher velocities it appears to be approximately proportional to the square or even higher powers of the velocity. In any practical situation the relationship between the resistance force and the velocity has to be obtained experimentally.

Without air resistance a parachute would be useless. When a man jumps from a plane with his open parachute, he is accelerated vertically until he reaches such a velocity that the air resistance force exactly equals the total weight of man and parachute. This maximum velocity is called the *terminal velocity* of the system. Suppose that the mass of the falling object is m and, at any particular velocity v, the air resistance force is mkv, where k is a constant of proportionality with dimensions $[T^{-1}]$. At the terminal velocity v_t, the resistance force is exactly equal to the weight:

$$mkv_t = mg$$

or

$$v_t = \frac{g}{k} \qquad (2.12)$$

For a downward velocity v less than v_t, the body has a downward acceleration dv/dt, and the equation of motion is, from Fig. 2.15,

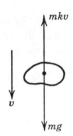

$$mg - mkv = m\frac{dv}{dt}$$

where the downward direction is taken as positive. Thus

$$\frac{dv}{g - kv} = dt$$

If we assume the acceleration of gravity g to be constant, the integration of this equation gives

$$-\frac{1}{k}\ln(g - kv) = t + C$$

Fig. 2.15 Motion against air resistance.

where ln is used for $\log_e$.

Suppose that the body starts from rest. Then, at $t = 0$, $v = 0$, and $C = -1/k \ln g$, so that

$$-\frac{1}{k}\ln\left(\frac{g - kv}{g}\right) = t$$

or

$$g - kv = ge^{-kt}$$

The velocity v at any time t is

$$v = \frac{g}{k}(1 - e^{-kt}) \qquad (2.13)$$

Thus the velocity starts from zero and increases until after an infinite time, when the exponential term becomes zero, the velocity reaches a constant or terminal value of v_t, as shown in Fig. 2.16.

Suppose that the body starts from some height h above the ground. The velocity v may be written as dy/dt, where y is measured downward with its origin at the point of fall. Thus from Eq. 2.13,

$$\frac{k}{g}dy = (1 - e^{-kt})dt$$

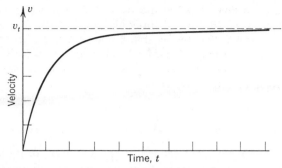

Fig. 2.16   Velocity plotted against time for motion in a resisting medium.

and by integration

$$\frac{k}{g} y = t + \frac{e^{-kt}}{k} + C$$

Since $y = 0$ at $t = 0$,

$$C = -\frac{1}{k}$$

and the distance y fallen through in time t is given by

$$\frac{k}{g} y = t + \frac{e^{-kt}}{k} - \frac{1}{k}$$

or

$$y = \frac{g}{k^2} (e^{-kt} + kt - 1) \tag{2.14}$$

If the body started from a height h above the ground and reached the ground in time T, then

$$h = \frac{g}{k^2} (e^{-kT} + kT - 1)$$

or the time of fall T is

$$T + \frac{e^{-kT}}{k} = \frac{kh}{g} + \frac{1}{k} \tag{2.15}$$

Possibly the simplest way to solve for the time T is to proceed by trial and error. Suppose that we wish to calculate the time of fall of a body from the height of 10,000 ft, with the value of the air resistance constant equal to 0.1 sec^{-1}. If there were no air resistance, the time of fall would be given by the equation

$$h = \tfrac{1}{2} g t^2$$

or

$$t^2 = \frac{2 \times 10^4}{32} \qquad \text{giving} \quad t = 25 \text{ sec}$$

With air resistance the time of fall would be larger. As a first guess, let us take the time T as 40 sec. Then $40 + 10e^{-4}$ should be equal to $(10^3/32) + 10$, or 41.25. From the table of exponentials the left-hand side of Eq. 2.15 is found to equal 40.18, which is less than the number of seconds required. Since the exponential term yields only a small number, then, for a second guess, let us take $T = 41.1$ sec. With this value for T the left-hand side is

$$41.1 + 10e^{-4.11} = 41.1 + 0.16 = 41.26$$

This is sufficiently close to the value of 41.25 on the right-hand side to be acceptable.

If the particle were projected vertically upward with an initial velocity V and were subjected to a resistance force of mkv, the equation of motion would be

$$m\frac{dv}{dt} = -mg - mkv \qquad \text{or} \qquad \ddot{y} = -g - k\dot{y}$$

where v is the velocity at any time t. In this example the origin is taken at the ground and the upward direction is taken as positive. This equation can be integrated if the initial conditions are known. This would give the maximum height attained by the particle and also the time taken to reach the maximum height. The result of a similar integration is given by Eq. 2.17.

Example of a projectile moving in a resisting medium. If a projectile is shot with a velocity V at an angle α with the horizontal and is subject to a retardation of kv due to air resistance, then it no longer follows a parabolic path. The equations of motion along the horizontal or X axis and the vertical or Y axis are

$$m\ddot{x} = -mk\dot{x} \qquad \text{or} \qquad \ddot{x} = -k\dot{x}$$

and

$$m\ddot{y} = -mg - mk\dot{y} \qquad \text{or} \qquad \ddot{y} = -g - k\dot{y}$$

Integrating the first of these equations with the conditions that, at time $t = 0$, $\dot{x} = V\cos\alpha$, we have

$$\int\frac{d\dot{x}}{\dot{x}} = \int -k\,dt + C$$

or

$$\ln\dot{x} = -kt + \ln(V\cos\alpha)$$

or

$$\dot{x} = (V\cos\alpha)e^{-kt}$$

The distance traveled in a time t is by integration

$$x = \frac{-V\cos\alpha}{k}e^{-kt} + C$$

If $x = 0$ at $t = 0$, then

$$C = \frac{V\cos\alpha}{k}$$

and

$$x = \frac{V \cos \alpha}{k}(1 - e^{-kt}) \qquad (2.16)$$

The elevation y of the projectile at any time t can be found by integration to be

$$y = \frac{1}{k^2}(g + kV \sin \alpha)(1 - e^{-kt}) - \frac{gt}{k} \qquad (2.17)$$

since the initial velocity is $V \sin \alpha$. Eliminating the time t between Eqs. 2.16 and 2.17, we obtain the equation of the path as

$$y = \frac{1}{k^2}(g + kV \sin \alpha)\left(\frac{kx}{V \cos \alpha}\right) - \frac{g}{k^2} \ln\left(\frac{V \cos \alpha}{V \cos \alpha - kx}\right)$$

or

$$y = \frac{x}{V \cos \alpha}\left(V \sin \alpha + \frac{g}{k}\right) + \frac{g}{k^2} \ln\left(1 - \frac{kx}{V \cos \alpha}\right)$$

If the right-hand side of this equation is expanded in powers of kx, then

$$y = \frac{x}{V \cos \alpha}\left(V \sin \alpha + \frac{g}{k}\right) + \frac{g}{k^2}\left(-\frac{kx}{V \cos \alpha} - \frac{k^2 x^2}{2V^2 \cos^2 \alpha} - \frac{k^3 x^3}{3V^3 \cos^3 \alpha} - \cdots\right)$$

or

$$y = x \tan \alpha - \frac{gx^2}{2V^2 \cos^2 \alpha} - \frac{gkx^3}{3V^3 \cos^3 \alpha} - \frac{gk^2 x^4}{4V^4 \cos^4 \alpha} - \cdots \qquad (2.18)$$

which, when $k = 0$, reduces to Eq. 2.10, the equation for the parabolic path of a projectile without air resistance. The path of the projectile given by Eq. 2.18 is no longer parabolic, since for each value of k there is a differently shaped path. It can be shown that the angle at which the projectile strikes the ground is greater than the angle of projection α. The slope of the path at any point is given by the value of dy/dx at that point. Thus it is readily seen that at the point of projection, $x = 0$,

$$\left(\frac{dy}{dx}\right)_{x=0} = \tan \alpha$$

Although the assumption that the resistance force is proportional to the velocity is not completely justified, in practice this analysis does give a first approximation to the path followed by a projectile in air.

2.14 Relativity of Mass

According to the relativity of Galileo and Newton for a coordinate system S' moving along the x axis with a constant speed v relative to system S, the following relationships hold:

$$x' = x - vt; \qquad t = t'; \qquad m = m'$$

If Newton's second law, $F = m\ddot{x}$, is valid in system S then the corresponding equation in system S' is $F' = m\ddot{x}'$ since $\ddot{x} = \ddot{x}'$ or Newton's second law is invariant in form in a Galilean transformation. That is, if S is an inertial system then S' is also an inertial system or there is no unique inertial system.

In Chapter 1 the Lorentz transformation equations gave the relationships between x and x' and t and t' in the relativity of Einstein. Another change, which Einstein gave in 1905, consisted in showing that the mass or inertia of an object increases rapidly as the speed of the object approaches the speed of light, the actual relationship being

$$m = m_0/\sqrt{1 - v^2/c^2} \qquad (2.19)$$

where m is the mass of the object moving with speed v and m_0, called the rest mass, is the mass when the object is at rest or moving with a speed $v \ll c$ the speed of light. Since about 1930 there is considerable evidence from high energy particle experiments that the mass of an object does vary according to Eq. 2.19.

In the Einstein relativity the linear momentum is given by $\mathbf{p} = m\mathbf{v}$ and the force by $\mathbf{F} = d\mathbf{p}/dt$. However, since m varies with $\mathbf{v}$ it is no longer valid to write the equation of motion as $\mathbf{F} = m\mathbf{a}$, although it is assumed that the law of conservation of momentum is obeyed in the mechanics of Einstein.

In order to establish Eq. 2.19 we shall analyze an impact by a method originally given by Tolman and Lewis.* Consider two coordinate systems S and S' with S' moving in the x, x' direction with velocity $\mathbf{v}$ relative to system S (Fig. 2.17a). Along the y and y' axes are two balls having equal rest masses m_0. Ball A has a velocity $\mathbf{u}$, measured in system S, along the y axis and ball B

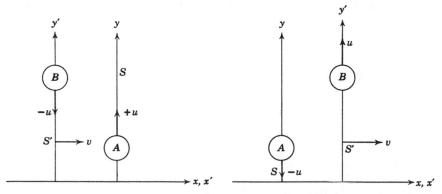

Fig. 2.17.   Relativistic collision of two similar balls. (a) Before collision. (b) After collision.

* "The Principle of Relativity, and Non-Newtonian Mechanics." G. N. Lewis and R. C. Tolman. *Phil. Mag.*, **18**, 510–523, 1909.

a velocity $-\mathbf{u}$, measured in system S', along the y' axis. We shall assume that $\mathbf{u}$ is small compared to the relative velocity $\mathbf{v}$. This thought or gedanken experiment is so arranged that when the y' axis of S' coincides with the y axis of S, the two balls collide elastically. Now an observer in the S coordinate system would see ball A travel up the y axis with velocity $+\mathbf{u}$ and rebound on collision down the y axis with velocity $-\mathbf{u}$, giving a total change in velocity of $-2\mathbf{u}$. Likewise an observer in the S' system would see ball B travel down the y' axis with velocity $-\mathbf{u}$ and rebound up the y' axis with velocity $+\mathbf{u}$, giving a total change of $+2\mathbf{u}$. That is, the collisions in S and S' are symmetrical. The velocity of ball B in S' is $-\mathbf{u}$ but this velocity when measured in S is $-\mathbf{u}\sqrt{1 - v^2/c^2}$ from Eq. 1.28. Similarly, after the collision the velocity of ball B in S', when measured in system S, is $+\mathbf{u}\sqrt{1 - v^2/c^2}$. Consider the collision as measured in S in which system the mass of ball A is m_0 but the mass of ball B measured in system S is some quantity m yet to be found. For this we assume the conservation of momentum at collision.

The momenta of balls A and B, before and after collision, when measured in system S, may be written as:

	Before Collision	After Collision
Momentum of ball A	$+m_0\mathbf{u}$	$-m_0\mathbf{u}$
Momentum of ball B	$-m\mathbf{u}\sqrt{1 - v^2/c^2}$	$+m\mathbf{u}\sqrt{1 - v^2/c^2}$

The momentum change of ball A is $-2m_0\mathbf{u}$
The momentum change of ball B is $2m\mathbf{u}\sqrt{1 - v^2/c^2}$

These momenta changes must be equal and opposite for the total change in momenta to be zero. Thus

$$m_0 = m\sqrt{1 - v^2/c^2} \qquad \text{or} \qquad m = \frac{m_0}{\sqrt{1 - v^2/c^2}} \qquad (2.19)$$

For a single particle the components of momentum are written as

$$p_x = \frac{m_0 v_x}{\sqrt{1 - v^2/c^2}} \; ; \qquad p_y = \frac{m_0 v_y}{\sqrt{1 - v^2/c^2}} \; ; \qquad p_z = \frac{m_0 v_z}{\sqrt{1 - v^2/c^2}} \qquad (2.20)$$

In the above it is stated that the velocity $\mathbf{u}$ of the balls should be small compared to the relative velocity $\mathbf{v}$ of the two coordinate systems S and S'. The reason for this is that we assumed that the mass of ball A in S is m_0, the rest mass of the ball, when actually the ball is moving with the velocity $\mathbf{u}$ relative to the coordinate system S. It is only when the velocity $\mathbf{u}$ is very small compared to the velocity of light c that the assumption is justified.

PROBLEMS

1. An Atwood machine has weights of 10 and 6 lbw respectively hanging on the ends of a cord passing over a pulley. Neglecting the masses of the cord and pulley and assuming no friction, find the acceleration of the system, the tension in the cord, and the force on the axle of the pulley. Give the physical reason for the force on the axle being less than 16 lbs.

2. A block of wood is started over a horizontal table with a speed of 40 ft/sec. If the coefficient of friction between the block and table is 0.4, find the distance the block moves before coming to rest.

3. A body of mass 0.8 kg is placed on an inclined plane of angle of inclination of 30°. A force of 0.8 kgw, at an angle of 60° with the plane, is pushing the body up the plane. (*a*) If the plane is frictionless, find the value of the acceleration up the plane. (*b*) Assuming a coefficient of friction of 0.2, find the acceleration. (*c*) If the force pushing the body up the plane is removed find the acceleration down the plane, assuming friction as in part *b*.

4. A periodic force *F* given in dynes varies in a manner shown in Fig. P4. Find

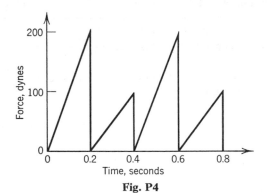

Fig. P4

the average force acting over a period of 0.4 sec. Interpret the result geometrically on the diagram.

5. A 6-lb body is placed on an inclined plane of angle of inclination of 30°. A force of 6 lb at an angle of 60° above the horizontal is pulling upwards on the upper edge of the body. (*a*) If the plane is frictionless, find the acceleration of the body. (*b*) If the coefficient of friction is 0.2, find the acceleration of the body up the plane.

6. A bullet is shot at an angle α with the horizontal along the line of maximum slope of a hill whose angle of inclination is β. If the bullet has an initial velocity of v, show that when air resistance is neglected the range of the bullet on the hill is

$$R = \frac{2v^2 \cos^2 \alpha}{g \cos \beta} (\tan \alpha - \tan \beta) = \frac{v^2}{g} \sec^2 \beta [\sin (2\alpha - \beta) - \sin \beta]$$

and that the range has a maximum value when $\alpha = 45° + \beta/2$, and that this maximum range is $v^2/[g(1 + \sin \beta)]$. If the bullet strikes the inclined plane at an angle ϕ with the horizontal, show that $\tan \phi = (dy/dx) = 2 \tan \beta - \tan \alpha$.

7. A bullet is shot at an angle of 60° with the horizontal along the line of maximum slope of a hill whose angle of inclination is 20°. If the bullet has an initial velocity of 1200 ft/sec, find the range of the bullet along the hill.

8. A bullet is shot with an initial speed of 1200 ft/sec at an angle of 60° with the horizontal from the top of a hill whose angle of inclination with the horizontal is 20°. If the bullet is shot along the line of greatest slope, find its range down the hill.

9. A block slides without friction from the top of a roof 18 ft long which is inclined at an angle of 30° with the horizontal. If the edge of the roof is 16 ft above the ground, find how far the block strikes the ground from the point directly below the edge of the roof.

10. A ball A is projected at an angle of 30° with the horizontal towards a ball B, and at the instant the ball A is projected the ball B starts to fall. Prove that the two balls will collide if the velocity of projection is large enough. If the initial distance between the balls is 4 ft and the initial velocity of A is 16 ft/sec at an angle of 30° with the horizontal, find (a) the length of time after projection when the balls collide; (b) the magnitude and direction of the velocity of ball A just before it collides with ball B.

11. Masses of 100 and 500 gm are connected by a cord which passes over a smooth peg, as shown in Fig. P11. Find the acceleration of the system and the tension in the cord if the coefficient of friction is 0.20.

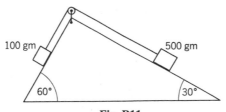

100 gm   500 gm

60°   30°

Fig. P11

12. A heavy rope of length L ft and weight w lb is hung over a small, smooth peg such that lengths $L/2$ hang over each side. One side of the rope is pulled down a small amount and let go so that the rope accelerates. Find the accelerating force when there is a length y longer on one side than on the other, and find the

expression for the displacement y as a function of the time t. (Assume a solution of the type $y = Ae^{\lambda t}$.)

13. A bullet whose mass is 2.5 gm has a velocity of 30,000 cm/sec and is horizontally fired into a block of wood of mass 500 gm lying at rest on a horizontal table. If the coefficient of friction between the block and the table is 0.40 and the bullet becomes embedded in the block, find (a) the velocity of the bullet and the block immediately after the bullet comes to rest in the block; (b) the distance moved by the block before it comes to rest. If the bullet penetrates 5 cm into the block of wood, find (c) the force exerted by the bullet on the block.

14. Suppose an object in space moving in the absence of any forces is collecting interplanetary debris at the rate of $kv = dm/dt$, where k is a constant. Show the deceleration of the object, assuming it moves in a straight line, is $-kv^2/m$.

15. In Fig. P15 the block of wood A of mass 2 kg is placed on a block B of mass 8 kg with block A connected by a light cord to a hanging mass of 4 kg. If the

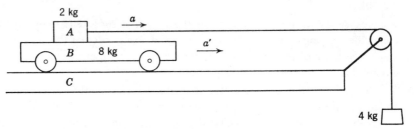

Fig. P15   Block A slides on block B.

coefficient of friction between A and B is 0.6 and zero friction between the block B and the table C, show that block A will slide along the block B. Find the accelerations of blocks A and B relative to C.

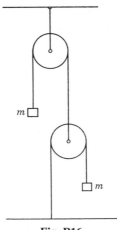

Fig. P16

16. A pulley system is set up as shown in Fig. P16 with equal masses on the ends of the cords. If each pulley has a negligible mass, show the acceleration of the mass on the left and the acceleration of the mass on the right are $g/5$ and $2g/5$ respectively.

17. If the movable pulley in problem 16 has the same mass as the hanging masses but negligible rotational inertia, show the accelerations of the mass on the left and right are $g/3$ and $2g/3$.

18. If the pulleys have negligible mass and there is no friction, show the accelerations of the 10- and 8-lb masses and the tension in the cord are $g/7$ and $2g/7$ and $T = 5.71$ lbw (Fig. P18).

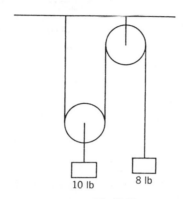

10 lb   8 lb

Fig. P18

19. A rocket shot vertically upwards in a uniform gravitational field, with air resistance neglected, uses up fuel at a constant rate such that the fuel is used up in 50 secs. The velocity of the escaping gases relative to the rocket is 7000 ft/sec and the ratio of the initial mass of fuel to the mass of the empty rocket is 300. Show the maximum velocity is attained in 50 sec and has a value of about 3.84×10^4 ft/sec.

20. A body is shot vertically upwards with an initial speed of u_0. If gravity is constant but there is a resistance force at speed v of mkv^2, show that the maximum height H which the body reaches is given by $2kH = \ln [1 + (ku_0^2/g)]$ and that the speed v_0 with which the body reaches the ground is given by $2kH = \ln [g/(g - kv_0^2)]$. Show that u_0 and v_0 are related by $1/v_0^2 = 1/u_0^2 + k/g$.

21. A bullet is shot vertically upward with an initial velocity of 1600 ft/sec. If it is subjected to an air resistance of mkv, where k has the value of 0.10 sec^{-1}, m is the mass, and v the velocity of the bullet, find (a) how long it takes to reach its maximum altitude; (b) how high it rises; (c) how long it takes to get from its maximum altitude to the ground; (d) the velocity with which it strikes the ground.

22. Repeat Problem 21 for a resistance force of mkv^2, where k is (a) 0.10 ft^{-1}; (b) 0.001 ft^{-1}.

23. A bullet is shot with an initial velocity of 1600 ft/sec at an angle of 30° with the horizontal. Assuming no air resistance, find (a) the range on a horizontal plane. If the air resistance force is mkv with $k = 0.1$ sec^{-1}, find (b) the maximum altitude attained; (c) the total time of flight; (d) the range on a horizontal plane.

24. Qualitatively explain how a simple pendulum motion involves the inertial and gravitation concepts of mass.

25. The system is balanced when the masses m_1, m_2 and M are at rest. If m_2 is larger than m_1, will the beam balance as m_2 descends? Find the relationship between the tension T and the total weight $(m_1 + m_2)g$ (Fig. P25).

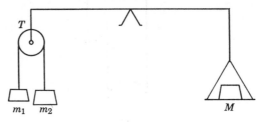

Fig. P25

26. Find the ratio of the masses m/m_0 corresponding to speed ratios of v/c equal to 0.1, 0.9, 0.99, 0.999.

27. Find the ratio v/c corresponding to mass ratios of m/m_0 of 1.04, 10, 100, 1000.

3

WORK AND ENERGY
OF PARTICLES

3.1 Introduction

The concept of energy is one of the most important in physics, with applications in every branch of science. We are so familiar with its manifestations and transformations in our daily life that they often evade our notice. Energy is now recognized in a great variety of forms—gravitational, electrical, chemical, thermal, nuclear, etc. The standard of living of man has always been dependent upon the extent to which he could utilize the various forms of energy. With the discovery by Einstein in 1905 that matter itself is a form of energy, science has progressed until our very civilization may be threatened by its uncontrolled use for military purposes.

It was in the 1840's that the principle of conservation of energy was first announced. This principle states that energy can never be destroyed; it can only be changed from one form to another. Though many new forms of energy have been recognized in the past 100 years, the principle of conservation of energy has never been invalidated.

The energy of a body is measured by the work the body is capable of doing. Work is done whenever a force moves a body; it is measured by the product of the force and the distance moved in the direction of the force. Though force and displacement are each vector quantities, work and energy are scalars with magnitude but not direction. Since we are not concerned with energy of rotation, we shall consider bodies to be particles in this chapter.

3.2   Definitions, Units, and Dimensions of Work

Suppose that a force F acts on a particle and moves it a distance dr in such a direction that the angle between the force and the displacement is θ (Fig. 3.1). Then the work W done by the force on the particle is

$$W = F \cos \theta \, dr \tag{3.1}$$

Since the quantity $F \cos \theta$ is the component of the force in the direction of the displacement dr, *the work done by any force is the product of the magnitude of the displacement and the magnitude of the component of the force in the direction of the displacement.*

Since $dr \cos \theta$ is the component of displacement in the direction of the force, Eq. 3.1 for work may be interpreted as the magnitude of the force multiplied by the component of the displacement in the direction of the force. If the force is at right angles to the displacement, no work is done by the force in this displacement. Thus no work is done by or against the force due to gravity when a body is moved in a horizontal direction. The work done by a force on a body can be positive, zero, or negative, according to whether the cosine term is positive, zero, or negative. Work is positive when the force has a component in the direction of the displacement, zero when the force is at right angles to the displacement, and negative when the force has a component in the direction opposite to the displacement. As an example of negative work, consider a body being moved along a rough horizontal plane. Work is done by the forward force against the backward force of friction. Thus the work done by the force of friction during the displacement is a negative quantity.

The common unit of work in the cgs system is the dyne centimeter which is called the *erg.* One erg of work is done on a body when a force of 1 dyne moves the body through a distance of 1 cm in the direction of the force. In the mks system, the unit of work is the newton meter which is called a *joule.* Thus 1 joule of work is done on a particle when a force of 1 newton moves the particle through a distance of 1 m in the direction of the force. Since 1 newton is equal to 10^5 dynes and 1 m is equal to 10^2 cm, it follows

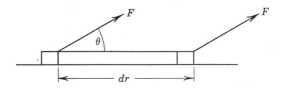

Fig. 3.1   Force F moving an object through a distance dr.

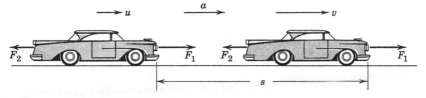

Fig. 3.2 Kinetic energy given to an automobile.

that 1 joule $= 10^7$ ergs. In the metric gravitational system, a unit of work is 1 gmw multiplied by 1 cm, or 1 gmw cm, where approximately 1 gmw cm = 980 ergs. In the British gravitational or engineering system, the unit of work is 1 lbw of force multiplied by 1 ft, 1 ft lbw. Thus 1 ft lbw of work is done on a particle when a force of 1 lbw moves the particle through a distance of 1 ft in the direction of the force.

The dimensions of work in the metric absolute system are $[MLT^{-2}L]$ or $[ML^2T^{-2}]$. In the engineering or gravitational system, force, length, and time are arbitrarily taken as fundamental, so the dimensions of work in this system are $[FL]$.

3.3   Kinetic and Potential Energy

Let us now consider the work done by a horizontal force moving a particle a horizontal distance. Since we are discussing linear and not rotational motion, consider this to be an automobile accelerating along a horizontal road. The forward force F is the resultant of a forward force F_1 and a backward resistance force F_2. Consider both of these forces to be horizontal so that

$$F = F_1 - F_2$$

Work is done by the forward force F_1 and against the background force F_2 during the displacement. The resultant force F gives the automobile an acceleration, and its speed is increased. Thus the work done by the resultant force goes into increasing the speed of the automobile or giving the automobile kinetic energy.

Suppose that the automobile has a mass m and an initial velocity u when the force F starts to act, so that the automobile has a constant acceleration a, and, after traveling a distance s, its velocity is v (Fig. 3.2). Assuming the motion is along the x direction, we may give the equation of motion as $F = m\ddot{x}$ and since $dx = \dot{x}\,dt$ it follows that

$$F\,dx = m\ddot{x}\dot{x}\,dt = \frac{m}{2}\frac{d}{dt}(\dot{x}^2)\,dt$$

Hence

$$Fs = \int_0^s F\,dx = \frac{m}{2}\int_u^v d(\dot{x}^2) = \tfrac{1}{2}mv^2 - \tfrac{1}{2}mu^2 \qquad (3.2)$$

The quantity $\tfrac{1}{2}mv^2$ is by definition the *linear* or *translational kinetic energy*, K.E., of a particle of mass m moving with a linear velocity v. Thus the work done by the accelerating force F on the car is equal to the increase in kinetic energy of the car. To stop the car, a braking or decelerating force is necessary. The work done by the braking force on the car is equal to its decrease in kinetic energy. To increase or decrease the kinetic energy of a particle, an accelerating or decelerating force must act through some distance.

The units and dimensions of kinetic energy are the same as those of work, as may be easily shown. In the British engineering system of units, the kinetic energy equation is written

$$Fs = \frac{1}{2}\frac{w}{g}(v^2 - u^2)$$

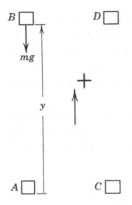

Fig. 3.3 Gravitational potential energy.

where w is the weight of the accelerated particle and F the resultant force acting on the particle. Both F and w are in the same units, namely, pounds weight.

Now a body at rest may possess energy owing to its position or configuration. A raised weight and an extended or compressed spring are examples of bodies possessing energy owing to their position; that is, they possess potential energy. Consider a body of mass m raised a vertical height y from position A to B (Fig. 3.3). The work done *against* the force of gravity in lifting the body from A to B is mgy, since the force due to gravity mg acts vertically downward and the upward vertical direction is taken as positive. In position B the mass m possesses gravitational potential energy mgy relative to position A. In other words, the difference in potential energy of the mass m in positions B and A is mgy. It is assumed that the body is at rest at A and B and that any displacement is made very slowly so that the kinetic energy is negligible. Gravitational potential energy increases upward while the force of gravity acts downward.

Let us now consider the work done against gravity when the mass m is taken along the rectangular path $ACDB$ from A to B. Along the horizontal paths AC and DB no work is done by or against the force of gravity since the displacements are horizontal. The only work done against gravity along the path $ACDB$ is mgy. No matter what path is chosen to take the mass m from A to B, it may be broken up into a series of horizontal and vertical paths.

The total vertical displacement will be AB or y, and the total work done against gravity mgy. When the mass m is moved from B back to A, an amount of work mgy is done by the force of gravity. There is a decrease in the potential energy of the mass m equal to mgy. If the work done against gravity is taken as positive, then the work done by gravity must be negative. Therefore the total work done against or by gravity when a body is taken around a closed path is zero. Any force of such a nature that the total work done against it in a closed path is zero is called a *conservative force*. The force of gravity is such a force. There are other conservative forces, a familiar one being the force exerted by an electrostatic field on an electrically charged body.

The change in potential energy of a body moved between two points in a conservative force field is the work done *against* the conservative force, or the negative of the work done by the conservative force, when the body is moved between the two points. Thus, if F_y is a conservative force, then, when a body is moved against this conservative force from position y_0 to y, the increase in potential energy V is given by

$$V = -\int_{y_0}^{y} F_y \, dy \tag{3.3}$$

and the conservative force F_y is

$$F_y = -\frac{dV}{dy} \tag{3.4}$$

In the gravitational case the difference in potential energy of a mass m moved a vertical distance y is

$$V = mgy$$

and

$$F_y = -\frac{dV}{dy} = -mg$$

where y is positive in an upward direction, while F_y which is equal to $-mg$ acts downward. This, of course, is in accord with Eq. 3.3.

3.4   The Energy Integral for Linear Motion

Let us assume that the accelerating force F on an object is a function only of the position of the object. That is, the force is independent of the velocity of the object or the time at which the object arrives at any position. The force depends only on the coordinate, the value of y, and to indicate this explicitly, the force F is written as $F(y)$. Thus:

$$m\frac{d^2y}{dt^2} = m\ddot{y} = F(y) \tag{3.5}$$

This equation may be integrated by first multiplying both sides by $\dot{y}\,dt = dy$, or

$$m\ddot{y}\dot{y}\,dt = F(y)\dot{y}\,dt$$

$$\frac{m}{2}\frac{d}{dt}(\dot{y}^2)\,dt = F(y)\,dy$$

Integration gives

$$\frac{m}{2}\int_{\dot{y}_0}^{\dot{y}} d(\dot{y}^2) = \int_{y_0}^{y} F(y)\,dy$$

where $\dot{y}_0$ is the velocity of the object at position y_0 and $\dot{y}$ is that at position y. Hence

$$\frac{m}{2}\dot{y}^2 - \frac{m}{2}\dot{y}_0^{\,2} = \int_{y_0}^{y} F(y)\,dy \tag{3.6}$$

In words, this equation states that the increase in kinetic energy of the mass m is equal to the work done by the force F on the mass m over the path from y_0 to y.

Let us now consider the potential energy $V(y)$ of the system. The force $-F(y)$ equal and opposite to the force $F(y)$ above. Then by Eq. 3.3 the work done by $-F(y)$ in moving the mass from y_0 to y is defined as the difference in potential energy $V(y) - V(y_0)$ between the points y_0 and y, that is:

$$V(y) - V(y_0) = -\int_{y_0}^{y} F(y)\,dy \tag{3.7}$$

Hence Eq. 3.6 may be written as

$$\tfrac{1}{2}m\dot{y}^2 - \tfrac{1}{2}m\dot{y}_0^{\,2} = -V(y) + V(y_0)$$

or

$$\tfrac{1}{2}m\dot{y}^2 + V(y) = \tfrac{1}{2}m\dot{y}_0^{\,2} + V(y_0) = E \tag{3-8}$$

where E is the mechanical energy, that is, the sum of the kinetic and potential energies and is a constant over the path. Thus from Eq. 3.8 we have the law of conservation of mechanical energy, and $F(y)$ is a conservative force, for its value depends only on the coordinate y. The velocity $\dot{y}$ may be written as:

$$\dot{y} = \sqrt{\frac{2}{m}[E - V(y)]} \tag{3.9}$$

Since the force $F(y)$ and the potential energy $V(y)$ are functions of the position y only, it is theoretically possible to have any shape for the potential energy

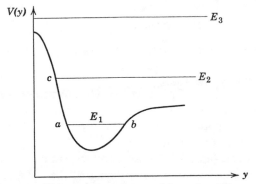

Fig. 3.4 Potential energy $V(y)$ plotted against position y for different values of the total energy E.

curve, but this shape must be unique for the particular problem. Suppose the potential energy curve is that shown in Fig. 3.4 in which $V(y)$ is plotted against y, and suppose the total mechanical energy E of the system is that represented by the horizontal line E_1. From Eq. 3.9 it follows that the velocity of the object is real or physically possible where E is larger than V. Hence for energy E_1 the object would oscillate between the values of y corresponding to the points a and b. At the points a and b, the value of the potential energy $V(y)$ is equal to the total energy E_1, so that from Eq. 3.9 the velocity of the object is zero at the points a and b. This type of motion is that associated with simple harmonic motion as, for example, a vibrating spring in which the points a and b represent the limits of the vibration.

For the total energy given by the line at height E_2, the object would be at rest only at the point c. If the object were coming in from the right with energy E_2, it would speed up near b where $(E - V)$ becomes larger, and then the object would rapidly slow down as the potential energy increases from a to c. At c the value of the potential energy $V(c)$ is equal to the total energy E_2, so the velocity of the object must be zero. The motion of the object might be reversed so that it might proceed to infinity to the right. One might think of this type of motion as that of a ball, uninfluenced by gravity, being thrown at a vertical wall and being reflected without any loss of energy. Can you say what the motion would be like if the energy were E_3?

3.5 Work and Kinetic Energy

A more general and sophisticated manner of deriving the relationship between work done and kinetic energy will now be given. If a particle of mass m, acted on by a resultant force $\mathbf{F}$, moves a distance $d\mathbf{r}$ in a direction making an angle θ with the force, then the work done by the force is by Eq. 3.1

$$W = F \cos \theta \, dr$$

This can be written in vector form, as is done in section 1.6. Thus

$$W = \mathbf{F} \cdot d\mathbf{r}$$

This is called the dot or scalar product of the vectors $\mathbf{F}$ and $d\mathbf{r}$. The expression $\mathbf{F} \cdot d\mathbf{r}$ is read as F dot dr and by definition has a magnitude equal to the product of the magnitudes of the two vectors and the cosine of the angle between their positive directions.

An analytical expression for the dot product can be written by expressing the force $\mathbf{F}$ and the displacement $d\mathbf{r}$ in terms of their respective components and the unit vectors along the X, Y, Z axes. If F_x, F_y, F_z are the respective magnitudes of the force $\mathbf{F}$ along the X, Y, Z axes, then from Eq. 1.4 it follows that

$$\mathbf{F} = \mathbf{i}F_x + \mathbf{j}F_y + \mathbf{k}F_z$$

Similarly, if dx, dy, dz are the magnitudes of the components of the displacement dr, then

$$d\mathbf{r} = \mathbf{i} \, dx + \mathbf{j} \, dy + \mathbf{k} \, dz$$

Thus, as in Section 1.6,

$$\mathbf{F} \cdot d\mathbf{r} = F_x \, dx + F_y \, dy + F_z \, dz$$

The quantity $F_x \, dx$ is the work done by the component F_x moving a distance dx. Since both these quantities are scalars, it follows that the work $F_x \, dx$ is also a scalar quantity. Thus the work $\mathbf{F} \cdot d\mathbf{r}$ is a scalar quantity.

Let us return now to the particle of mass m acted on by a resultant force $\mathbf{F}$. Its acceleration is given by

$$\mathbf{F} = \frac{m \, d^2\mathbf{r}}{dt^2} = m\ddot{\mathbf{r}}$$

The work done by the force $\mathbf{F}$ in the displacement $d\mathbf{r}$ is

$$W = \mathbf{F} \cdot d\mathbf{r} = m\ddot{\mathbf{r}} \cdot d\mathbf{r}$$

From the identity

$$d\mathbf{r} = \frac{d\mathbf{r}}{dt} dt = \dot{\mathbf{r}} \, dt$$

the acceleration equation may be changed into a velocity equation. Thus,

$$\mathbf{F} \cdot d\mathbf{r} = m\ddot{\mathbf{r}} \cdot \dot{\mathbf{r}} \, dt$$

$$= \frac{m}{2} \frac{d(\dot{\mathbf{r}} \cdot \dot{\mathbf{r}})}{dt} \, dt$$

$$= \frac{m}{2} d(\dot{r}^2)$$

If the velocity of the particle is u at time t_0 and position $\mathbf{r}_0$ and is v at a later time t and position $\mathbf{r}$, then by integration

$$\int_{\mathbf{r}_0}^{\mathbf{r}} \mathbf{F} \cdot d\mathbf{r} = \int_{u}^{v} \frac{m}{2} d(\dot{r}^2) = \frac{m}{2} v^2 - \frac{m}{2} u^2 \tag{3.10}$$

Thus the work done by the resultant force $\mathbf{F}$ during the time $(t - t_0)$ in which there is a displacement of $(\mathbf{r} - \mathbf{r}_0)$ is equal to the change in kinetic energy of the particle. The question now arises whether Eq. 3.10 will lead to the conservation of energy if the force $\mathbf{F}$ depends only on the coordinate, as it did for motion in a straight line.

3.6 Condition for Conservative Forces for Motion in Two or Three Dimensions

The law of conservation of mechanical energy may be derived in a more general manner. Consider that a conservative force $\mathbf{F}$ moves a body from position A to position B (Fig. 3.5). If V_A, V_B are the respective potential energies of the body at A and B, then by Eq. 3.3 it follows that the work done by the force $\mathbf{F}$ is

$$\int_{A}^{B} \mathbf{F} \cdot d\mathbf{r} = \int_{A}^{B} -dV = V_A - V_B \tag{3.11}$$

Also, if the conservative force $\mathbf{F}$ is the resultant force acting on the mass m and the velocity at A is v_A and at B is v_B, then by Eq. 3.10

$$\int_{A}^{B} \mathbf{F} \cdot d\mathbf{r} = \tfrac{1}{2}mv_B{}^2 - \tfrac{1}{2}mv_A{}^2$$

Thus

$$V_A - V_B = \tfrac{1}{2}mv_B{}^2 - \tfrac{1}{2}mv_A{}^2$$

or

$$V_A + \tfrac{1}{2}mv_A{}^2 = V_B + \tfrac{1}{2}mv_B{}^2$$

If the symbol T indicates kinetic energy, then the law of conservation of

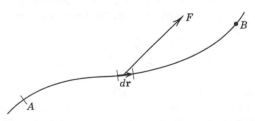

Fig. 3.5 Work done by a conservative force in moving a body between two points is independent of the path.

mechanical energy may be written

$$V_A + T_A = V_B + T_B$$

or

$$V + T = \text{a constant} \tag{3.12}$$

At any point along the path of a particle moving in a conservative force field the sum of the kinetic and potential energies of the body is a constant.

In the above we have assumed that the force $\mathbf{F}$ and its line integral given by Eq. 3.11 can be expressed as the difference between the potential energies at the end points, A and B of the path. In other words, the work given by the integral of $\mathbf{F} \cdot d\mathbf{r}$ does not depend on the path between A and B, but only on the end points A and B. That is, the line integral is independent of the path.

For the motion along a straight line, given in Section 3.4, the conservation of mechanical energy was realized for the condition that the force depended only on its position or coordinate. Such is not necessarily the case for motion in a plane or in space, as the following examples show.

Consider the force whose components F_x, F_y, and F_z are:

$$F_x = x^2 \qquad F_y = 0 \qquad F_z = 0$$

Such a force acts only along the X axis and has a magnitude of 1, 4, 9, etc., at values of x of 1, 2, 3, etc., respectively. The work done by this force in moving from the point x_0, y_0, z_0 to the point x, y, z is

$$\int_{x_0 y_0 z_0}^{x,y,z} x^2 \, dx = \frac{x^3}{3} - \frac{x_0^3}{3}$$

Thus, in this case, the work depends only on the end points and not at all on the type of path between x_0, y_0, z_0 and x, y, z. For such a force a potential function can be set up.

As a second example of a force which depends on the coordinates consider the one with the components:

$$F_x = y \qquad F_y = 0 \qquad F_z = 0$$

Again this force acts only along the X axis but its magnitude is governed by the value of the y coordinate. The work done by the force between the points x_0, y_0, z_0 and x, y, z is given by

$$\int_{x_0 y_0 z_0}^{x,y,z} y \, dx$$

and this is the equation for the area under a curve connecting the two points. The value of the integral depends on the shape of the curve, so that it does not depend only on the values of the end points and no potential function can be set up for this force.

We must now find the necessary condition for the force to be conservative or for the conservation of mechanical energy to hold. This is done by assuming that a potential function exists for the force and finding the relationship between the components of the force F_x, F_y, F_z and the coordinates x, y, z.

Suppose now that the components F_x, F_y, F_z of the force **F** are functions only of the coordinates x, y, z of the point at which the force is being considered, so that

$$F_x \, dx + F_y \, dy + F_z \, dz = -dV(x, y, z)$$

where dV is a perfect differential. If dV is a perfect differential, it follows that

$$dV = \frac{\partial V}{\partial x} \, dx + \frac{\partial V}{\partial y} \, dy + \frac{\partial V}{\partial z} \, dz$$

Thus

$$F_x = -\frac{\partial V}{\partial x} \qquad F_y = -\frac{\partial V}{\partial y} \qquad F_z = -\frac{\partial V}{\partial z} \tag{3.13}$$

Since the function V is the potential energy of the particle, it follows that *the component of the conservative force in any direction is equal to the space rate of decrease of potential in that direction.* Partial derivatives are employed since the function V depends on the three coordinates x, y, z, and in taking the derivative with respect to one coordinate the other two coordinates are considered constant. A scalar field, such as potential energy V or temperature T, is one in which the quantity V or T is given by a single number and depends only on its position in space.

This result may become somewhat clearer if we consider a contour map, Fig. 3.6, on which the closed curves represent places of equal elevation above sea level. They are curves of equal gravitational potential energy or equipotential curves. If you have carried such a map when walking through mountainous country, you have probably realized, sometimes painfully, that the hill is steepest where the contour lines are closest together. The steepest direction of ascent or descent is in the direction perpendicular to the contour lines or to the equipotential curves. If a direction s is chosen perpendicular to the contour lines, or s is in the direction of maximum slope, then along this direction the rate of change of potential energy dV/ds is a maximum. Along an

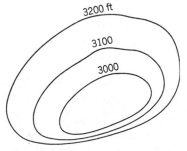

Fig. 3.6 A contour map showing places of equal elevation above sea level.

equipotential curve in which V is constant, the rate of change of the potential energy is zero; that is, there is no gravitational component of force along the horizontal equipotential curves. In any arbitrary direction, as for instance along the X axis, the rate of change of potential energy $\partial V/\partial x$, or the steepness of the hill in this direction is equal to the negative of the gravitational force in this direction.

The vector having the components given in Eq. 3.13 is called the gradient of the potential function V, usually abbreviated grad V, so that the force **F** can be written as $\mathbf{F} = -\text{grad } V$.

Largely for the sake of abbreviation, a vector differential operator del $\boldsymbol{\nabla}$ is sometimes used. This operator is defined as

$$\boldsymbol{\nabla} = \mathbf{i}\frac{\partial}{\partial x} + \mathbf{j}\frac{\partial}{\partial y} + \mathbf{k}\frac{\partial}{\partial z}$$

A conservative force **F** may be written in terms of this operator and the potential energy V. From Eq. 3.13

$$\mathbf{F} = -\mathbf{i}\frac{\partial V}{\partial x} - \mathbf{j}\frac{\partial V}{\partial y} - \mathbf{k}\frac{\partial V}{\partial z} = -\boldsymbol{\nabla}V$$

This operator adds nothing to the analysis here, though it is used extensively in vector analysis.

Using Eq. 3.13, let us now differentiate F_x with respect to y and F_y with respect to x, noting that the second partial differentiation does not depend on the order of the differentiation. Since the potential function V depends on x, y, z, or $V = V(x, y, z)$ it follows that:

$$\frac{\partial F_x}{\partial y} = -\frac{\partial}{\partial y}\left(\frac{\partial V}{\partial x}\right) = -\frac{\partial^2 V}{\partial y\,\partial x} = \frac{\partial F_y}{\partial x}$$

or

$$\frac{\partial F_x}{\partial y} - \frac{\partial F_y}{\partial x} = 0 \tag{3.14}$$

By proceeding in a similar manner for the other pairs of variables it follows that

$$\frac{\partial F_y}{\partial z} - \frac{\partial F_z}{\partial y} = 0; \qquad \frac{\partial F_z}{\partial x} - \frac{\partial F_x}{\partial z} = 0 \tag{3.15}$$

The three quantities in Eqs. 3.15 are the components of a vector† called the

† To demonstrate that a cross product of two vectors is also a vector requires a more thorough discussion than that given here. In such a discussion the definition of a vector is given in terms of equations relating the components of a vector displacement **r** in two coordinate systems having a common origin but whose axes are rotated with respect to each other. For an interesting development of this topic, see *The Feynman Lectures on Physics* by R. P. Feynman, R. B. Leighton, and M. Sands. Addison-Wesley, Reading, Mass. Vol I, Chapter 11 and Vol II, Chapter 2.

curl of the force **F**. Thus the necessary condition for a conservative force, or for the existence of a potential function whose negative gradient gives the value of the force, is that the curl of the force must be zero, i.e., the condition is the curl $\mathbf{F} = 0$.

In actual use one must always go back to the components of the force given in Eqs. 3.15 to test whether each one of the components for the curl is zero. Though we have stated that the necessary condition for a force to be conservative is that its curl be zero, we have not shown that this is a sufficient condition. This may be done by making use of Stokes' theorem, which relates the integral of the curl of a vector over an area to the line integral of the vector completely around the edge of the area. If the curl of **F** is zero, then the line integral, $\int \mathbf{F} \cdot d\mathbf{r}$, around a closed path is also zero. Thus the line integral, $\int \mathbf{F} \cdot d\mathbf{r}$, between any two points depends only on the coordinates of the two points and not on the path joining them, and the force is conservative and derivable from a potential function. Hence the curl $\mathbf{F} = 0$ is a necessary and sufficient condition for the existence of a potential function. (A proof of Stokes' theorem can be found in more advanced texts of mathematical physics.)

The name "curl" appears to have come from the Scottish game of curling, which is played on ice, and involves some twisting or rotation. The rotation produced by the curl of a vector can easily be visualized for the simple case in which the rate of change of the component of the force with the coordinate is a constant. Let us consider the component $\partial F_y / \partial x - \partial F_x / \partial y$. Suppose that F_y increases uniformly with x, then the rate of change of F_y with x, $\partial F_y / \partial x$ is given by the tangent of the angle θ in Fig. 3.7 and represents rotation about the positive direction of the Z or **k** axis. In a similar manner, if F_x increases uniformly with y, then $\partial F_x / \partial y$ is given by the tangent of the angle ϕ and represents rotation about the negative **k** axis. Thus the component $(\partial F_y / \partial x - \partial F_x / \partial y)$ represents the rotation of the force about the positive Z axis. When the curl is zero, there is no rotation.

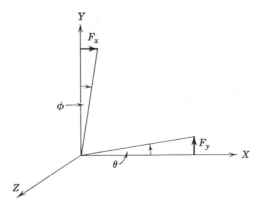

Fig. 3.7 Rotation in a curl of a vector.

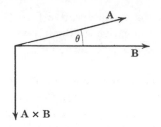

Fig. 3.8 Cross product of two vectors.

Frequently the curl **F** is written as $\nabla \times \mathbf{F}$ (del cross **F**). By definition, the cross product of two vectors is a vector whose magnitude is the product of the magnitudes of the two vectors and the sine of the angle between their positive directions, and whose direction is that of the progress of a right-handed screw turned from the first vector into the second. Thus, the cross product of the two vectors **A** and **B**, $\mathbf{A} \times \mathbf{B}$, has a magnitude of $AB \sin \widehat{AB}$ or $AB \sin \theta$ and a direction downwards as shown in Fig. 3.8. The cross product of two vectors is illustrated in the discussion of torques, given in Section 6.11. From that discussion it may be seen that $\nabla \times \mathbf{F}$ can be obtained from the following determinant, or the curl $\mathbf{F} = \nabla \times \mathbf{F}$

$$\nabla \times \mathbf{F} = \begin{vmatrix} \mathbf{i} & \mathbf{j} & \mathbf{k} \\ \dfrac{\partial}{\partial x} & \dfrac{\partial}{\partial y} & \dfrac{\partial}{\partial z} \\ F_x & F_y & F_z \end{vmatrix} = \mathbf{i}\left(\frac{\partial F_z}{\partial y} - \frac{\partial F_y}{\partial z}\right)$$

$$+ \mathbf{j}\left(\frac{\partial F_x}{\partial z} - \frac{\partial F_z}{\partial x}\right) + \mathbf{k}\left(\frac{\partial F_y}{\partial x} - \frac{\partial F_x}{\partial y}\right) \quad (3.16)$$

which are the three components given in Eqs. 3.14, 3.15. Notice that the **j** and **k** components can be written from the **i** component by cyclical rotation of the coordinates $(x \to y \to z \to x)$.

As an example of a conservative force, consider the force of attraction of two particles m_1 and m_2, separated by a distance r, where the force F is given by

$$F = \frac{-Gm_1m_2}{r^2}$$

where G is the constant of gravitation. The negative sign indicates an attractive force. In Cartesian coordinates, $r^2 = x^2 + y^2 + z^2$, so that

$$\frac{\partial r}{\partial x} = \frac{2x}{2r} = \frac{x}{r} \qquad \frac{\partial r}{\partial y} = \frac{y}{r} \qquad \frac{\partial r}{\partial z} = \frac{z}{r}$$

Also $F^2 = F_x^2 + F_y^2 + F_z^2$, so that $\partial F/\partial F_x = F_x/F = x/r$, since the force **F** lies along the vector **r**, Fig. 3.9. Hence $F_x = (xF)/r = -Gm\,m_2(x/r^3) = -a(x/r^3)$, where $a = Gm_1m_2$. Similarly, $F_y = -a(y/r^3)$, $F_z = -a(z/r^3)$.

To show that this gravitational force is conservative or derivable from a potential, we must show that its curl is zero. Taking the **k** component, it is necessary to show that;

$$\partial F_y/\partial x - \partial F_x/\partial y = 0$$

Now

$$\frac{1}{a}\frac{\partial F_y}{\partial x} = -\frac{\partial}{\partial x}\left(\frac{y}{r^3}\right) = -y\frac{\partial}{\partial x}(x^2 + y^2 + z^2)^{-3/2} = y\frac{3x}{r^5}$$

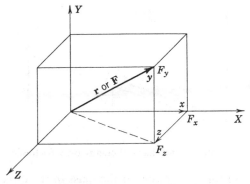

Fig. 3.9   Showing the vector r lies along F.

Similarly

$$\frac{1}{a}\frac{\partial F_x}{\partial y} = x\frac{3y}{r^5}$$

so that

$$\frac{\partial F_y}{\partial x} - \frac{\partial F_x}{\partial y} = 0$$

and corresponding results hold for the other two components, so that the

$$\text{curl } F = 0 \qquad \text{or} \qquad \nabla \times F = 0$$

Thus a potential function V may be set up which is

$$V(r) - V(r_\infty) = V = -\int_\infty^r F \cdot dr = Gm_1m_2\int_\infty^r \frac{dr}{r^2} = \frac{-Gm_1m_2}{r}$$

where V at infinity is taken as zero. Note that the force in the direction of r, that is, along the line joining m_1 and m_2, is given by

$$F(r) = \frac{-\partial V}{\partial r} = \frac{-Gm_1m_2}{r^2}$$

It should be noticed that it is always a potential difference that is found, and giving the potential at any point implies some arbitrary zero of potential.

3.7   An Example of Conservation of Energy

Let us consider a particle of mass m placed on a frictionless inclined plane of height h and inclination θ, Fig. 3.10. The potential energy of the mass m at any height y is $V = mgy$ relative to its potential energy on the horizontal X axis. If the distance s is measured from the origin along the incline, then $y = s\sin\theta$ and $V = mgs\sin\theta$.

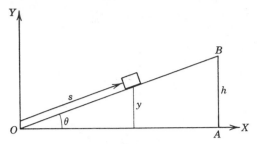

Fig. 3.10 Conservation of mechanical energy on a frictionless inclined plane.

The conservative force F_s acting on the mass in the direction of increasing s is

$$F_s = -\frac{\partial V}{\partial s} = -\frac{\partial (mgs \sin \theta)}{\partial s} = -mg \sin \theta$$

Thus the force along the plane acting on the mass m is $mg \sin \theta$ in a direction down the plane. This same result was previously obtained by resolving the vertical force mg into components along and at right angles to the plane.

Since we have assumed that the plane is frictionless, the system in Fig. 3.10 is a conservative one, and the sum of the potential and kinetic energies of the mass as it moves down the plane is a constant. If the particle starts from rest at the top of the plane, its energy in this position is all potential and has a value of mgh. This potential energy is equal and opposite to the work done by the force F_s in moving the particle down the plane a distance BO. The work done is

$$F_s \frac{h}{\sin \theta} = mg \sin \theta \frac{h}{\sin \theta} = mgh$$

If v is the velocity of the particle at the bottom of the plane, then by the law of conservation of energy it follows that

$$\tfrac{1}{2}mv^2 = mgh \qquad \text{or} \qquad v^2 = 2gh$$

This is the same result as that obtained for a particle starting from rest and falling through a vertical height h with a constant acceleration g. It could also be obtained by considering the motion of a particle down the plane where the acceleration is $g \sin \theta$ and the distance traveled is $h/\sin \theta$.

3.8   An Elastic Collision in the Center of Mass Coordinate System

Elastic collision theory is of importance in nuclear reactors in which fast neutrons are slowed down by colliding with stationary nuclei. In the first operating chain reactor of December 1942, the uranium that underwent fission was embedded in blocks of graphite. When fission takes place, neutrons of high energy are emitted, and these have to be slowed down or moderated in order to cause further fission. This slowing down is accomplished by the elastic collisions between the neutrons and the carbon nuclei.

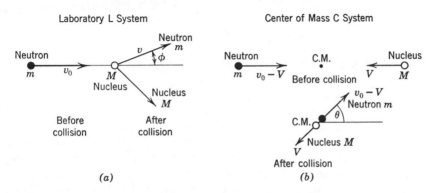

Fig. 3.11 Collision of a neutron with a nucleus in the laboratory and center of mass systems.

As may be shown by working Problem 2 at the end of this chapter, there is conservation of kinetic energy at an elastic collision. The fractional loss of kinetic energy of the neutron at each collision can be readily calculated from the laws of conservation of momentum and energy. However, the analysis is simplified if it is calculated with respect to a coordinate system attached to the center of the mass of the colliding particles. In order to illustrate the procedure, we shall essentially repeat some of the material given in Chapter 2 on elastic collisions.

Up to the present we have employed a coordinate system which is at rest with respect to the moving particles, or one in which the target nucleus is at rest before the collision. This system is frequently called the laboratory or L system inasmuch as this is the system in which all measurements are made. In contrast to this, let us consider a collision in the center of mass or C system.* Suppose that a neutron of mass m moving to the right with a velocity v_0 collides with a stationary nucleus of mass M. After the collision the neutron and nucleus are scattered, as shown in Fig. 3.11a for the laboratory system. In the C system, an observer attached to the center of mass would see the collision, as shown in Fig. 3.11b. In the C coordinate system the center of mass is at rest. In the L system the velocity V of the center of mass to the right is given from the law of conservation of

* If you are not familiar with the center of mass, you will find a discussion in Section 6.2.

momentum as

$$(M + m)V = mv_0 \quad \text{or} \quad V = \frac{mv_0}{(M + m)}$$

In the C system in which the center of mass is considered to be at rest, the neutron is moving to the right with a velocity of

$$v_0 - V = \frac{Mv_0}{(M + m)}$$

and the nucleus is moving to the left with a velocity of

$$V = \frac{mv_0}{(M + m)}$$

In the C system the total momentum before collision is zero, as is readily seen from the above equations, and from the law of conservation of momentum it is zero after the collision. Since there is conservation of kinetic energy at the elastic collision, the kinetic energy before impact must be equal to the kinetic energy after impact. This must mean that in the C system the speeds of the neutron and the nucleus must be unchanged at the collision, for a change in speeds would imply a change in the total kinetic energy. In the C system, after the collision, the two particles go off in opposite directions with their speeds unchanged, or the total effect of the collision is to change the direction but not the magnitudes of the velocities, as shown in Fig. 3.11b. This is not true in the L system, in which both the magnitude and the direction of the velocities are changed. Since experimental measurements are made in the L system, it is necessary to change the velocities from the C to the L system. This is done by the vector velocity triangle, Fig. 3.12. The velocity v of the neutron after the collision is obtained in the L system by adding the velocity $(v - V)$, of the same magnitude as $v_0 - V$, in the C system to the velocity V with which the center of mass moves in the L system relative to the C system. From Fig. 3.12 and the law of cosines and substituting for V and $(V - v_0)$, we obtain

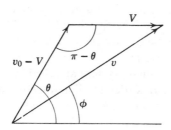

Fig. 3.12   The velocity v of the neutron after the collision in the L system determined from the velocity $v_0 - V$ in the C system and the velocity V of the center of mass.

$$v^2 = (v_0 - V)^2 + V^2 + 2V(v_0 - V) \cos \theta$$

$$= \frac{v_0^2 M^2}{(M + m)^2} + \frac{v_0^2 m^2}{(M + m)^2} + \frac{2v_0^2 mM \cos \theta}{(M + m)^2}$$

The ratio of the kinetic energy E of the neutron after the collision to its kinetic energy E_0 before collision is given in the laboratory system as

$$\frac{E}{E_0} = \frac{mv^2/2}{mv_0^2/2} = \frac{M^2 + m^2 + 2mM \cos \theta}{(M + m)^2}$$

It follows that, the larger the mass of the nucleus, the smaller the amount of energy which the neutron loses on impact.

3.9   General Law of Conservation of Energy

In the foregoing examples we have neglected the force of friction, though in practice it can never be entirely eliminated. Let us consider the first example dealing with a particle sliding down an inclined plane. If we now include friction, then with a coefficient of fraction μ between the particle and the plane, the force down the plane would be $(mg \sin \theta - \mu mg \cos \theta)$. In this case the kinetic energy of the particle at the bottom of the plane would not equal the potential energy at the top of the plane since some work would be done against the force of friction as the particle slides down the plane. The system is no longer a conservative one since the work done in taking the mass around a closed path, as for instance from the top to the bottom and back to the top of the plane, is not zero. When the mass moves up or down the plane, work is done against friction, and this goes into an equivalent amount of heat energy.

About the middle of the last century Joule, an English physicist, carried out a large number of experiments in which the heat and the work against the force of friction were carefully measured. These experiments showed that 4.187 joules of work done against friction would heat 1 gm of water 1°C, that is, would produce 1 calorie of heat. It was not until after quite a controversy that heat became recognized as another form of energy. However, the results of the experiments of Joule and others could not be denied, and it was on the basis of these experiments that the general law of conservation of energy was enunciated. According to this law, the total energy in a closed or isolated system is a constant. If the isolated system taken is our universe, it follows that the total energy of it is a constant independent of time. Although energy is continually changing from one form to another, the total energy of the universe must remain constant.

Since the time of Joule many other forms of energy have been recognized. Possibly the most important discovery in our time is that matter itself is a form of energy. This discovery was made by means of theoretical reasoning by Einstein, who showed that any piece of matter of mass m grams has an energy equivalent to E ergs according to the relation

$$E = mc^2 \tag{3.17}$$

where c is the velocity of light, approximately 3×10^{10} cm/sec. Thus each gram of matter is equivalent to approximately 9×10^{20} ergs of energy. The practical possibility of changing matter into energy did not come until the early 1930's when research with nuclear accelerators was started. Until the advent of the atomic bomb the amount of matter ever transformed into energy was exceedingly minute.

All the various forms in which energy can exist are included in the general law of conservation of energy, making this a law of universal application. This law in a sense disposes of the law of conservation of mass inasmuch as the latter is included in the conservation of energy law. However, classical physics deals with such relatively small amounts of energy that their mass equivalents are negligible, so we shall continue to use the law of conservation of mass.

3.10   Kinetic Energy in Relativity Theory

The following is a simple derivation of the relation between mass and energy, assuming the equation given by Einstein for the change in mass with velocity. Starting with Newton's second law of motion:

$$F = \frac{d}{dt}(mv) = m\frac{dv}{dt} + v\frac{dm}{dt}$$

The kinetic energy T may be written in terms of the work $F\,dx$ as;

$$dT = F\,dx = m\frac{dv}{dt}\,dx + v\frac{dm}{dt}\,dx$$

$$= mv\,dv + v^2\,dm$$

since $dx/dt = v$.

From Section 2.14 where the relationship for change of mass with velocity is given,

$$m = m_0(1 - v^2/c^2)^{-\frac{1}{2}} \quad\text{then}\quad dm = \frac{m_0}{c^2}\frac{v\,dv}{(1 - v^2/c^2)^{\frac{3}{2}}} = \frac{mv\,dv}{(c^2 - v^2)}$$

where m_0 is the rest mass of the particle and c is the velocity of light. Hence the change in kinetic energy may be written as

$$dT = mv\,dv + v^2\,dm = (c^2 - v^2)\,dm + v^2\,dm = c^2\,dm$$

$$= \frac{m_0 v\,dv}{(1 - v^2/c^2)^{\frac{3}{2}}} = \frac{m_0\,d(v^2)}{2[1 - (v^2/c^2)]^{\frac{3}{2}}}$$

By integration the kinetic energy in relativistic form is given by

$$T = \int_0^T dT = \frac{m_0}{2} \frac{2c^2}{(1 - v^2/c^2)^{1/2}} \bigg|_0^v = m_0 c^2 \left[\frac{1}{\sqrt{1 - v^2/c^2}} - 1 \right]$$

$$= mc^2 - m_0 c^2$$

or

$$mc^2 = T + m_0 c^2 = E \tag{3.18}$$

This is the famous equation given by Einstein for the equivalence of mass and energy. The total energy E of a free particle is given by the rest energy $m_0 c^2$ plus the kinetic energy T or $E = mc^2 = m_0 c^2/\sqrt{1 - v^2/c^2}$. The relativistic equation for the momentum p is given as

$$p = mv = \frac{m_0 v}{\sqrt{1 - v^2/c^2}} \tag{3.19}$$

If the velocity of the mass is small compared to the velocity of light, then the relativistic equation for kinetic energy may be written, by the binomial theorem, as

$$T = m_0 c^2 \left(1 + \frac{v^2}{2c^2} \cdots - 1 \right)$$

$$= \tfrac{1}{2} m_0 v^2$$

when terms in v^4/c^4 and higher powers are neglected. Thus we are justified in using the expression $m_0 v^2/2$ for the kinetic energy of bodies such as moving automobiles or airplanes. The Newtonian or classical theory is the limit of the relativistic theory when the velocities of the moving bodies are small compared to the velocity of light.

As an example, let us consider how the relativity theory affects the mass and kinetic energy of an airplane traveling at the very large speed of 1000 mph. The ratio of the velocity of the plane to the velocity of light (186,000 mi/sec) is

$$\frac{v}{c} = \frac{1}{186} \times \frac{1}{3600} = 1.493 \times 10^{-6}$$

Hence

$$\frac{v^2}{c^2} = 2.229 \times 10^{-12}$$

and

$$\frac{1}{\sqrt{1 - v^2/c^2}} = 1 + 1.115 \times 10^{-12}$$

Thus the increase in mass is quite negligible, and the relativistic kinetic energy is negligibly different from that obtained by the classical theory.

A further relationship between the energy $E = mc^2$ and the momentum $p = mv$ is the following:

$$E^2 - p^2c^2 = m^2c^4 - m^2v^2c^2 = m^2c^4\left(1 - \frac{v^2}{c^2}\right) = m_0^2c^4 \qquad (3.20)$$

Though the energy E and momentum p are different when measured in different coordinate systems, it is somewhat surprising that the difference between the two quantities as given in Eq. 3.20 is a constant in all coordinate systems moving with uniform velocity relative to one another. This statement is expressed in relativity by considering that energy and momentum form a four component vector, called a world vector, with components p_x, p_y, p_z, and a fourth component iE/c where $i = \sqrt{-1}$. These components are such that

$$p_x^2 + p_y^2 + p_z^2 + \frac{i^2E^2}{c^2} = p^2 - \frac{E^2}{c^2} = -m_0^2c^2 \qquad (3.21)$$

where $i^2 = -1$.

3.11 Power

Frequently one is interested not only in the amount of work done by a force but also in the time taken to do this work. The time rate of doing work is called power. If an amount of work ΔW is done by a force in a time Δt, the average power developed $\bar{P}$ is

$$\bar{P} = \frac{\Delta W}{\Delta t}$$

The power P developed at any instant is given by the limit of this expression as Δt approaches zero or

$$P = \underset{\Delta t \to 0}{\text{limit}} \frac{\Delta W}{\Delta t} = \frac{dW}{dt} = \dot{W} \qquad (3.22)$$

If a force $\mathbf{F}$ moves a particle through a displacement $d\mathbf{r}$ in a time dt, the work done is

$$dW = \mathbf{F} \cdot d\mathbf{r}$$

and the instantaneous power developed by the force is

$$\frac{dW}{dt} = \mathbf{F} \cdot \frac{d\mathbf{r}}{dt} = \mathbf{F} \cdot \mathbf{v} \qquad (3.23)$$

where $\mathbf{v}$ is the instantaneous velocity of the particle.

The dimensions of power in the absolute system of units are $[ML^2T^{-3}]$ and in the gravitational system are $[FLT^{-1}]$. In the cgs system the unit of power is the erg per second, which is very small for practical purposes, so the unit of the mks system is used instead. This is the joule per second, also called a *watt* (w). Energy and work are frequently measured in units of power multiplied by time, as for example in kilowatt-hours (kwh). In the British engineering system of units the unit of power is the foot pound weight per second (ft lbw/sec). Another unit is the horsepower (hp), which is equal to 550 ft lbw/sec or to 746 w.

An example of energy and power.   A train is traveling at a constant speed of 60 mph and is picking up water from a trough between the rails. If it picks up 15,000 lb of water in 400 yd, let us determine the exact force and horsepower required.

The speed of the engine is 60 mph or 88 ft/sec. It goes 400 yd in $(400 \times 3)/88$ sec so that the time rate of picking up water is $(15{,}000 \times 88)/(400 \times 3) = 1100$ lbw/sec. The force F which the engine must exert to pick up this water is given by Newton's second law as

$$F = \frac{d(mv)}{dt} = v\frac{dm}{dt} = \frac{v\,dw}{g\,dt} = \frac{88}{32} \times 1100 = 3025 \text{ lbw}$$

since the velocity v is constant. The additional horsepower which the engine must develop to maintain its constant speed and pick up the water is

$$\frac{Fv}{550} = \frac{3025 \times 88}{550} = 484 \text{ hp}$$

Now it might be thought that the additional force necessary could be found from the relationship between the work done by the force F and the increase in kinetic energy of the water. An amount of kinetic energy

$$\text{K.E.} = \frac{1}{2}\frac{w}{g}v^2 = \frac{1}{2} \times \frac{15000}{32} \times 88 \times 88 = 1.815 \times 10^6 \text{ ft lbw}$$

is picked up in a distance of 1200 ft, so the force required for this is

$$F = \frac{1.815 \times 10^6}{1200} = 1512.5 \text{ lbw}$$

This is exactly half the value obtained by using the rate of change of momentum. The discrepancy lies in the fact that the engine has done more work than that required to give the water kinetic energy. Some energy goes into heat and, as illustrated in Problem 18, is exactly equal to the kinetic energy received. Thus the total amount of energy received by the water in the 400 yd is double that given by the kinetic energy, or 3.63×10^6 ft lbw. With this value for the energy the force comes out correctly. For the situation in which a mass in motion captures another object, or whenever one mass is added to or subtracted from another, the force exerted cannot be obtained directly from the relationship between the work done and the increase in kinetic energy.

PROBLEMS

1. A bullet with a mass of 0.025 kg and a velocity of 300 m/sec is fired into a block of wood whose mass is 10 kg and which is suspended so as to swing as a pendulum. The bullet penetrates 0.03 m into the block of wood before coming to rest. (*a*) Find the vertical height through which the block and bullet are raised. (*b*) Find the amount of heat produced in the block and bullet. (*c*) Find the average resistance force exerted by the block on the bullet.

2. Two spheres *A* and *B* moving in the same direction impact along their line of centers. The mass of *A* is 10 gm, and its initial velocity is 100 cm/sec. The mass of *B* is 20 gm, and its initial velocity is 40 cm/sec. Find the loss energy at impact for coefficients of restitution of (*a*) 0; (*b*) 1.0; (*c*) 0.5.

3. A body whose weight is 20 lbw is pulled up a frictionless inclined plane whose angle of inclination is 30° and whose length is 4 ft by a force of 20 lb making an angle of 60° with the horizontal. If the body starts from rest at the bottom of the plane, find its kinetic energy and velocity at the top of the plane. Show that the work done by the force in pulling the body up the plane is equal to the increase in mechanical energy.

4. If the plane in problem 3 is rough with a coefficient of friction of 0.30, find the kinetic energy and velocity of the body at the top of the plane. Show that the difference between the work done by the force and the increase in mechanical energy is equal to the work done against the frictional force.

5. A block having a weight of 10 lb is placed on a horizontal surface and is attached to a horizontal spring whose force constant *k* is 3 lbw/in. of compression. Assume that the force of compression is proportional to the distance *x* through which the spring is compressed, that is, $F = kx$, and that the coefficient of friction between the block and the horizontal surface is 0.20. A bullet whose weight is 0.125 lb and whose velocity is 1600 ft/sec is fired into the block and comes to rest in it. (*a*) Show that the work done in compressing the spring a distance *x* is $kx^2/2$. (*b*) Find the distance the block moves before it comes to rest. (*c*) Assuming that the force constant for extension is the same as that for compression, find the second position of rest. (*d*) Describe qualitatively the subsequent motion of the block (Fig. P5).

Fig. P5

6. A particle weighing 3 lb is acted on by a force F_x in pounds weight given by $F_x = 5 - x - 2x^2$, where x is measured in feet. (a) Plot a graph of F against x. (b) Show that F is a conservative force. (c) Find the work done by the force when the particle moves from $x = 0$ to $x = 2$ ft. (d) Assuming that the particle starts from rest at origin, find the velocity of the particle at $x = 2$ ft. (e) Show that the particle comes to rest at $x = 2.39$ ft. (f) Describe the motion of the particle.

7. An electrically charged body A is attracted by another fixed charged body B with a force of $F = -10^4/r^2$ dynes, where r is the distance between A and B and is measured in centimeters. (a) Show that a potential function exists for F and find the function. (b) Find the work done against the force when the charged body A is moved from $r = 1$ to $r = 5$ cm. (c) What is the shape of the equipotential surfaces about the charged body B? (d) If the potential energy is considered to be zero at an infinite distance, find its potential energy at $r = 20$ cm.

8. If a continuous function f of x, y, z is such that $f = \ln (x^2 y^3 z^4)$, show that grad $f = 2i/x + 3j/y + 4k/z$.

9. Show grad $(1/r) = -r/r^3$ where $r = (x^2 + y^2 + z^2)^{1/2}$ and $r = ix + jy + kz$.

10. Expand the expression for the curl F, $\nabla \times F$, using the definition of a cross product and show that this gives the components in Eq. 3.16.

11. Suppose a force has the components $F_x = y$, $F_y = -x$, $F_z = 0$. Prove that the force is tangent to circles in the XY plane and find the line integral completely around a circle of radius r. Show that the vector field is not conservative, that is, the curl of F is not zero.

12. If ϕ is a scalar function of position (similar to the potential energy V) show $\nabla\phi \cdot d\mathbf{r} = d\phi$. If displacement $d\mathbf{r}$ is along a direction in which ϕ is constant, then $d\phi = 0$. From this show $\nabla\phi$ is at right angles to the surface $\phi = $ constant.

13. If potential energy $V = 3xz/y$, find the force F associated with this potential. Repeat this for $V = -ax^2 + bxy - cy^2$.

14. Prove the curl of the gradient of ϕ is zero or $\nabla \times \nabla\phi = 0$.

15. Find the curl of the force having the following components and set up the potential function where possible. (a) $F_x = y^2/\sqrt{(x^2 + y^2)}$, $F_y = x^2/\sqrt{(x^2 + y^2)}$, $F_z = 0$; (b) $F_x = F_1(x)$, $F_y = F_2(y)$, $F_z = F_3(z)$, where the functions F_1, F_2, F_3 represent arbitrary but different functions; (c) $F_x = xF(r)$, $F_y = yF(r)$, $F_z = zF(r)$, where $F(r)$ is some function of the distance r from the origin.

16. Find the horsepower developed by a fire engine which can project 120 lb of water per second with a nozzle speed of 132 ft/sec.

17. A uniform flexible cord having a weight of 8 lb/ft is drawn from a heaped-up coil with a constant speed of 16 ft/sec along a frictionless floor. Find the force necessary and the horsepower used.

18. A number of 1-gm masses are placed in a straight line at equal intervals of 2 cm apart on a frictionless horizontal table. A 10-kg mass is pushed along the table so as to make inelastic impacts with each of the 1-gm masses, thus adding to its own mass. If the 10-kg mass is pushed along the table at a constant speed of 30 cm/sec, find (*a*) the kinetic energy received by the 1-gm masses each second; (*b*) the mechanical energy lost at each impact and the energy lost per second owing to the inelastic impacts; (*c*) the force necessary to maintain the constant velocity both from the total work done per second and from the time rate of change of momentum.

19. (*a*) Find the speed with which a 1500-hp engine can pull a train weighing 500 tons along a horizontal track if the resistance forces are constant at 30 lbw/ton. (*b*) Find the instantaneous acceleration of the train when its speed is 30 mph. (*c*) Find the maximum speed with which the engine could pull the train up an incline of 1 in 200 against the same resistance forces.

20. It appears that the scattering of neutrons by nuclei is spherically symmetric in the center of mass coordinate system for relatively light nuclei and neutron energies of a few thousand electron volts. This implies that all values of cos θ in Fig. 3.11 are equally probable. Show that this also means approximately spherically symmetric scattering in the laboratory L system for nuclei in which M/m is large but not for hydrogen nuclei for which $M = m = 1$.

21. Show that the maximum fractional loss in energy of a neutron in an elastic collision between a neutron and a stationary nucleus occurs when there is a head-on collision, and find this fractional loss for a collision with a deuteron, mass 2, and a beryllium nucleus, mass 9. Take the mass of the neutron as unity. Make these calculations in both the laboratory and the center of mass coordinate systems.

22. Find the energy equivalent of 1 μgm (1 μgm $= 10^{-6}$ gm) of mass. At the price of 4 cents/kwh, how much is this energy worth?

23. If the mass of a body at rest is 100 gm, find the mass of the body relative to an observer at rest when the body has a speed of (*a*) 3×10^9 cm/sec; (*b*) 2×10^{10} cm/sec; (*c*) 2.99×10^{10} cm/sec. (*d*) Compare the kinetic energy of a body moving with a speed of 2.99×10^{10} cm/sec as given by classical mechanics with that obtained from relativistic mechanics.

24. Suppose **F** is a conservative force such that $\mathbf{F} = -\nabla V$, and a particle of mass m moves in this force field from point A to point B. If the velocity of the particle at A is v_A and at B is v_B show there is conservation of energy such that $V(A) + \frac{1}{2}mv_A^2 = V(B) + \frac{1}{2}mv_B^2$. [*Hint.* Calculate $\int \mathbf{F} \cdot d\mathbf{r} = \int m\ddot{\mathbf{r}} \cdot \dot{\mathbf{r}} \, dt$.]

NEWTON'S LAW
OF GRAVITATION AND
SOME OF ITS CONSEQUENCES

4.1 The Law of Gravitation

During the great plague in 1666 the English universities closed and Newton, who was then an undergraduate at Cambridge, returned to his home in the midlands of England. When Newton was about 73 years old he wrote of this early period: "In the same year (1666) I began to think of gravity extending to the orb of the moon ... From Kepler's rule of the periodical times of the planets ... I deduced the forces which keep the planets in their orbs must be reciprocally as the squares of their distances from the centres about which they revolve: and thereby compared the force requisite to keep the moon in her orb with the force of gravity·at the surface of the earth, and found them answer pretty nearly. All this was in the plague years of 1665 and 1666, for in those days I was in the prime of my age for invention, and minded mathematics and philosophy more than in any time since."

The rule of the inverse square force was extended into what has become known as *Newton's law of gravitation*, which may be stated as follows.

Every particle in the universe attracts every other particle with a force which is proportional to the product of their masses and inversely proportional to the square of their distance apart.

As we shall see, one of the important contributions made by this law is in the interpretation of the motions in our solar system.

Mathematically the law of gravitation may be stated in the following manner. If m_1, m_2 are the masses of two particles separated by a distance r, the force of attraction F along the line joining the particles is

$$F \propto \frac{m_1 m_2}{r^2}$$

This equation may also be written

$$F = G \frac{m_1 m_2}{r^2} \tag{4.1}$$

where G, the constant of proportionality, is called the *constant of gravitation*. The mass involved here is called the gravitational or passive mass, and this should be compared with a totally different property of mass: its inertial or active mass.* Newton, by simple experiments with pendulums, concluded that the two masses are equal to one part in 300; later, R. von Eötvös showed their equality to 1 part in 10^6; and still later, Professor R. H. Dicke at Princeton University showed the equality to 1 part in 10^{11}.

The earliest experiments for determining G were made in the eighteenth century by the Reverend Mitchell and by Lord Cavendish. These experiments were repeated in the nineteenth century by Professors Poynting and Boys with improved apparatus. The present accepted value of G obtained by Dr. Heyl† at the National Bureau of standards in Washington is

$$(6.673 \pm 0.003) \times 10^{-8} \frac{\text{dyne cm}^2}{\text{gm}^2}$$

Essentially all these determinations were carried out by the method indicated in Fig. 4.1. In this method a light rod has a mass m placed on each end. The rod and masses are suspended horizontally by a vertical fine fiber. Close to each mass m is a large mass M. When the large masses are in the positions A and B, the small masses are attracted and take the equilibrium position at $m'm'$. When the large masses M are moved to the position $A'B'$, the rod and light masses move to a position indicated at $m''m''$. If the masses and their distances apart, the torsion constant of the fiber, and the angles of torsion are known, it is possible to calculate G. Such a calculation is called for in Problem 15, Chapter 6. Since the force of attraction is very small, the

---

* If the same size wood and iron balls are dropped, they fall with the same downward acceleration although the inertia of the iron ball is, say, k times that of the wooden ball. This is because the force on the iron ball, due to gravity, is also k times that on the wooden ball. This equality of the gravitational and inertial masses is by no means a trivial fact. For example, a proton and an alpha particle would fall in a gravitational field with the same acceleration, but in an electrc field they would experience very different accelerations.

† P. R. Heyl and P. Chrzanowski, *J. Res. Nat. Bur. Standards*, **29**, July 1942.

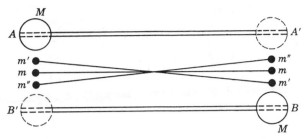

Fig. 4.1   Cavendish method for measuring the gravitational constant *G*.

fibers must have a very small torsion constant. Professor Boys made a significant improvement when he developed a method for drawing very fine quartz fibers.

In the statement of Newton's law of gravitation, the masses are considered to be particles, and, since the earth is anything but a particle from the point of view of a falling apple, the question arises as to what is the appropriate distance between the earth and the apple. To solve this problem, Newton is said to have invented the calculus. All this he did as an undergraduate while on vacation!

4.2   The Gravitational Attraction between a Uniform Sphere and a Particle

We shall now prove that the earth attracts an external particle as though all the mass of the earth were concentrated in a point at its center. To do this we must first consider the attraction between a uniform spherical shell and an external particle, since a solid sphere may be thought of as built up from a large number of concentric thin shells.

Let R be the radius of a thin shell whose mass per unit area is σ. The problem is to find the force of gravitational attraction between this shell and a particle of mass m at P located at a distance l from the center of the shell, as shown in Fig. 4.2. In order to find this force, we first divide the shell into a number of strips formed by the intersection of the shell with spheres, having centers at P, of radii r and $r + dr$. The particular strip shown in Fig. 4.2 has a width $R \, d\alpha$. The area of the strip is $2\pi R \sin \alpha R \, d\alpha$, and its mass is $\sigma 2\pi R^2 \sin \alpha \, d\alpha$. Every element on this strip is, to a first approximation, at the same distance r from the particle at P. The force of attraction between any element in this strip and the particle at P is along a straight line drawn between the element and P. Thus the vectors representing the forces at P form a cone with its vertex at P. Two of these vectors are shown in the plane

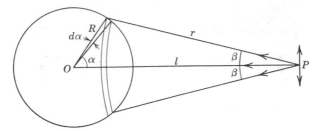

Fig. 4.2 Attraction of a uniform shell and an external particle.

of the paper. The forces at P can be resolved into components along and at right angles to PO. It is easily seen that the components at right angles to OP cancel and have a resultant of zero, whereas the components along PO add together. The resultant force between the mass m at P and the strip is along the line PO and has a magnitude of

$$dF = G\frac{(2\pi\sigma R^2 \sin\alpha\, d\alpha)}{r^2} m\cos\beta \tag{4.2}$$

Since the values of r, α, and β are different for each strip, it is not possible to integrate Eq. 4.2 directly to find the force between the shell and the particle. We must eliminate some of the variables. From the cosine formula we have

$$r^2 = R^2 + l^2 - 2Rl\cos\alpha$$

and by differentiation

$$2r\, dr = 2Rl\sin\alpha\, d\alpha$$

since R and l are constants. Thus

$$\sin\alpha\, d\alpha = \frac{r\, dr}{lR}$$

Similarly

$$\cos\beta = \frac{r^2 + l^2 - R^2}{2rl}$$

Substituting for $\sin\alpha\, d\alpha$ and $\cos\beta$ in Eq. 4.2, we have

$$dF = G\frac{\sigma\pi m R}{l^2}\left(1 + \frac{l^2 - R^2}{r^2}\right)dr$$

Since the total mass of the shell m_s is $4\pi R^2\sigma$, the force dF may be given as

$$dF = G\frac{m_s m}{4Rl^2}\left(1 + \frac{l^2 - R^2}{r^2}\right)dr$$

This equation now contains only the variable r so that the resultant force F between the shell and the particle can be obtained by integrating this equation between the limits $l - R$ and $l + R$. By integration

$$F = \frac{Gm_sm}{4Rl^2}\left[r - \frac{l^2 - R^2}{r}\right]_{l-R}^{l+R}$$

$$= \frac{Gm_sm}{4Rl^2}\left[2R + \frac{2R(l^2 - R^2)}{l^2 - R^2}\right]$$

$$= \frac{Gm_sm}{l^2} \tag{4.3}$$

This is the value of the force of attraction between two particles of mass m_s and m respectively, separated by a distance l. Thus, as far as the attraction of the shell and an external particle is concerned, the shell behaves as though all its mass were concentrated at its center. Since a uniform solid sphere may be considered to be built up of an infinite number of concentric shells, it follows that a uniform solid sphere attracts an external particle as though all the mass of the sphere were concentrated at a point at its center. If the earth is a uniform sphere or made up of uniform concentric shells, it attracts an external particle such as the apple as though all the mass of the earth were concentrated at a point at its center.

In Problem 1 at the end of the chapter you are asked to show that the resultant force on a particle inside a closed spherical shell is zero. There is no gravitational force inside a closed shell, and the gravitational potential within the shell is constant and equal to the value on its surface. This is analogous to a similar conclusion made in electrostatics, namely that the electric field inside a closed charged conductor is zero and the electric potential inside the shell is constant and equal to the value at its surface. These results are a consequence of the inverse square law which holds for point electric charges and for particles.

It follows from these results that the force on a particle within a solid sphere is due to the mass enclosed by the concentric spherical surface on which the particle lies, since the portion of the sphere outside the particle exerts no force on the particle.

4.3 Gauss' Law

Another manner of stating the inverse square law, which is useful in solving problems involving symmetrical figures, was first given by the

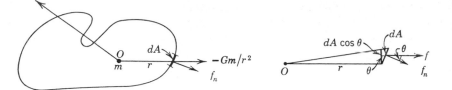

Fig. 4.3   Gauss' law for a closed surface about a mass m.

German scientist Gauss and is called Gauss' law. This law may be stated in the following manner.

The total outward gravitational flux through any closed surface is equal to $-4\pi G$ times the total mass enclosed by the surface.

If f_n is the normal component of the gravitational force on a unit mass at an element of area dA, then $f_n\, dA$ is defined as the gravitational flux through dA, and Gauss' law may be written

$$\int_{\substack{\text{closed}\\ \text{surface}}} f_n\, dA = -4\pi GM$$

where M is the total mass enclosed within the surface. To prove this law, let us consider a mass m at O (Fig. 4.3) with a closed surface drawn around it. If dA is a small element of area in the surface at a distance r from O, then the gravitational force exerted on a unit mass at dA by the mass m is Gm/r^2. This force is directed along r from dA to O, and the normal component of this force is f_n in an outward direction where

$$f_n = -\frac{Gm\cos\theta}{r^2}$$

The gravitational flux through dA is

$$f_n\, dA = -\frac{Gm\cos\theta\, dA}{r^2}$$

But $dA\cos\theta$ is the component of the area dA perpendicular to the radius r, and $(dA\cos\theta)/r^2$ is the element of solid angle $d\Omega$ subtended by dA at O. Hence

$$f_n\, dA = -Gm\, d\Omega$$

and integrating this expression over any closed surface gives

$$\int_{\substack{\text{closed} \\ \text{surface}}} f_n \, dA = -Gm \int_{\substack{\text{closed} \\ \text{surface}}} d\Omega = -4\pi Gm$$

If the surface drawn about the mass has re-entrant portions as shown in the figure, then there must always be an odd number of times in which the lines from elementary cones drawn from O to the outside must cut the surface so that the expression given holds for any surface enclosing a mass. When the surface encloses a number of separate masses m_1, m_2, etc., whose total mass is M, then Gauss' law becomes

$$\int_{\substack{\text{closed} \\ \text{surface}}} f_n \, dA = -4\pi G(m_1 + m_2 \cdots) = -4\pi GM$$

As an example, let us consider the gravitational force on a unit mass placed outside a homogeneous sphere of mass M (Fig. 4.4). Since there is symmetry, we shall choose as the closed surface a sphere having the same center as the homogeneous mass. This closed surface, shown dotted in the figure, has a radius R larger than that of the mass. Since f_n is everywhere the same on the surface, the total gravitational flux over the closed surface $\int f_n \, dA$ is in this case $4\pi R^2 f_n$. Thus from Gauss' law we have

$$4\pi R^2 f_n = -4\pi GM \quad \text{or} \quad f_n = \frac{-GM}{R^2}$$

Hence the gravitational force on a unit mass at a distance R from the center of a homogeneous sphere of mass M is $-GM/R^2$, where the negative sign indicates that there is attraction between the mass M and the unit mass. This is essentially the result given in Eq. 4.3. It should be stated that Gauss' law is of great importance in electrostatics where the inverse square holds between point charges.

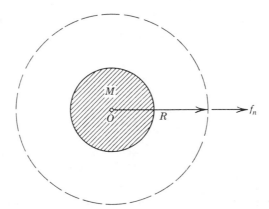

Fig. 4.4   Gravitational force at a distance R from a homogeneous sphere of mass M.

4.4   The Acceleration Due to Gravity and the Law of Gravitation

The force the earth exerts on a mass m near its surface is its weight mg and is given as

$$mg = G\frac{Mm}{R^2} - (mR\cos\theta)\omega^2 \tag{4.4}$$

where M is the mass, R the radius, ω the angular velocity of rotation of the earth, θ the latitude at which the mass m is located, and g the acceleration due to gravity.[*] Since the centrifugal term has a maximum value of 0.3% of

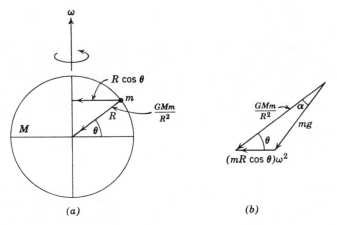

Fig. 4.5   (a) Earth mass M, radius R. (b) Forces acting on particle of mass m; α is the angle which a plumb-line direction mg deviates from line from m to center of earth.

the first term at the equator $(\theta = 0)$, Fig. 4.5, the approximate value of the acceleration resulting from gravity g is given by

$$mg = G\frac{Mm}{R^2}$$

Hence

$$g = G\frac{M}{R^2} \tag{4.5}$$

[*] It should be noted that the earth is not a perfect sphere; the equatorial radius is larger than the polar radius by about 21 km, the mean radius being 6371 km. Thus the shape of the earth is slightly elliptical and is such that the direction of the vector g, or of a stationary plumb line, is perpendicular to the surface of the earth at the point being considered. This is apart from local variations of mass near the surface of the earth.

If g and R are known from independent experiments, the measurement of G, the gravitational constant, provides a means of obtaining the mass of the earth, which is about 6.0×10^{27} gm or 6.7×10^{21} tons.

4.5   The Variation of the Acceleration Due to Gravity with Elevation

From the previous discussion it follows that the acceleration due to gravity at any point varies as the inverse square of the distance from the center of the earth to the point. The change in this acceleration is negligibly small for objects falling from points near the earth, but this is not so where the distance of fall is an appreciable fraction of the radius of the earth, 4000 miles.

Let O be the center of the earth of radius R and mass M, Fig. 4.6. A particle of mass m is placed at P at a distance r from the center of the earth. In order to locate the point P, let O be the origin of the vector $\mathbf{r}$ which is equal to OP. The force on the mass m is

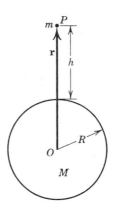

$$\mathbf{F} = -G \frac{Mm}{r^2} \mathbf{r}_1$$

where the negative sign indicates that the force $\mathbf{F}$ is in the opposite direction to the vector $\mathbf{r}$ or the unit vector $\mathbf{r}_1$. By Newton's second law, the equation of motion of the particle at P is

$$m\ddot{r}\mathbf{r}_1 = -G \frac{Mm}{r^2} \mathbf{r}_1$$

or

$$\ddot{r} = -G \frac{M}{r^2} \qquad (4.6)$$

Fig. 4.6   Variation of gravity with elevation.

To integrate Eq. 4.6 or to change from an acceleration to a velocity, we use the identity $\dot{r}\, dt = dr$. Thus

$$\ddot{r}\dot{r}\, dt = -G \frac{M}{r^2}\, dr$$

or

$$\frac{1}{2} \frac{d}{dt} (\dot{r}^2)\, dt = \tfrac{1}{2}\, d(\dot{r}^2) = GM\, d\left(\frac{1}{r}\right) \qquad (4.7)$$

For the initial conditions, suppose that the mass m starts from rest at a height h above the surface of the earth, or, at $t = 0$, $\dot{r} = 0$, and $r = R + h$.

Then the velocity v with which the mass strikes the ground is given by

$$\tfrac{1}{2}[\dot{r}^2]_0^v = GM\left[\frac{1}{r}\right]_{R+h}^{R}$$

Thus

$$v^2 = 2GM\left(\frac{1}{R} - \frac{1}{R+h}\right) = \frac{2GMh}{Rh + R^2} \tag{4.8}$$

If the particle starts at a very large distance from the earth so that h is much larger than R, then the velocity v_∞ with which it strikes the earth when falling from an infinite distance is

$$v_\infty = \sqrt{\frac{2GM}{R}} \tag{4.9}$$

This is likewise the vertical velocity a particle must have if it is to escape permanently from the earth's gravitational force. This velocity of escape from the earth, v_∞, is about 7 mi/sec.

If the body falls from a height h which is small relative to the radius of the earth, show from Eq. 4.8 that its velocity v on reaching the earth is given by $v^2 = 2gh$.

We shall now determine the time a body takes to fall from a large elevation h to the earth. For convenience let $R + h = R_1$. Then the velocity $\dot{r}$ at any distance r from the center of the earth is obtained from Eq. 4.8.

$$\dot{r}^2 = 2GM\left(\frac{1}{r} - \frac{1}{R_1}\right)$$

Thus

$$\frac{dr}{dt} = -\sqrt{2GM}\sqrt{\frac{1}{r} - \frac{1}{R_1}} \tag{4.10}$$

The square root with the minus sign has been chosen because the velocity dr/dt is positive in the upward direction of increasing r whereas the right-hand side of the equation decreases as r increases.

In order to integrate Eq. 4.10, let us change the variable from r to θ by making the substitution

$$r = R_1 \cos^2 \theta$$

Then

$$dr = -2R_1 \cos \theta \sin \theta \, d\theta$$

Substituting these values in Eq. 4.10, we have

$$\sqrt{2GM} \, dt = \sqrt{\frac{R_1^2 \cos^2 \theta}{R_1(1 - \cos^2 \theta)}} \, 2R_1 \cos \theta \sin \theta \, d\theta$$

or

$$\sqrt{2GM} \, dt = 2R_1^{3/2} \cos^2 \theta \, d\theta = R_1^{3/2}(1 + \cos 2\theta) \, d\theta$$

Integrating

$$\sqrt{2GM}\ t = R_1^{3/2}\left(\theta + \frac{\sin 2\theta}{2}\right) + C$$

$$= R_1^{3/2}\left[\cos^{-1}\sqrt{\frac{r}{R_1}} + \sqrt{\frac{r(R_1 - r)}{R_1^2}}\right] + C$$

If we again assume the same initial conditions as above, namely, that at $t = 0$, $r = R_1$, then the constant of integration C is zero. The time T taken by the body to reach the earth is found by putting $r = R$, where R is the radius of the earth.

$$T = \sqrt{\frac{R_1}{2GM}}\left[R_1 \cos^{-1}\sqrt{\frac{R}{R_1}} + \sqrt{RR_1 - R^2}\right] \qquad (4.11)$$

4.6   Gravitational Potential Due to the Earth

In Section 3.6 it is shown that the attractive force between two particles is a conservative force and that the potential energy V at a distance r between the particles is $V = (-Gm_1m_1)/r$, where the potential is considered to be zero when r is infinite. If one of the masses is considered to be that of the earth and r is measured from the center of the earth, then for the earth of mass M and a unit mass, the change in potential energy dV per unit mass for a displacement dr is

$$dV = \frac{GM}{r^2}\ dr$$

If the potential is V_0 at a distance r_0 from the center of the earth and V at a distance r, then by integration

$$\left[V\right]_{V_0}^{V} = \left[-\frac{GM}{r}\right]_{r_0}^{r}$$

or

$$V - V_0 = GM\left(\frac{1}{r_0} - \frac{1}{r}\right)$$

If the gravitational potential is taken as zero at an infinite distance, or $V_0 = 0$ when $r_0 = \infty$, the potential at any distance r from the center of the earth is

$$V = -\frac{GM}{r} \qquad (4.12)$$

The presence of the negative sign sometimes gives trouble in Eq. 4.12. This need not be so if we bear in mind the definition of potential difference. Since the gravitational force is directed toward the earth, the potential increases away from the earth or upward. If the potential is taken as zero at an infinite distance, then it must become increasingly negative in a direction toward the earth. The smaller the value of r, the smaller, that is to say the more negative, the potential becomes.

4.7   Kepler's Laws

During the seventeenth century there was considerable interplay between the sciences of astronomy, mathematics, and physics. Late in the sixteenth century a Polish monk, Copernicus, had accounted for the motion of the planets and stars by assuming a central sun with the earth and other planets revolving around it. This was called the heliocentric or Copernican theory in contrast to the earlier geocentric or Ptolemaic theory which assumed a central stationary earth with the sun and planets revolving around it. Ptolemy's theory had been accepted for about fifteen centuries, but during this time there had developed an increasing number of discrepancies between the observations and the theory.

On the observational side Tycho Brahe, the last great astronomer to make observations without the aid of a telescope, had carefully collected data on the position of the planet Mars. These data were given to his student Kepler, who worked nearly twenty years before he obtained any satisfactory explanation of the observations. His results are summarized in what have become known as Kepler's three laws. They are:

1. *The law of elliptical orbits: The planets move in ellipses about the sun which is located at one of the foci.*

2. *The law of areas: The line joining any planet to the sun sweeps out equal areas in equal intervals of time.*

3. *The harmonic law: The square of the period of any planet about the sun is proportional to the cube of its mean distance from the sun.*

These are empirical laws inasmuch as they were derived directly from observational data without any interpretation by physical theory. Newton was able to show that Kepler's laws could be derived from the laws of motion and the law of gravitation. This was a great triumph for the Newtonian theory and had much to do with the subsequent rapid increase in knowledge in the physical sciences. We shall now repeat Newton's analysis, though not in the form given in *The Principia*.

4.8   Motion as Governed by Newton's Law of Gravitation

Let us consider the motion of a particle of small mass m which is attracted by a particle of large mass M. The force of gravitational attraction between these masses is along the line joining them, so the resulting motion is often called *motion under a central force*. The acceleration of the large mass M will be much smaller than that of the small mass m, so that without making too much error we may consider M to be at rest with m moving about it.

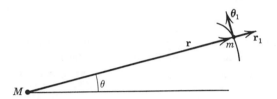

Fig. 4.7 Motion of a small mass m under the gravitational field of a large mass M.

Any motion under a central force takes place in a plane which we shall here assume to be the plane of this page. Suppose the origin of the coordinates to be at M and the position of m to be given by the vector $\mathbf{r}$ joining M to m and making an angle θ with some fixed direction (Fig. 4.7). The coordinates r, θ are the polar coordinates of m with respect to M.

By Newton's law of gravitation the force $\mathbf{F}$ exerted by M on m is

$$\mathbf{F} = -\frac{GMm}{r^2}\mathbf{r}_1$$

where $\mathbf{r}_1$ is the unit vector in the direction of $\mathbf{r}$, namely, from M to m. The general expression for the acceleration of a particle moving in a plane is given in Chapter 1. Thus the equation of motion of mass m is

$$\mathbf{F} = -\frac{GMm}{r^2}\mathbf{r}_1 = \mathbf{r}_1 m(\ddot{r} - r\dot{\theta}^2) + \boldsymbol{\theta}_1 m(r\ddot{\theta} + 2\dot{r}\dot{\theta}) \qquad (4.13)$$

From this it follows that, since the component of force along $\boldsymbol{\theta}_1$, or perpendicular to $\mathbf{r}_1$, is zero, the component of acceleration in this direction must be zero. Hence for *any* central force

$$r\ddot{\theta} + 2\dot{r}\dot{\theta} = 0$$

This may be written

$$\frac{1}{r}\frac{d}{dt}(r^2\dot{\theta}) = 0$$

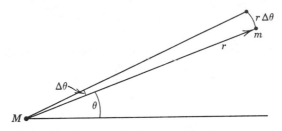

Fig. 4.8   Area swept out by the vector **r**.

which gives by integration

$$r^2\dot\theta = \text{constant} = k \qquad (4.14)$$

The quantity $mr^2\dot\theta$ is called the angular momentum or moment of momentum, mvr, of the moving mass m about the origin or the position of M. From the discussion it follows that the angular momentum of a planet moving about the sun is a constant.

Let us consider the significance of the equation $r^2\dot\theta = k$. Let the vector **r** in Fig. 4.7 rotate through a small angle $\Delta\theta$, as shown in Fig. 4.8. The area ΔA swept out by the vector **r** in turning through $\Delta\theta$ would then be

$$\Delta A = \frac{r^2\,\Delta\theta}{2}$$

If the angle $\Delta\theta$ is turned through in a time Δt, then the area ΔA is also swept out in this time.

$$\frac{\Delta A}{\Delta t} = \frac{r^2\,\Delta\theta}{2\,\Delta t}$$

and in the limit

$$\frac{dA}{dt} = \frac{r^2\,d\theta}{2\,dt}$$

Thus from Eq. 4.14

$$r^2\dot\theta = 2\frac{dA}{dt} = k \qquad (4.15)$$

or the rate at which the small mass m sweeps out an area as it moves about M is a constant. This is Kepler's law of areas in which the small mass represents a planet and the large mass the sun. From what has been said it follows that the law of areas is valid for any type of central force whether of the inverse square type or not.

In order to find the equation of the path in which m moves it is necessary to eliminate the time t and to obtain an equation between r and θ. The equations from which t is to be eliminated are

$$r^2 \dot{\theta} = k$$

and from Eq. 4.13

$$-\frac{GMm}{r^2} = m(\ddot{r} - r\dot{\theta}^2) \tag{4.16}$$

A considerable simplification in the elimination of the time from these equations can be introduced by changing the variable to u where $r = 1/u$. By differentiation

$$\frac{dr}{d\theta} = -\frac{1}{u^2}\frac{du}{d\theta}$$

and

$$\dot{r} = \frac{dr}{dt} = \frac{dr}{d\theta}\frac{d\theta}{dt} = -\frac{1}{u^2}\frac{du}{d\theta}\frac{k}{r^2} = -k\frac{du}{d\theta}$$

Similarly

$$\ddot{r} = \frac{d\dot{r}}{dt} = -k\frac{d}{dt}\left(\frac{du}{d\theta}\right) = -k\frac{d}{d\theta}\left(\frac{du}{d\theta}\right)\frac{d\theta}{dt} = -k^2 u^2 \frac{d^2 u}{d\theta^2}$$

Substituting these values in Eq. 4.16, we have

$$-GMu^2 = -k^2 u^2 \frac{d^2 u}{d\theta^2} - \frac{k^2 u^4}{u}$$

or

$$\frac{d^2 u}{d\theta^2} + u = \frac{GM}{k^2} \tag{4.17}$$

A solution of this equation, as may be verified by substitution, is

$$u = C\cos(\theta + \alpha) + \frac{GM}{k^2}$$

where C and α are the arbitrary constants of integration. Thus the equation of the path is

$$\frac{1}{r} = C\cos(\theta + \alpha) + \frac{GM}{k^2}$$

The constant α may be eliminated by measuring θ from a new axis, making an angle $-\alpha$ with the old axis. In terms of this new axis the equation of the path is

$$\frac{1}{r} = C\cos\theta + \frac{GM}{k^2} \tag{4.18}$$

This is the equation of a conic section: an ellipse, a parabola, or a hyperbola.

Let us consider the properties of an ellipse, Fig. 4.9, with the two foci marked F, F'. The fundamental property of an ellipse is that the sum of the

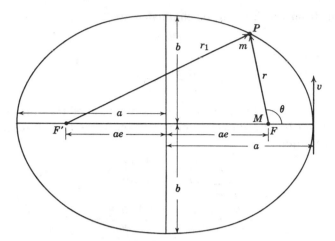

Fig. 4.9   An ellipse, $r_1 + r = 2a$.

distances from each of the foci to any point on the ellipse is a constant. Thus if r_1 and r are the distances from the foci to any point P on the ellipse, then $r_1 + r$ is equal to a constant* which, as can be seen from the figure, is equal to $2a$, the length of the major axis. The distance between the two foci is $2ae$, where e is the eccentricity of the ellipse and is less than unity. Notice as the two foci come closer together, the eccentricity becomes smaller, so that, when the two foci coincide, $e = 0$, the ellipse becomes a circle. From the cosine formula for the triangle FPF', Fig. 4.9, it follows that

$$r_1{}^2 = r^2 + (2ae)^2 + 2(2ae)r \cos \theta$$

Also

$$r_1 + r = 2a, \text{ so } r_1 = 2a - r \text{ and } r_1{}^2 = r^2 + 4a^2 - 4ar$$

By subtraction:

$$4a^2(1 - e^2) = 4ar(1 + e \cos \theta)$$

The equation for an ellipse is

$$\frac{1}{r} = \frac{1 + e \cos \theta}{a(1 - e^2)} = \frac{e \cos \theta}{a(1 - e^2)} + \frac{1}{a(1 - e^2)} \tag{4.19}$$

* You can readily draw an ellipse by placing two pins at the foci and tying a loose loop of string around them. Place a pencil within the loop and, keeping the string taut, trace out the ellipse. For a hyperbola it is the difference of the two distances from the foci which is constant, or $r_1 - r = $ constant, and the equation of a hyperbola can be shown to be $\dfrac{1}{r} = \dfrac{e \cos \theta' - 1}{a(e^2 - 1)}$, with the eccentricity e larger than unity. (See Fig. 4.12 for θ' or Problem 28.)

Since the ellipse (or circle) is the only closed figure among the conic sections, it follows that the planets must move in ellipses about the sun, with the sun at one of the foci. Comparing Eq. 4.18 with Eq. 4.19, it follows that:

$$C = \frac{e}{a(1 - e^2)} \quad \text{and} \quad \frac{GM}{k^2} = \frac{1}{a(1 - e^2)}$$

or

$$a = \frac{k^2}{GM(1 - e^2)} \quad \text{and} \quad e = \frac{Ck^2}{GM} \tag{4.20}$$

The area of the ellipse may be shown to be πab, and the semimajor and semiminor axes are related by $b = a\sqrt{1 - e^2}$ (see Fig. 4.9). If P is the period of revolution of the mass m, then the area πab is swept out in time P. From Eq. 4.15, the rate of sweeping out of this area is $k/2$. Thus

$$\frac{dA}{dt} = \frac{\pi ab}{P} = \frac{k}{2}$$

Hence

$$P = \frac{2\pi ab}{k} = \frac{2\pi a^2 \sqrt{1 - e^2}}{k}$$

Using Eq. 4.20, we have

$$P = \frac{2\pi a^{3/2}}{\sqrt{GM}}$$

or

$$P^2 = \frac{4\pi^2 a^3}{GM} \tag{4.21}$$

Thus the square of the period of revolution is proportional to the cube of the semimajor axis or the cube of the mean distance of m from M, which is Kepler's harmonic law. By means of this law, the mass of the sun can be determined if the period and mean distance of one of the planets are known. Similarly, if a planet has a satellite whose period and mean distance are known, then the mass of the planet can be determined.

The harmonic law as given in Eq. 4.21 is only approximately correct since we have assumed the large mass M to be stationary. The acceleration of mass M or of the sun toward a planet of mass m is Gm/r^2 and of the planet toward the sun is GM/r^2, where r is their distance apart (Fig. 4.10). If the acceleration

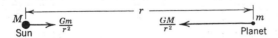

Fig. 4.10   Relative accelerations of the sun M and planet m.

of the sun is to be made zero, then it must be given an acceleration of Gm/r^2 away from the planet. The acceleration of the planet relative to the sun is then

$$\frac{GM}{r^2} + \frac{Gm}{r^2} = \frac{G}{r^2}(M + m)$$

The corrected form of Kepler's harmonic law is then

$$P^2 = \frac{4\pi^2 a^3}{G(M + m)} \tag{4.22}$$

Thus the law correctly stated should be: The square of the periodic time of a planet about the sun is proportional to the cube of its mean distance from the sun and inversely proportional to the sum of the masses of the planet and sun. Actually this correction is very small for the solar system, as may be seen from the values of the relative masses: earth 1, moon $\frac{1}{81}$, jupiter 320, sun 330,000.

4.9   The Energy of a Planet Moving about the Sun

Assuming the sun to be at rest and the planet to move in an elliptical path about the sun, the energy of the planet consists of its kinetic and potential energies. This total energy E can be written as

$$E = \frac{m\dot{r}^2}{2} + \frac{mr^2\dot{\theta}^2}{2} - \frac{GmM}{r} \tag{4.23}$$

where the first two terms on the right give the linear and angular kinetic energies respectively, and the last one gives the potential energy which is intrinsically negative for an attractive force. This energy E is constant for the motion of a planet about the sun, and we may therefore calculate its value at convenient positions, such as at the ends of the major axis where $\dot{r} = 0$ and $\theta = 0$ or π (Fig. 4.9). Also for this motion, Eq. 4.14 gives $r^2\dot{\theta} = k$, so that $r^2\dot{\theta}^2 = k^2/r^2$.

When $\theta = 0$, r has its minimum value, r_m, which is given from Eq. 4.19 as

$$\frac{1}{r_m} = \frac{1 + e}{a(1 - e^2)} = \frac{GM}{k^2}(1 + e) \tag{4.24}$$

using also Eq. 4.20. For $\theta = 0$, $\dot{r}_m = 0$, and Eq. 4.23 can be written as

$$E = \frac{mk^2}{2r_m^2} - \frac{GmM}{r_m} = \frac{mk^2 G^2 M^2(1 + e)^2}{2k^4} - \frac{GMm(1 + e)GM}{k^2} \tag{4.25}$$

$$= \frac{G^2 M^2 m}{k^2}\left[\frac{1 + 2e + e^2}{2} - (1 + e)\right] = -\frac{G^2 M^2 m}{2k^2}(1 - e^2) \tag{4.26}$$

Since the eccentricity e is less than unity for an ellipse, it follows that the energy of a planet in its orbit is negative as it must be for an attractive force as between a planet and the sun. The situation is comparable to that of the potential due to the earth, Eq. 4.12, where the negative sign for an attractive force is discussed.

The expression for the energy E, given in Eq. 4.26, can be rewritten using Eq. 4.20 to give

$$E = -\frac{G^2M^2m}{2GMa} = -\frac{GMm}{2a}$$ (4.27)

Thus the energy of a planet, going about the sun in an elliptical orbit of major axis $2a$, depends only on the size of the major axis and not on the eccentricity e of the orbit. From Eq. 4.21 a similar statement holds for the period of a planet about the sun. Hence all elliptical orbits having the same major axes have the same energy and period.

The expression for the energy E, Eq. 4.23, can be written, using the fact that $r^2\dot\theta = k$, as

$$E = \frac{m\dot{r}^2}{2} + \frac{mk^2}{2r^2} - \frac{GMm}{r}$$ (4.28)

In Eq. 4.28 the first term on the right-hand side is the linear kinetic energy of the planet, whereas the remaining terms depend on $1/r^2$ and $1/r$ respectively, and may be looked on as an effective potential energy U where

$$U = \frac{mk^2}{2r^2} - \frac{GMm}{r}$$ (4.29)

(Check the dimensions of this equation.) Notice that the first term on the right arises from the kinetic reaction commonly called the centrifugal force. It may be readily shown that $-\partial U/\partial r$ gives $m\ddot{r}$ by using Eq. 4.16.

As a qualitative discussion of the effective potential energy U, Eq. 4.29, consider the potential energy function $U' = 1/r^2 - 1/r$, which is similar to the function U in its dependence on r but has different numerical constants. A plot of U' against r is shown in Fig. 4.11a. For the region of positive energies such as at E_1, no oscillatory motion is possible. Notice that if a mass m starts out from rest at an infinite distance and is attracted by the mass M, then its gain in kinetic energy at any position is equal to its loss in potential energy so that its total energy is zero. For the positive energy E_1 the mass must have been supplied with some outside energy and would come up to the distance r_a of mass M and stop and then reverse its direction and travel outward. At negative energies such as E_2, there are two values of r which have the energy E_2, and these values of r correspond to points at the ends of the major axis of the ellipse. The minimum of the negative portion of

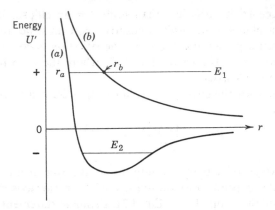

Fig. 4.11   Variation of potential energy U' with distance r for (a) force of attraction; (b) force of repulsion.

the curve at which there is a single value of r corresponds to a circular orbit for the mass m about M.

If there is a repulsive force between the two masses, as in the case of an alpha particle shot at a stationary heavy atomic nucleus, then the potential energy is intrinsically a positive quantity. This situation may be qualitatively discussed in terms of a potential energy function such as $1/r^2 + 1/r$. This is shown in Fig. 4.11b. In this case, a negative energy such as at E_2 is physically impossible, whereas for a positive energy such as at E_1, the particle would come up from infinity to the minimum distance r_b and be repelled and return to infinity.

Another manner of examining the energy of the system is by using Eq. 4.26. Rewriting this equation in terms of the eccentricity e gives

$$e = \sqrt{1 + \frac{2Ek^2}{mM^2G^2}} \qquad (4.30)$$

This shows that for elliptical motion for which e is less than 1 the energy E must be negative, whereas for parabolic motion in which e is equal to 1 the energy must be zero, and for hyperbolic motion, e greater than 1, the energy E must be positive.

4.10   Repulsive Forces and the Nuclear Atom

Repulsive forces are not encountered in gravitation but do appear between like electric charges. Historically, it was the bombardment of very thin gold leaf by alpha particles that led Rutherford to the concept of the nuclear atom. The observed deflection or scattering of the alpha particles was interpreted in terms of the repulsion between the positive charge q of the moving alpha particle of mass m, and the stationary positive charge Q of mass M associated with the nucleus of the atom of gold.

It was Coulomb who showed that the force between point electric charges varies inversely as the square of the distance between the charges. This is similar to the law of gravitation for masses so that a considerable amount of theory concerning the planets and the sun can be used in solving this problem. While there is gravitational attraction between the masses m and M, this is negligibly small compared to the repulsion force between the charges. The repulsive force is taken into account by substituting $-Qq$ for GMm in Eq. 4.17. Thus the equation of motion of mass m is

$$\frac{d^2u}{d\theta^2} + u = \frac{-Qq}{mk^2} \tag{4.31}$$

The general solution of this equation is

$$u = \frac{1}{r} = C \cos(\beta - \theta) - \frac{Qq}{mk^2} \tag{4.32}$$

where $k = r^2\dot{\theta}$. The charges Q and q must be measured in electrostatic units and all other quantities in Eq. 4.31 must be expressed in the cgs system if numerical values are used with the equation.

With a repulsive force, for which the potential energy is $+Qq/r$, the total energy E is positive and the orbit is hyperbolic. The equation for a hyperbolic orbit (see footnote in Section 4.8 and problem 28 in this chapter) is

$$\frac{1}{r} = \frac{e \cos(\beta - \theta) - 1}{a(e^2 - 1)} \tag{4.33}$$

where e is the eccentricity of the orbit and θ' has been replaced by $(\beta - \theta)$. Comparing Eqs. 4.32 and 4.33, it is seen that

$$C = \frac{e}{a(e^2 - 1)} \quad \text{and} \quad \frac{1}{a(e^2 - 1)} = \frac{Qq}{mk^2} \, ; \quad \text{hence } C = \frac{eQq}{mk^2}$$

Now the eccentricity e can be expressed in terms of the energy and other constants by Eq. 4.30, and when $-Qq$ is substituted for GMm this gives

$$e = \sqrt{1 + \frac{2k^2mE}{Q^2q^2}} \quad \text{and} \quad C = \frac{Qq}{mk^2}\sqrt{1 + \frac{2k^2mE}{Q^2q^2}}$$

Thus Eq. 4.32 can be written as

$$\frac{1}{r} = \frac{Qq}{mk^2}\left[\sqrt{1 + \frac{2k^2mE}{Q^2q^2}}\cos(\beta - \theta) - 1\right]$$ (4.34)

This is the equation for the hyperbolic path taken by a positive charge q moving under the influence of a fixed positive charge Q.

The charge q is shown in Fig. 4.12 coming in from infinity along the direction of the asymptote SC with velocity v_0 and moving away to infinity in the direction of the other asymptote CS' due to the repulsive force from Q situated at the exterior focus F of the hyperbola. In the figure the alpha particle q is shown at some instant of time with the corresponding distance r or Fq. The angle θ associated with r is the angle between a line FT, through the focus F and parallel to the asymptote CS, and the line Fq or r.

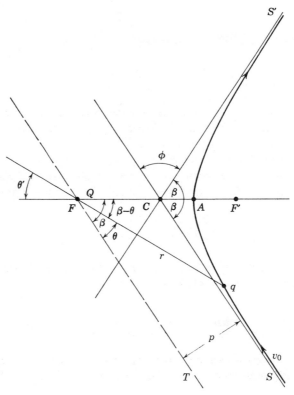

Fig. 4.12 Alpha particle coming in along the asymptotic SC with velocity v_0 is scattered in hyperbolic orbit by force due to Q at exterior focus F ($FC = F'C = ae$, $CA = a$, $\theta' = \beta - \theta$).

When the alpha particle starts out with velocity v_0 from a very large distance, $r = \infty$, the angle θ is very small, and in the limit $r = \infty$, the angle $\theta = 0$. From Eq. 4.34 it follows that r has its minimum value, or $1/r$ its maximum value (*FA* in Fig. 4.12) when $\cos(\beta - \theta) = 1$, that is, $\beta - \theta = 0$ or $\theta = \beta$. The charge q moves in from infinity corresponding to $\theta = 0$ and then moves out to infinity along asymptote CS' corresponding to $\theta = 2\beta$. This latter statement may be seen from Eq. 4.34, since $r = \infty$ when $\theta = 0$, the cosine factor becomes $\cos(\beta)$; so also when $\theta = 2\beta$ the cosine factor becomes $\cos(-\beta)$, and this corresponds to $r = \infty$, since $\cos(\beta) = \cos(-\beta)$. It is in this manner that the angle 2β is drawn between the two asymptotes CS and CS'.

Now the alpha particle comes in from infinity along asymptote SC and moves out to infinity along CS' so that the angle of deflection or scattering, between the positive directions SC and CS', is angle ϕ. Thus from Fig. 4.12 it is seen that

$$\phi + 2\beta = \pi$$

Since $r = \infty$ when $\theta = 0$, it follows from Eq. 4.34 that $r = \infty$ when

$$\sqrt{1 + \frac{2k^2mE}{Q^2q^2}} \cos\beta = 1 \quad \text{or} \quad \cos\beta = \frac{1}{\sqrt{1 + \frac{2k^2mE}{Q^2q^2}}}$$

From a right triangle it is readily found that

$$\tan\beta = \frac{k}{Qq}\sqrt{2mE} \tag{4.35}$$

and since $\beta = (\pi/2 - \phi/2)$, then $\phi/2$ is the other angle in the right triangle so that

$$\tan\beta = \cot(\phi/2)$$

Now the distance p, called the impact parameter, is the perpendicular distance between the asymptote along which the alpha particle comes in ($\theta = 0$) and a parallel line FT through the scattering center. Now $k = r^2\dot\theta = pv_0$ and the total energy $E = mv_0^2/2$, so by substitution in Eq. 4.35 it follows that

$$\tan\beta = \cot\frac{\phi}{2} = \frac{pv_0}{Qq}\sqrt{2mE} = \sqrt{\frac{2E}{m}}\frac{p}{Qq}\sqrt{2mE} = \frac{2Ep}{Qq}$$

or

$$p = \frac{Qq}{2E}\cot(\phi/2) \tag{4.36}$$

To make the connection of this equation with that given by Rutherford for the scattering of alpha particles, we must first investigate the meaning of

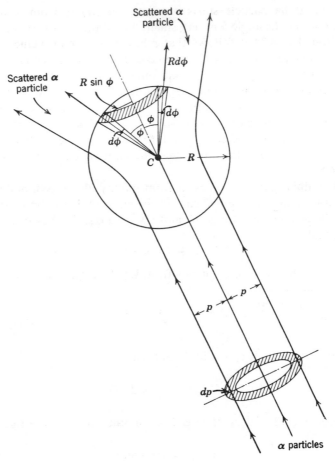

Fig. 4.13 Alpha particles between impact parameters p and $p + dp$ scattered into solid angle $d\Omega$ lying between ϕ and $\phi + d\phi$ by force from charge Q at C. The alpha particles are considered coming from infinity along the asymptote SC in Fig. 4.12.

cross section for scattering. Consider an incident beam of alpha particles, in which the intensity I of the beam is the number of particles crossing unit area perpendicular to the beam per second. This beam of positively charged particles is deflected or scattered by a fixed positive charge Q situated at the center C of a spherical surface (Fig. 4.13). The force exerted by charge Q is spherically symmetric about C.

The scattering cross section σ in direction ϕ is defined so that the number of alpha particles which are scattered into a solid angle $d\Omega$ in direction ϕ per second is $I\sigma \, d\Omega$. A solid angle $d\Omega$ is defined as the area on a sphere divided by the square of the radius of the sphere, so that the element of solid angle subtended by the area between ϕ and $\phi + d\phi$ is

$$d\Omega = \frac{2\pi R \sin \phi (R \, d\phi)}{R^2} = 2\pi \sin \phi \, d\phi$$

The number of alpha particles scattered between angles ϕ and $\phi + d\phi$ is the number between the impact parameters p and $p + dp$, hence

$$2\pi I p \, dp = -2\pi I \sigma \sin \phi \, d\phi \tag{4.37}$$

It should be noted that less particles are scattered as p increases because the repulsive force is less, that is, as p increases ϕ decreases; hence the negative sign. From Eq. 4.37 the cross section for scattering σ in direction ϕ is given as

$$\sigma(\phi) = - \frac{p}{\sin \phi} \frac{dp}{d\phi} \tag{4.38}$$

From Eq. 4.36, by differentiation,

$$\frac{dp}{d\phi} = \frac{-Qq}{2E2 \sin^2 \phi/2}$$

Hence

$$\sigma(\phi) = \frac{1}{2}\left(\frac{Qq}{2E}\right)^2 \frac{\cos \phi/2}{\sin \phi/2} \times \frac{1}{\sin \phi} \times \frac{1}{\sin^2 \phi/2}$$

$$= \frac{1}{4}\left(\frac{Qq}{2E}\right)^2 \times \frac{1}{\sin^4 \phi/2} \tag{4.39}$$

Thus the number of alpha particles scattered per second at the angle ϕ is inversely proportional to the fourth power of the sine of the half angle of scattering. This result was experimentally confirmed by Rutherford and his associates.[*] A partial table of the data, given by Geiger and Marsden in Table II on page 610 of the article on the variation of scattering with angle, is given below.

Angle of deflexion ϕ	Number N of scattered α observed in gold at angle ϕ
150°	33.1
120°	51.9
60°	477
30°	7,800
15°	105,000

* H. Geiger and E. Marsden, "On the Laws of Deflexion of α Particles through Large Angles," *Phil. Mag.* **25**, 1913.

Since σ is proportional to N, it follows from Eq. 4.39 that $N \sin^4 \phi/2$ should be constant. As an exercise, show that within experimental error, this is the case.

In the same article, Geiger and Marsden report on experiments using α particles of different speeds scattered in a fixed direction. From Eq. 4.39 it is seen that, for this case, $E^2\sigma$ or $v^4\sigma$ should be constant. The data given on page 620 is as follows.

Relative values of $1/v^4$	Number N scattered per minute
1.0	24.7
1.21	29.0
1.91	44
4.32	101

As an exercise show these data agree, within experimental error, with the theory as given by Eq. 4.39.

PROBLEMS

1. Show that the force on a particle inside a closed spherical shell is zero, using the method of Section 4.2 and also Gauss' law. Show, as in Section 4.2, that the potential inside a closed spherical shell is a constant equal to its value at the surface of the shell. If the potential at infinity is taken as zero, find the potential at the surface.

2. Find the velocity a body must have if it is to escape from the earth's gravitational field. Also find the velocity of escape from the moon, supposing that the moon subtends an angle of 0.5° at the earth, is 3.84×10^8 m away from the earth, and has a mass of $\frac{1}{81}$ of the mass of the earth.

3. Show by Gauss' law that the force exerted on a mass, placed in a deep hole in the earth, is that due to the matter in the spherical portion below the hole. Find the force on 1 kg at a depth of $\frac{1}{4}$ and $\frac{1}{2}$ of the radius of the earth, assuming that the density of the earth is constant.

4. Find the speed of a satellite going around the earth in a circular path at an elevation of 2000 mi.

5. Show that, as a satellite encounters air resistance and comes closer to the earth, its speed increases. Discuss this increase in speed from an energy standpoint since mechanical energy is lost as heat.

6. Show that for free fall in the earth's gravitational field from infinity results in the same speed at the surface of the earth as that which would be achieved by a free fall from a height above the earth equal to its radius if this took place under a constant acceleration of gravity, 9.8 m/sec².

7. A hole is bored through a diameter of the earth, and assuming the earth to be of constant density, show that the force on a particle in the hole at a distance r from the center is proportional to r and in the direction towards the center. Show that $\ddot{r} = -gr/R$, where R is the radius of the earth and g is the acceleration of gravity at the surface of the earth. Assume a solution $r = R \cos 2\pi t/T$, where T is the period of the motion, and show that the time a particle dropped from the surface of the earth into the hole takes to reach the center is $\sqrt{\pi^2 R/4g}$. Show that the speed of the particle at the center of the earth is $\sqrt{gR}$ and the speed with which the particle emerges from the other side of the earth is zero.

8. What is the approximate per cent error made in calculating the period of the earth about the sun, if one assumes that the sun is fixed relative to the earth rather than the two rotate about their common center of mass.

9. Assuming the mass of the earth to be 5.95×10^{24} kg, find the mass of the moon if its mean distance from the earth is 239,000 mi or 3.84×10^8 m and its period about the earth is 27.3 days.

10. Find the mass of the sun in kilograms, assuming that the mass of the earth is negligible. The period of the earth about the sun is approximately 365.25 days, and the mean distance of the earth from the sun is 93,000,000 mi (1 mi = 1.608×10^3 m).

11. A body falls from a height of 1000 mi toward the earth. If air resistance is neglected but the variation of gravity with height is taken into consideration, find (a) the velocity with which the body reaches the ground; (b) the time taken to reach the ground; (c) the time taken in falling the last 500 mi to the earth. Assume that the radius of the earth is 4000 mi.

12. A bullet is shot vertically upward with a velocity of 5 mi/sec. If air resistance is neglected but the variation of gravity with height is considered, find (a) the height above the earth's surface to which the bullet rises, in miles; (b) the time the bullet takes to reach its maximum height; (c) the velocity with which the bullet strikes the ground, in miles per second.

13. A particle of mass m moves under a central repulsive force of mb/r^3 and is initially moving at a distance a from the origin of the force with a velocity V at

right angles to a. Show that the equation of the path of the particle is

$$r \cos (p\theta) = a$$

where $p^2 = (b/a^2 V^2) + 1$.

14. If the mean distance of Mars from the sun is 1.524 times that of the earth from the sun, find the time of revolution in years of the planet Mars about the sun.

15. The first check Newton made on the law of gravitation was to compare the acceleration of an apple toward the earth with that of the moon toward the earth. He assumed the path of the moon about the earth to be circular and the distance to the moon to be sixty times the mean radius of the earth. Show that the acceleration due to the earth's gravity at the moon is approximately equal to that of the moon in its orbit, each being equal to about 0.27 cm/sec².

16. A particle of mass m is attracted by a central force ma/r^3 toward a fixed point. Initially the particle is at a distance c from the fixed point and has a velocity of $\sqrt{a/c}$ in a direction at 45° to c and away from the fixed point. Show that the path of the particle is an equiangular spiral whose equation is $r = ce^\theta$.

17. The period of the planet Jupiter is 11.86 years, and its mean distance from the sun is 483×10^6 mi. A satellite of Jupiter, analogous to our moon, is 261,000 mi from Jupiter, and the satellite has a period of 1 day 18.5 hours. From these data show that the ratio of the mass of the sun to the mass of Jupiter is a little larger than 1000.

18. From the data in Problem 17, find the mass of the planet Jupiter.

19. Two particles having masses of m and M respectively attract each other according to the law of gravitation. Initially they are at rest at an infinite distance apart. Show that their relative velocity of approach is

$$\sqrt{\frac{2G(M + m)}{a}}$$

where a is their separation and G is the constant of gravitation.

20. The maximum and minimum velocities of a planet revolving about the sun are 20 and 18 kilometers per second respectively. Find the eccentricity of the orbit.

21. A particle of mass m is attracted by a central force of ma/r^5 toward a fixed point. Initially the particle is at a distance c from the origin and has a velocity of $\sqrt{a/2c^4}$ in a direction at right angles to c. Show that the equation of the orbit is $r = c \cos \theta$.

22. Show that, if the force between two masses m and M is that given by the repulsive inverse square law, as for similarly charged particles, the orbit of m about M is hyperbolic. Assume that M is very much greater than m. (Consider the total energy of the system.)

23. If the eccentricity of a planet's orbit about the sun is 0.4, find (a) the ratio of the lengths of the major to the minor axes of the orbit of the planet; (b) the ratio of the velocities of the planet when it is at the ends of the major axis of its elliptical orbit.

24. Suppose that a particle of very small mass is at rest at an infinite distance from a large mass M. If the particle m is pulled in toward M by gravitational attraction, show that the orbit of m about M is parabolic. Show that, if the orbit of m about M is to be an ellipse, then its velocity in its orbit must be less than that acquired in falling from infinity.

25. Find the angle α which the force mg makes with the line to the center of the earth (Fig. 4.5) for latitudes 0°, 90°, and 45°.

26. Plot the graph representing the energy, $V = 1/x^2 - 1/x$ for an attractive force, and on the same graph one for a repulsive force whose energy function is $V = 1/x^2 + 1/x$.

27. An alpha particle emitted from radium has a velocity of about 3×10^9 cm/sec and is shot out in such a direction as to make a head-on collision with the nucleus of a gold atom. The charge on the alpha particle is 9.6×10^{-10} electrostatic units, and its mass is 6.6×10^{-24} gm. The charge on the nucleus of the gold atom is $79 \times 4.8 \times 10^{-10}$ electrostatic units. The positive charge on the alpha particle and the positive charge on the nucleus of the gold atom repel each other according to the inverse square law of force. (a) Find the closest distance of approach of the alpha particle to the nucleus of the gold atom in the head-on collision. (b) Show that, in general, the alpha particle would move in a hyperbolic path about the nucleus.

28. Prove the equation for a hyperbola (see Fig. 4.12) is $1/r = (e \cos \theta' - 1)/a(e^2 - 1)$, where $\theta' = (\beta - \theta)$.

5

FREE AND FORCED
HARMONIC OSCILLATIONS

5.1 Introduction

A great many of the phenomena we observe in everyday life have a periodic motion. Our pulse or our heartbeat, a pendulum clock, many kinds of machinery as well as the seasons and night and day are examples of phenomena repeated at regular intervals of time. In many cases the periodic motion is maintained by a periodic driving force. Without this the oscillations would gradually die out, owing to dissipative forces in the system. For these so-called forced oscillations the amplitude or maximum displacement depends on the frequency of the driving force. There is usually some frequency of the driving force at which the oscillations have their maximum amplitude. The vibrating system and the periodic driving force are then in resonance. This phenomenon of resonance plays an important role in almost every branch of physics. It is encountered in mechanical vibrations, in musical instruments, in the dispersion of light and x-rays, and in alternating-current circuits.

The development of the analysis of periodic phenomena gives one an insight into a portion of the development of physics. It shows how relatively complicated phenomena in light, electricity, etc., were explained in terms of simple mechanical ideas. Perhaps from what is presented here a partial picture of this unity in the realm of physics may be obtained. As an introduction to the analysis of periodic phenomena, we shall first discuss the vibrations of an idealized spring.

5.2   Simple Harmonic Motion of a Mass Hung on an Idealized Spring

By an idealized spring is meant one for which it may be assumed that the mass of the spring is negligible, that there are no dissipative forces tending to decrease the motion of the spring, and that the restoring force exerted by the spring is strictly proportional to its extension. Although these assumptions are never completely justified in practice, nevertheless, for a light spring in which the extensions are kept small and the oscillations are not prolonged over a long time, the errors introduced are not large. At this point one may well ask what is meant by a light spring, a small extension, and a long time. It should be recognized that these terms are all relative. By a light spring is meant one whose mass is small compared to the mass hung on its lower end; a small extension is one that is small compared to the length of the spring; a long time is one long compared to the time of one oscillation.

Consider the spring (Fig. 5.1a) with its upper end rigidly supported and

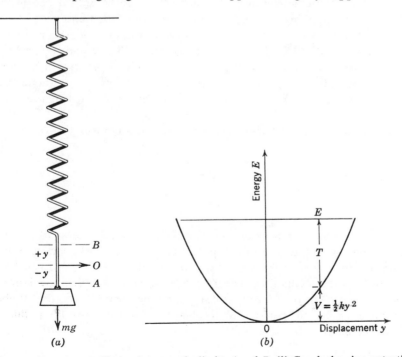

Fig. 5.1   (a) A spring oscillating between the limits A and B. (b) Graph showing potential energy V where $V + T = E$ in simple harmonic motion.

with a mass m or load mg hung on its lower end. The equilibrium position of the spring with the load mg is indicated by the position O. Let O be the origin of the Y coordinates with displacements upward from O being considered positive and below O negative. Suppose that the load is pulled downward a distance $-y$ by an applied force $-F$. There is then an upward restoring force of $+F$ exerted by the spring. If the spring obeys Hooke's law, i.e., the restoring force is proportional to the displacement, then

$$F = -ky \tag{5.1}$$

where k is a constant called the force constant of the spring. This force constant k can be measured in dynes per centimeter, newtons per meter, or pounds per foot. The negative sign in Eq. 5.1 indicates that the restoring force exerted by the spring on the mass m is in the opposite direction to the displacement. Actually the maximum displacement must not go beyond the unstretched length of the spring, otherwise the force is not given by Eq. 5.1.

Suppose that the downward applied force $-F$ is removed. Then the upward restoring force $+F$ accelerates the mass m upward from A to O. Since the restoring force decreases as the mass approaches the equilibrium position at O, the acceleration of the mass decreases until it becomes zero at O. At the equilibrium position O, the velocity, momentum, and kinetic energy of the mass m have their maximum values, and the mass moves upward to position B where the displacement OB is equal to the original displacement OA. Although at position B the kinetic energy of the mass is zero, its potential energy is at maximum. There is a constant interchange of potential energy into kinetic energy and vice versa as the spring oscillates up and down.

By Newton's second law the acceleration $\ddot{y}$ at any instant of time is

$$F = m\ddot{y}$$

Substituting in Eq. 5.1, it follows that

$$\ddot{y} = -\frac{ky}{m} \tag{5.2}$$

This is the fundamental equation of *undamped simple harmonic motion*. It shows that the acceleration is proportional to and in the opposite direction from the displacement. The maximum displacement OB or OA is called the *amplitude* of the oscillations. The time taken for one complete oscillation, for instance from A to B and back to A, is called the *period P*, and the number of complete oscillations per second is called the *frequency f* of the oscillations. From these two definitions it follows that

$$P = \frac{1}{f}$$

To obtain the displacement at any instant of time, we must solve Eq. 5.2. This may be done by multiplying both sides of the equation by $2\dot{y}$, giving

$$2\dot{y}\ddot{y} = - \frac{2k}{m}\dot{y}y$$

or

$$\frac{d}{dt}(\dot{y}^2) = - \frac{k}{m}\frac{d}{dt}(y^2)$$

Thus

$$d(\dot{y}^2) = - \frac{k}{m}d(y^2)$$

By integration

$$\dot{y}^2 = - \frac{k}{m}y^2 + C$$

where C is a constant which may be determined if the velocity $\dot{y}$ is known at some displacement y. Hence

$$\frac{dy}{\sqrt{C - (k/m)y^2}} = \pm dt$$

Integration gives

$$\sin^{-1}\sqrt{\frac{k}{mC}}\,y = \pm\sqrt{\frac{k}{m}}\,t + \alpha$$

where α is a constant which may be determined if the displacement y is known at some time t. Thus the displacement at any time is given by

$$y = \sqrt{\frac{mC}{k}}\sin\left(\pm\sqrt{\frac{k}{m}}\,t + \alpha\right)$$

This solution of the second-order differential equation, Eq. 5.2, contains two arbitrary constants C and α, and may be written in the equivalent form

$$y = A\sin\sqrt{\frac{k}{m}}\,t + B\cos\sqrt{\frac{k}{m}}\,t \qquad (5.3)$$

in which A and B are the two arbitrary constants and are obtained from the initial or boundary conditions.

A less direct but simpler method of solving the equation of motion, Eq. 5.2, is by assuming a solution of the form which when differentiated twice with respect to time gives the original form of the solution multiplied by a negative constant. Solutions using the sine or cosine or the exponential are suitable. Consider the solution with two arbitrary constants, A and α:

$$y = A\cos(\omega t + \alpha) \qquad (5.4)$$

in which ω is an angular velocity in radians per second and ωt is an angle in radians. From the second differentiation of Eq. 5.4 with respect to time it follows that

$$\ddot{y} = -\omega^2 A \cos(\omega t + \alpha) = -\omega^2 y$$

Comparing this expression for $\ddot{y}$ with that given in the equation of motion, Eq. 5.2, it is seen that ω^2 must equal k/m or $\omega = \sqrt{k/m}$. This result may be obtained directly from Eq. 5.3. The period of the simple harmonic motion is the period of the sine or cosine functions in Eqs. 5.3 and 5.4. If the period is denoted by P, then

$$\omega P = 2\pi \quad \text{or} \quad P = \frac{2\pi}{\omega} = 2\pi\sqrt{\frac{m}{k}} \tag{5.5}$$

If the spring is started from the position $-Y$, then the initial conditions may be stated as, at $t = 0$, $y = -Y$ and $\dot{y} = 0$. Using Eq. 5.4 and $y = -Y$ at $t = 0$ gives

$$-Y = A \cos\alpha$$

Also $\dot{y} = 0$ at $t = 0$ gives $-A\omega \sin\alpha = 0$, hence α must equal zero, for setting A equal to zero would lead to a trivial solution. The solution with the above boundary conditions is given as

$$y = -Y \cos\omega t = -Y \cos\sqrt{\frac{k}{m}}\, t \tag{5.6}$$

As we have seen, there are several equivalent solutions of the original second-order differential equation of motion. These general solutions must contain two arbitrary constants. For any given set of initial conditions the displacement-time relationship must be unique since it corresponds to a particular physical situation. As an exercise show that the initial conditions $t = 0$, $y = -Y$, $\dot{y} = 0$ as given above, when applied to Eq. 5.3, yield Eq. 5.6.

It can be readily verified that the quantity $\sqrt{m/k}$ has the dimensions of time. The force constant k has the dimensions of force divided by distance or $[MLT^{-2}L^{-1}]$ or $[MT^{-2}]$. The dimensions of m/k are $[T^2]$.

5.3   Kinetic and Potential Energies in Simple Harmonic Motion

The kinetic energy of the oscillating mass at any time is $m\dot{y}^2/2$. Taking the same initial conditions of motion used in deriving Eq. 5.6, the kinetic energy of the mass m is

$$\tfrac{1}{2}m Y^2\omega^2 \sin^2\omega t = \tfrac{1}{2}k Y^2 \sin^2\omega t \tag{5.7}$$

since in this problem $\omega^2 = k/m$. The maximum value for the kinetic energy occurs when $\sin^2 \omega t = 1$. With T the symbol for kinetic energy, it follows that

$$T_{max} = \tfrac{1}{2}k\,Y^2 \qquad (5.8)$$

The potential energy V at any displacement y is the work done on the mass m in giving it this displacement. From Eq. 3.3

$$dV = -F\,dy$$

and since the restoring force is

$$F = -ky$$

it follows that

$$V = \int_0^V dV = \int_0^y ky\,dy = \frac{ky^2}{2} \qquad (5.9)$$

Thus the potential energy at any time t is

$$V = \frac{k}{2}\,Y^2 \cos^2 \omega t$$

and has a maximum value of

$$V_{max} = \tfrac{1}{2}k\,Y^2 \qquad (5.10)$$

Since we have assumed that there are no dissipative forces present, the total mechanical energy of the system should be constant and independent of the time. This is easily proved for

$$T + V = \tfrac{1}{2}k\,Y^2 \sin^2 \omega t + \tfrac{1}{2}k\,Y^2 \cos^2 \omega t$$

$$= \tfrac{1}{2}k\,Y^2 \qquad (5.11)$$

The quantity $\tfrac{1}{2}k\,Y^2$ is also the maximum value of the kinetic or potential energy, for at the time when the kinetic energy has its maximum value the potential energy is zero, and vice versa. The relationship $V = \tfrac{1}{2}ky^2$ is plotted in Fig. 5.1b, giving a parabolic curve. For a total energy E, drawn as a horizontal line perpendicular to the energy axis, the value of the kinetic energy T and potential energy V can be readily determined for any displacement y.

The average value of the kinetic energy over a complete oscillation will now be considered. In Fig. 5.2 this average value of the kinetic energy $\bar{T}$ is the ordinate of the rectangle whose area is equal to that enclosed by the sine-squared curve. Mathematically this may be expressed as

$$\bar{T} = \frac{1}{P}\int_0^P T\,dt = \frac{1}{P}\int_0^P \frac{k}{2}\,Y^2 \sin^2 \omega t\,dt$$

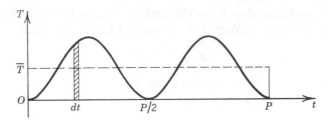

Fig. 5.2   Average value of kinetic energy $\bar{T}$ over a complete period P.

Since $2 \sin^2 \omega t = 1 - \cos 2\omega t$, this equation may be written

$$\bar{T} = \frac{kY^2}{4P} \int_0^P (1 - \cos 2\omega t)\, dt$$

$$= \frac{kY^2}{4P} \left[t - \frac{\sin 2\omega t}{2\omega} \right]_0^P = \frac{kY^2}{4}$$

since $\omega P = 2\pi$.

It can be readily verified in a manner similar to that given above that the average value of $\sin^2 \theta$ or $\cos^2 \theta$ over a complete cycle or angle of 2π radians is one-half and the average value of $\sin \theta$ or $\cos \theta$ over a complete cycle is zero. The time average of the potential energy V over a complete oscillation is similarly given by $kY^2/4$. The sum of these averages is

$$\bar{T} + \bar{V} = \frac{kY^2}{2} \tag{5.12}$$

which is the total energy at any instant of time.

5.4   Effect of the Mass of the Spring

So far we have assumed that the mass of the spring was negligible compared to the oscillating mass hung on its lower end. We shall now take the mass of the spring into account by calculating the total kinetic energy of the system when oscillating. Suppose that the spring is uniformly wound and that its length is l and its mass M. The mass per unit length σ is then given by

$$\sigma = \frac{M}{l}$$

Consider a small portion of the spring of length dy at a height y above the equilibrium position, Fig. 5.3. The mass of this small portion is $\sigma\, dy$. Suppose that the velocity of the lower end of the spring at any instant of time is v. Since the upper end of the spring is always at rest and, if we assume that the velocity at any point on the spring is proportional to its distance from the upper end, it follows that the velocity v_y at any point y is given by

$$v_y = \frac{v(l - y)}{l}$$

The kinetic energy of the small portion at position y is

$$\tfrac{1}{2}\sigma v_y{}^2 \, dy = \frac{1}{2} \frac{\sigma v^2 (l - y)^2}{l^2} \, dy$$

The total kinetic energy of the spring and the mass m is

$$\tfrac{1}{2}mv^2 + \frac{1}{2} \frac{\sigma v^2}{l^2} \int_0^l (l - y)^2 \, dy = \tfrac{1}{2}mv^2 + \frac{1}{2} \frac{\sigma v^2}{l^2} \left[l^2 y + \frac{y^3}{3} - \frac{2ly^2}{2} \right]_0^l$$

$$= \tfrac{1}{2}mv^2 + \tfrac{1}{2}\sigma v^2 \frac{l}{3} = \tfrac{1}{2}v^2 \left(m + \frac{M}{3} \right) \quad (5.13)$$

Thus the fraction of the mass of the spring partaking in the oscillation is one-third.

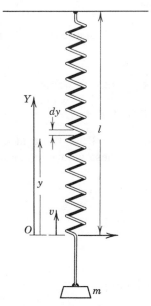

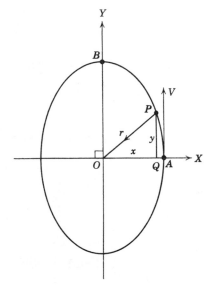

Fig. 5.3 Kinetic energy of spring and mass m.

Fig. 5.4 The path of a particle of mass m attracted to a point O by a force proportional to its distance from O.

5.5   Composition of Simple Harmonic Motions

Let us consider the problem of finding the path of a particle attracted toward a fixed point by a force proportional to its distance from the fixed point. Depending on the initial direction of the velocity of the particle, the path may be a straight line, a circle, or an ellipse. Let O be the center of attraction and A be the point of projection of the particle with a velocity V. Take OA as the direction of the X axis and OB parallel to V as the direction of the Y axis, Fig. 5.4.

At any point P, a distance of r from O, the equation of motion for the particle of mass m is

$$m\frac{d^2\mathbf{r}}{dt^2} = -k\mathbf{r} \tag{5.14}$$

where k is the force constant.

If PQ is drawn parallel to the Y axis and P has the coordinates x, y, then the force $-k\mathbf{r}$ along PO is equivalent to the vector sum of the forces $-kx$ and $-ky$ along QO and PQ respectively. Thus

$$m\ddot{x} = -kx$$

$$m\ddot{y} = -ky$$

The solution of these equations, with $\omega = \sqrt{k/m}$, may be written

$$x = X\sin(\omega t + \alpha) \qquad y = Y\sin(\omega t + \beta)$$

The initial conditions of motion assumed in Fig. 5.4 are those at $t = 0$: $x = OA = a$, $y = 0$, $\dot{x} = 0$, $\dot{y} = V$. Substituting these values in the above equations gives

$$a = X\sin\alpha \qquad 0 = \omega X\cos\alpha$$

Hence

$$\alpha = \frac{\pi}{2} \quad \text{and} \quad a = X$$

and

$$0 = Y\sin\beta \qquad V = \omega Y\cos\beta$$

Hence

$$\beta = 0 \quad \text{and} \quad V = \omega Y \quad \text{or} \quad Y = \frac{V}{\omega}$$

Thus we have

$$x = a \sin\left(\sqrt{\frac{k}{m}}\,t + \frac{\pi}{2}\right) = a \cos\left(\sqrt{\frac{k}{m}}\,t\right)$$

$$y = \frac{V}{\omega} \sin\left(\sqrt{\frac{k}{m}}\,t\right)$$

Hence by squaring and adding

$$\frac{x^2}{a^2} + \frac{y^2}{V^2/\omega^2} = 1 \tag{5.15}$$

The path of the moving particle is therefore an ellipse referred to OX and OY as conjugate diameters.

The figures produced by two sources of simple harmonic vibrations such as tuning forks or pendulums were first studied by Lissajous about the middle of the last century. A modern simple method of producing these so-called Lissajous' figures is by means of an oscilloscope. In such an instrument the moving particles, the electrons, have a mass of about 9×10^{-28} gm. These small electrically charged particles can be made to oscillate in mutually perpendicular directions by applying alternating voltages to appropriate deflecting plates. By applying voltages of the same frequency and varying the initial conditions, it is possible to produce a straight line, a circle, or an ellipse on the viewing screen. If the voltages applied to the X and Y deflecting plates have different frequencies, then very interesting figures may result.

We shall now investigate the case in which the amplitudes of the mutually perpendicular X and Y vibrations are equal and the frequencies are $\omega/2\pi$ and $(\omega + \gamma)/2\pi$ respectively. If the particle is started with maximum displacement along the two axes, then the displacements along the X and Y directions are given by

$$x = A \cos \omega t$$
$$y = A \cos (\omega + \gamma)t$$

Expanding and eliminating ωt gives

$$y = x \cos \gamma t - \sqrt{A^2 - x^2} \sin \gamma t$$

Thus

$$x^2 + y^2 - 2xy \cos \gamma t = A^2 \sin^2 \gamma t \tag{5.16}$$

This may be regarded as the equation of an ellipse whose axes rotate with time. It is easily shown that the ellipse will repeat itself after some time only if the ratio of ω to γ is an integer. Suppose that $(\omega + \gamma)/\omega = 7/6$ or $\gamma = \omega/6$. Then, for one complete vibration along the X direction, there is 7/6 of a vibration along the Y direction, or, for each vibration in the X direction, the vibration in the Y direction gains a phase difference of 60°. Thus, when there have been six complete cycles in the

TABLE 5.1   *x* and *y* Values for Different Values of ω*t*

ω*t*	(ω + γ)*t*	*x*	*y*	Position on Fig. 5.5
0°	0°	A	A	a
60°	70°	0.5A	0.34A	b
360°	360° + 60°	A	0.5A	c
2 × 360°	2 × 360° + 120°	A	−0.5A	d
3 × 360°	3 × 360° + 180°	A	−A	e
4 × 360°	4 × 360° + 240°	A	−0.5A	f
5 × 360°	5 × 360° + 300°	A	0.5A	g
6 × 360°	6 × 360° + 360°	A	A	h

X direction, there have been seven complete cycles in the *Y* direction, and the two motions are back at their starting points. This is not possible unless ω/γ is an integer.

To plot the path of the particle it is possible to eliminate *t* in the above equations, but in practice it is much simpler to take the equations for *x* and *y* given above and proceed to solve for *x* and *y* by brute force. Table 5.1 is a partial table of the *x* and *y* values for different values of ω*t*. The values show how (ω + γ)*t* increases by 60° for each 360° of ω*t*.

A graph of this motion is shown in Fig. 5.5 where the path of the particle may be traced.

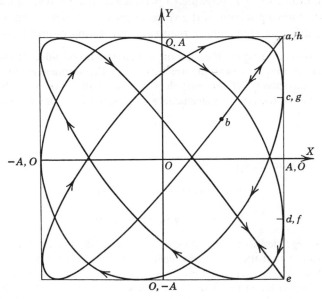

Fig. 5.5 Composition of two simple harmonic motions at right angles and having frequencies in the ratio of 7/6.

5.6   The Simple Pendulum

A simple pendulum consists of a small mass suspended by a light inextensible cord from a fixed support. We shall assume that there are no damping forces so that theoretically once the pendulum is oscillating it continues indefinitely with the same amplitude. Let O in Fig. 5.6 be the fixed point of support, OP the length of the pendulum, and m the mass of the particle at P. The point O is taken as the origin of the radial vector $\mathbf{r}$ which at time t is along OP and makes an angle θ with the vertical through O. We shall now set up the equations of motion of the particle m along and perpendicular to the radius vector OP.

From Eq. 1.14 the force along the radius vector F_r is

$$F_r = m(\ddot{r} - r\dot{\theta}^2)$$

and this is equal to the components of the applied forces acting on mass m in the direction OP. Thus

$$m(\ddot{r} - r\dot{\theta}^2) = mg\cos\theta - T$$

where T is the tension in the cord at the angle θ. Similarly, for the transverse force F_θ in the direction of increasing θ, we have from Eq. 1.15

$$F_\theta = m(2\dot{r}\dot{\theta} + r\ddot{\theta}) = -mg\sin\theta$$

Since the string is assumed to be inextensible, the vector $\mathbf{r}$ has a constant magnitude equal to l, and it follows that

$$\dot{r} = 0 \qquad \text{and} \qquad \ddot{r} = 0$$

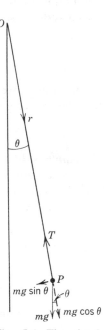

Fig. 5.6 The simple pendulum.

The equations of motion become

$$-ml\dot{\theta}^2 = mg\cos\theta - T \tag{5.17}$$

$$ml\ddot{\theta} = -mg\sin\theta \tag{5.18}$$

To make Eq. 5.18 linear, let us assume that θ is so small that $\sin\theta$ may be set equal to θ in radian measure. This really assumes that θ^3 and higher powers of θ are negligibly small compared to θ, since the expansion of $\sin\theta$ in radian measure is

$$\sin\theta = \theta - \frac{\theta^3}{3!} + \frac{\theta^5}{5!}\cdots$$

Thus for $\theta = 3°$, $\sin 3° = 0.05234$ and $3° = 0.05236$ radian. If therefore the maximum angular amplitude of the pendulum is a few degrees, we may write Eq. 5.18 approximately as

$$\ddot{\theta} = -\frac{g}{l}\theta \qquad (5.19)$$

This is a second-order differential equation similar to Eq. 5.2 with the difference that here the variable is an angular displacement rather than a linear one. Following the discussion given earlier, we may say that Eq. 5.19 represents a system oscillating with simple harmonic motion whose period is

$$P = 2\pi\sqrt{\frac{l}{g}} \qquad (5.20)$$

The tension T exerted by the cord on the particle depends on the angular displacement as given by Eq. 5.17, namely,

$$T = mg\cos\theta + ml\dot{\theta}^2$$

where $ml\dot{\theta}^2$ is the centrifugal or fictitious force. Thus, when $\theta = 0$, the tension has a maximum value equal to the weight of the particle plus the centrifugal force. Since we have assumed no dissipative force, the system is a conservative one, and there is conservation of mechanical energy.

We shall now drop the assumption of a small angular amplitude and investigate the period for relatively large amplitudes. To do this, we must integrate Eq. 5.18 or

$$\ddot{\theta} = -\frac{g}{l}\sin\theta$$

Again we employ the trick of multiplying the respective sides of this equation by the identity

$$\dot{\theta}\,dt = \frac{d\theta}{dt}\,dt = d\theta$$

giving

$$\ddot{\theta}\dot{\theta}\,dt = -\frac{g}{l}\sin\theta\,d\theta$$

or

$$\frac{1}{2}d(\dot{\theta}^2) = \frac{g}{l}d(\cos\theta)$$

Integrating and using the initial conditions that the pendulum has a maximum angular amplitude of α or $\dot{\theta} = 0$ when $\theta = \alpha$

$$\dot{\theta}^2 = 2\frac{g}{l}(\cos\theta - \cos\alpha) \qquad (5.21)$$

Thus

$$\frac{d\theta}{\sqrt{\cos\theta - \cos\alpha}} = \sqrt{\frac{2g}{l}}\, dt$$

The time taken by the pendulum to swing from $+\alpha$ to $-\alpha$ is half of the period of oscillation P_α for the angular amplitude α or

$$\int_{-\alpha}^{+\alpha} \frac{d\theta}{\sqrt{\cos\theta - \cos\alpha}} = \sqrt{\frac{2g}{l}}\int_0^{(P_\alpha)/2} dt = \sqrt{\frac{2g}{l}}\,\frac{P_\alpha}{2}$$

Using the transformation

$$\cos\theta = 1 - 2\sin^2\frac{\theta}{2}$$

the integral becomes

$$\int_{-\alpha}^{+\alpha} \frac{d\theta}{\sqrt{\sin^2\alpha/2 - \sin^2\theta/2}} = \sqrt{\frac{g}{l}}\,P_\alpha \tag{5.22}$$

To evaluate this integral we shall make the substitution

$$\sin\frac{\theta}{2} = \sin\frac{\alpha}{2}\sin\phi \tag{5.23}$$

This is a legitimate substitution inasmuch as the limiting value of θ is α. Hence the limits of $\pi/2$ and $-\pi/2$. Differentiating Eq. 5.23, we have

$$\frac{1}{2}\cos\frac{\theta}{2}\,d\theta = \sin\frac{\alpha}{2}\cos\phi\,d\phi$$

or

$$d\theta = \frac{2\sin(\alpha/2)\cos\phi\,d\phi}{\sqrt{1 - \sin^2(\alpha/2)\sin^2\phi}}$$

Substituting in Eq. 5.22 for the period, it follows that

$$\sqrt{\frac{g}{l}}\,P_\alpha = \int_{-\pi/2}^{+\pi/2} \frac{2\,d\phi}{\sqrt{1 - \sin^2(\alpha/2)\sin^2\phi}}$$

$$= \int_{-\pi/2}^{+\pi/2} 2\,d\phi\left(1 - \sin^2\frac{\alpha}{2}\sin^2\phi\right)^{-\frac{1}{2}}$$

$$= \int_{-\pi/2}^{+\pi/2} 2\,d\phi\left(1 + \tfrac{1}{2}\sin^2\frac{\alpha}{2}\sin^2\phi + \tfrac{3}{8}\sin^4\frac{\alpha}{2}\sin^4\phi\cdots\right)^\dagger$$

$$= 2\left[\pi + \tfrac{1}{2}\sin^2\left(\frac{\alpha}{2}\right)\frac{\pi}{2} + \tfrac{3}{8}\sin^4\left(\frac{\alpha}{2}\right)\frac{3\pi}{8}\cdots\right]$$

$\dagger\ \int_{-\pi/2}^{\pi/2}\sin^2\phi\,d\phi = \int_{-\pi/2}^{\pi/2}\frac{(1-\cos 2\phi)\,d\phi}{2} = \frac{\pi}{2}$

Hence

$$P_\alpha = 2\pi \sqrt{\frac{l}{g}} \left(1 + \tfrac{1}{4} \sin^2 \frac{\alpha}{2} + \tfrac{9}{64} \sin^4 \frac{\alpha}{2} \cdots \right)$$

If the period for an angular amplitude of approximately zero is called P_0, where

$$P_0 = 2\pi \sqrt{\frac{l}{g}}$$

then the period P_α for an angular amplitude α is

$$P_\alpha = P_0 \left(1 + \tfrac{1}{4} \sin^2 \frac{\alpha}{2} + \tfrac{9}{64} \sin^4 \frac{\alpha}{2} \cdots \right) \tag{5.24}$$

From the three terms in the expansion given in Eq. 5.24, we obtain for an angular amplitude of 90° a period of

$$P_{90°} = P_0(1 + \tfrac{1}{8} + \tfrac{9}{256})$$

$$= 1.16 P_0$$

Even for this amplitude another term in the expansion should be added if real accuracy is desired. However, in most experiments, amplitudes as large as 90° are not encountered. For an amplitude of 20° the contribution of the $\sin^2 \alpha/2$ term is 0.0075 and of the $\sin^4 \alpha/2$ term is 0.00013. If the amplitude is smaller than 20°, then the $\sin^4 \alpha/2$ term may be neglected without making an appreciable error. A further simplification can be made by replacing $\sin^2 \alpha/2$ by $\alpha^2/4$. Thus for $\alpha = 20°$ the value of $\sin^2 10°$ is approximately 0.0308 and $\alpha^2/4$ is approximately 0.0305 in radian measure. With these limitations, we may write the period

$$P_\alpha = P_0 \left(1 + \frac{\alpha^2}{16} \right)$$

and by the binomial theorem to the same degree of accuracy it follows that

$$P_0 = P_\alpha \left(1 - \frac{\alpha^2}{16} \right)$$

If the angular amplitude is 4° or 0.0698 radian, then

$$P_0 = P_{4°}(1 - 0.0003)$$

or

$$P_{4°} = P_0(1 + 0.0003) = 1.0003 P_0$$

The periods differ by three parts in 10,000. The tension in the cord can now be evaluated for any angular displacement. From Eq. 5.17 the tension in the cord is

$$T = mg \cos \theta + ml\dot\theta^2$$

and, substituting for $\dot\theta^2$ from Eq. 5.21, we have

$$T = mg(3 \cos \theta - 2 \cos \alpha)$$

5.7 Simple Harmonic Motion with Damping

We shall now drop the assumption that there are no dissipative forces acting on the particle that is executing simple harmonic motion. The amplitude of a simple pendulum continuously decreases with time, largely owing to air friction, though if the pendulum were placed in a vacuum, there would still be some damping or decrease in amplitude due to the lack of perfect rigidity of the supports.

For relatively small velocities of the moving particle it is a fairly accurate assumption that the damping force is proportional to the velocity of the particle. This damping force always opposes the motion or is in the opposite direction to the velocity. Thus, if the velocity of the moving particle is v, the damping force may be set equal to $-Rv$ where R is a constant, namely, the damping force per unit velocity.

If at some instant of time the mass m has a displacement y from its equilibrium position and a velocity $\dot{y}$, then by Newton's second law the equation of motion of the mass m in Fig. 5.1 is

$$m\ddot{y} = -ky - R\dot{y}$$

where $-ky$ is the restoring force due to the spring and $-R\dot{y}$ is the damping force opposing the motion. Thus

$$m\ddot{y} + R\dot{y} + ky = 0 \tag{5.25}$$

This is a linear homogeneous differential equation of the second order with constant coefficients. It is a *linear equation* because it contains no powers of y higher than the first; it is *homogeneous* because it contains no term independent of y or its derivatives; it is of the second order because its highest differential is of the second order. Such equations can be solved relatively simply by assuming an exponential relationship between the variables y and t such as

$$y = Ce^{\lambda t} \tag{5.26}$$

where C and λ are constants depending on the particular problem. By differentiation of Eq. 5.26,

$$\dot{y} = \lambda Ce^{\lambda t} \quad \text{and} \quad \ddot{y} = \lambda^2 Ce^{\lambda t}$$

Substituting these values in the equation of motion, Eq. 5.25, we have the auxiliary or *characteristic equation*

$$m\lambda^2 + R\lambda + k = 0$$

Thus the solution of this equation is

$$\lambda = \frac{-R \pm \sqrt{R^2 - 4mk}}{2m} = -\frac{R}{2m} \pm \sqrt{\frac{R^2}{4m^2} - \frac{k}{m}}$$

The displacement of the mass m at any time t is

$$y = e^{-(R/2m)t}[Ae^{+\sqrt{(R^2/4m^2-k/m)}t} + Be^{-\sqrt{(R^2/4m^2-k/m)}t}] \qquad (5.27)$$

where A and B are the constants of integration required to satisfy a second-order differential equation.

Equation 5.27 presents three distinct cases according to whether $R^2/4m^2$ is greater than, equal to, or less than k/m. These correspond physically to the following situations.

(a)
$$\frac{R^2}{4m^2} > \frac{k}{m}$$

In this situation the damping force is large compared to the restoring force. This results in the displaced mass coming slowly to the equilibrium position but not passing it. The system is said to be *overdamped*.

(b)
$$\frac{R^2}{4m^2} = \frac{k}{m}$$

In this the displaced mass comes to its equilibrium position in a minimum time and does not pass beyond the equilibrium position. The system is said to be *critically damped*.

(c)
$$\frac{R^2}{4m^2} < \frac{k}{m}$$

Here the damping force is relatively small compared to the restoring force. The motion of the mass m is oscillatory with the amplitude decreasing with each oscillation. This is called *damped oscillatory motion*.

We shall now discuss each of these cases.

5.8   Overdamped Motion: $R^2 > 4km$

For this condition the radical is real and the solution is that given by Eq. 5.27. The constants of integration A and B are determined from the initial or boundary conditions. Let us assume these to be that at time $t = 0$, $y = y_0$ and $\dot{y} = 0$. That is, the mass m starts from rest with an initial upward displacement of $+y_0$. Substituting these values in Eq. 5.27 yields

$$y_0 = A + B$$

Differentiating Eq. 5.27 and for convenience replacing

$$\sqrt{(R^2/4m^2) - (k/m)}$$

by α, we have

$$\dot{y} = -\frac{R}{2m} e^{-(R/2m)t}(Ae^{\alpha t} + Be^{-\alpha t}) + e^{-(R/2m)t}(A\alpha e^{\alpha t} - B\alpha e^{-\alpha t})$$

Substituting $\dot{y} = 0$ at $t = 0$ gives

$$0 = -\frac{R}{2m}(A + B) + \alpha(A - B)$$

or

$$\frac{R}{2m}\frac{y_0}{\alpha} = A - B$$

Thus the quantities $(A + B)$ and $(A - B)$ can be determined so that A and B can be found. Rather than carry this out with symbols we shall take a numerical example.

Example of overdamped motion.   Suppose that a spring has a force constant of 2500 dynes/cm, a damping force constant R of 1000 dyne sec/cm, and a mass of 25 gm hung on the lower end. The problem is to find the displacement y at any time t if the mass is started from rest 5 cm below the equilibrium position. First we must check whether the motion is overdamped or whether R^2 is greater than $4km$. Substituting these numerical values gives $R^2 = 10^6$ and $4km = 2.5 \times 10^5$ so that the motion is overdamped and Eq. 5.27 is directly applicable. From the data

$$\frac{R}{2m} = 20 \text{ sec}^{-1} \qquad \frac{k}{m} = 100 \text{ sec}^{-2}$$

$$\sqrt{\frac{R^2}{4m^2} - \frac{k}{m}} = \sqrt{300} = 17.3 \text{ sec}^{-1} \quad \text{(approximately)}$$

Substituting these numerical values in Eq. 5.27 gives

$$y = e^{-20t}(Ae^{17.3t} + Be^{-17.3t})$$
$$= Ae^{-2.7t} + Be^{-37.3t}$$

The second term is negligible for anything but exceedingly small values of t. Nevertheless we cannot neglect it at present. From the initial condition that $y = -5$ cm at $t = 0$

$$-5 = A + B$$

Also

$$\dot{y} = -2.7Ae^{-2.7t} - 37.3Be^{-37.3t}$$

For $\dot{y} = 0$ at $t = 0$

$$0 = -2.7A - 37.3B$$

These equations give

$$A = 5.39 \text{ cm} \qquad B = 0.390 \text{ cm}$$

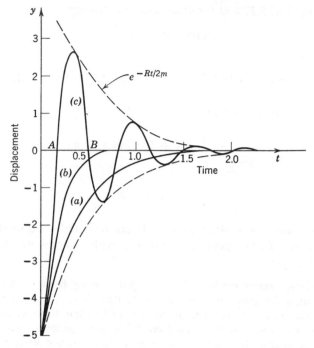

Fig. 5.7   Graph of displacement of mass m on the end of the spring plotted against time. (*a*) Overdamped motion. (*b*) Critically damped motion. (*c*) Damped oscillatory motion.

Thus the displacement at any time t is

$$y = -5.39e^{-2.7t} + 0.390e^{-37.3t}$$

The graph of this equation is shown in Fig. 5.7*a*.

5.9   Critically Damped Motion: $R^2 = 4km$

In this case Eq. 5.27 for the displacement reduces to

$$y = e^{-(R/2m)t}(A + B)$$

Since the two constants A and B are multiplied by the same time factor, they are really equivalent to one constant and, as such, this equation does not satisfy the second-order differential equation requirements. The solution for this condition found in most texts on differential equations is

$$y = (A + Bt)e^{-(Rt/2m)} \tag{5.28}$$

That this is a solution may be readily verified by substituting it in the original equation of motion, Eq. 5.25. The constants of integration, A and B, are obtained from the initial conditions as seen in the following example.

Example of critically damped motion. Let us take the data of the previous example, namely, $k = 2500$ dynes/cm, $m = 25$ gm, and, at $t = 0$, $\dot{y} = 0$ and $y = -5$ cm. First we must find the value of R_c required for critical damping. Substituting these values in the equation $R_c^2 = 4km$, we have

$$R_c^2 = 25 \times 10^4$$

or

$$R_c = 500 \text{ dyne sec/cm}$$

and

$$\frac{R_c}{2m} = 10 \text{ sec}^{-1}$$

Thus the displacement at any time t is

$$y = (A + Bt)e^{-10t}$$

Substituting the values $y = -5$ cm at $t = 0$ sec gives

$$A = -5 \text{ cm}$$

Also

$$\dot{y} = Be^{-10t} - 10(A + Bt)e^{-10t}$$

and using $\dot{y} = 0$ at $t = 0$ gives

$$0 = B - 10A$$

Thus $A = -5$ cm and $B = -50$ cm/sec. The displacement y at any time t for the critically damped motion is

$$y = -5e^{-10t}(1 + 10t)$$

The graph of this equation is given in Fig. 5.7b. Comparison of graphs a and b in Fig. 5.7 shows that when the motion is critically damped the mass m reaches its equilibrium position in a shorter time than when the damping is larger. Although theoretically, in both cases the equilibrium position is not reached until after an infinite time, nevertheless for all practical purposes the displacement becomes very close to zero in a finite time. For any damping constant less than 500 dyne sec/cm in our particular example the motion becomes oscillatory. This type of motion we shall now investigate.

5.10 Damped Oscillations: $R^2 < 4km$

In this case the radical $\sqrt{R^2/4m^2 - k/m}$ is the square root of a negative quantity and is therefore imaginary. To overcome this difficulty, we shall make use of the $\sqrt{-1}$ represented by the symbol j. Thus Eq. 5.27 for the displacement becomes, where $\hat{y}$ is a complex number,

$$\hat{y} = e^{-Rt/2m}[Ae^{j\sqrt{(k/m-R^2/4m^2)}t} + Be^{-j\sqrt{(k/m-R^2/4m^2)}t}] \qquad (5.29)$$

For convenience we shall put

$$\sqrt{\frac{k}{m} - \frac{R^2}{4m^2}} = \omega'$$

The appropriateness of the symbol ω' will appear later when it will be shown to be an angular velocity. Now by Euler's theorem (see Mathematical Appendix)

$$e^{j\omega't} = \cos \omega't + j \sin \omega't$$
$$e^{-j\omega't} = \cos \omega't - j \sin \omega't$$

With these relationships Eq. 5.29 becomes

$$\hat{y} = e^{-Rt/2m}[(A + B) \cos \omega't + j(A - B) \sin \omega't]$$

The displacement $\hat{y}$ is complex and is made up of a real and an imaginary portion such that $\hat{y} = y_1 + jy_2$, where y_1 and y_2 are both real quantities. Thus

$$y_1 = e^{-Rt/2m}(A + B) \cos \omega't \qquad \text{and} \qquad y_2 = e^{-Rt/2m}(A - B) \sin \omega't$$

are both solutions of the equation of motion, Eq. 5.25. Since this equation is linear, the sum of these solutions is also a solution, so that the complete solution, where y is a real quantity, may be written as:

$$y = e^{-Rt/2m}(C \cos \omega't + D \sin \omega't)$$

or

$$y = Ke^{-Rt/2m} \cos (\omega't - \phi) \qquad (5.30)$$

where

$$(A + B) = C \qquad (A - B) = D$$
$$K \cos \phi = C \qquad K \sin \phi = D$$

The two constants of integration determined from the initial conditions are A and B, or C and D, or K and ϕ. Which pair is used is immaterial, though Eq. 5.30 with K and ϕ is often the most convenient. From this equation it may be seen that the motion is oscillatory, having a period of $P = 2\pi/\omega'$. Note that the period of oscillation with damping is larger than that without damping. The amplitude of the motion is continuously decreasing owing to the factor $e^{-Rt/2m}$ which is called the damping factor. In practice it is found that the time for the displacement to reach zero is shorter for slightly under-damped oscillations than for the theoretical critically damped case. The principal features of this motion are given in the following example.

Example of damped oscillation.   Again we take the data of the previous problem: $k = 2500$ dynes/cm, $m = 25$ gm; but now the damping constant R is equal to 100 dyne sec/cm. The initial conditions are: at $t = 0$, $y = -5$ cm and $\dot{y} = 0$. To be sure that these data correspond to oscillatory motion, we must show that R^2 is less than $4km$. Since $R^2 = 10^4$ and $4km = 2.5 \times 10^5$, the motion is oscillatory and Eq. 5.30 is applicable. From the data

$$\omega' = \sqrt{\frac{k}{m} - \frac{R^2}{4m^2}} = \sqrt{96} = 9.80 \text{ sec}^{-1} \quad \text{(approximately)}$$

and

$$\frac{R}{2m} = 2 \text{ sec}^{-1}$$

Thus Eq. 5.30 becomes

$$y = Ke^{-2t} \cos (9.80t - \phi)$$

Substituting the initial conditions $y = -5$ cm at $t = 0$,

$$K \cos \phi = -5$$

Also

$$\dot{y} = -2Ke^{-2t} \cos (9.80t - \phi) - 9.80Ke^{-2t} \sin (9.80t - \phi) \qquad (5.31)$$

At $t = 0$, $\dot{y} = 0$. Hence

$$0 = -2K \cos \phi + 9.80K \sin \phi$$

Thus

$$\tan \phi = \frac{2}{9.8} = 0.204$$

and

$$\phi = 11° 32' \text{ or } 191° 32'$$

Also

$$K = -\frac{5}{\cos \phi}$$

If $\phi = 11° 32'$, then $K = -5/0.980 = -5.10$ cm, whereas if $\phi = 191° 32'$, then $K = -5/-0.980 = +5.10$ cm. Either value of ϕ can be substituted, but the appropriate sign for K must also be used. Thus we see that the displacement can be written either

$$y = -5.10e^{-2t} \cos (9.80t - 11° 32')$$

or

$$y = 5.10e^{-2t} \cos (9.80t - 191° 32')$$

If the angle ϕ is stated in degrees, then it is probably more convenient for purposes of calculation to express the $9.80t$ in degrees than in radians.

$$1 \text{ radian} = 57° 17' 45'' = \text{approximately } 57.3°$$

and

$$9.80t \text{ radians} = 561.5°t$$

The displacement y at any time t is

$$y = -5.10e^{-2t} \cos (561.5°t - 11.54°) \qquad (5.32)$$

A simple check on this equation is to see whether the displacement is -5 cm at $t = 0$. To plot the displacement curve shown in Fig. 5.7c, it is well if one determines the period P of the oscillations:

$$P = \frac{2\pi}{\omega'} = \frac{2\pi}{9.80} = 0.641 \text{ sec}$$

We can then determine at what times the displacement is zero. These times are obtained by setting the cosine function equal to zero. Thus $y = 0$ when $561.5°t - 11.54° = 90°, 270° \cdots (2n + 1) 90°$ where n is any integer. From this it follows that at times of approximately 0.181, 0.501, 0.822, 1.14 sec the displacement is zero. Since displacements are zero every half-period, once the first time of zero displacement has been determined, successive times are found by adding 0.3204 sec successively.

It might appear at first sight that maximum values of the displacement would be obtained by setting the cosine function equal to unity. However, this is not correct, for the maximum values of the displacement occur when the velocity $\dot{y}$ is zero. From Eq. 5.31 for the velocity $\dot{y}$ at any time t, it follows that $\dot{y} = 0$ when

$$\tan (9.80t - \phi) = -\frac{2}{9.80} = -0.204$$

The angles having a tangent of -0.204 are $-11°\,32', 180° - 11°\,32', 360° - 11°\,32'$, etc. Hence the times of maximum displacement occur when the term $(561.5°t - 11°\,32')$ is equal to one of the above angles, as when

$$561.5t = 0, 180, 360, \text{ etc.}$$

or

$$t = 0, 0.32, 0.64, 0.96, \text{ etc.}$$

Notice that these times are not exactly mid-way between the positions of zero displacement as would be given by setting the cosine function equal to unity. The values of y for these times can now be calculated, and a graph may be filled in free hand with a fair degree of accuracy. A partial table of these values is given below.

t	$\cos (561.5°t - 11.54°)$	e^{-2t}	y
0.000	0.985	1.000	-5.000
0.181	0.000		0.000
0.321	0.979	0.527	$+2.630$
0.501	0.000		0.000
0.641	0.978	0.278	-1.385

The graph of this damped oscillatory motion is shown in Fig. 5.7c. It is not a true cosine curve partly because of the negative exponential factor $e^{-Rt/2m}$. For this particular example the larger the value of R, up to the critical value of 500 dyne sec/cm, the more warped the curve becomes.

5.11 The Logarithmic Decrement

The amplitude of the oscillations decreases with time owing to the exponential or damping factor $e^{-Rt/2m}$. This decrease in amplitude is often expressed in terms of the logarithmic decrement δ, which is defined as the natural logarithm ($\ln \equiv \log_e$), of the ratio of two maxima a period apart.

$$\delta = \ln \frac{y_t}{y_{t+P}} = \ln \left\{ \frac{Ke^{-Rt/2m} \cos(\omega't - \phi)}{Ke^{-(R/2m)(t+P)} \cos[\omega'(t+P) - \phi]} \right\}$$

$$= \ln e^{RP/2m} = \frac{RP}{2m} \tag{5.33}$$

For the example given above, the value of the logarithmic decrement δ is

$$\delta = \frac{RP}{2m} = \frac{100}{50} \times 0.641 = 1.282$$

This value may also be obtained by taking the natural logarithm, or 2.303 times the logarithm to the base 10, of two maxima a period apart.

5.12 The Energy Equation with Damping and Q Value

Since the amplitude decreases with time, it follows that the mechanical energy of the system is being converted into heat energy through the agency of the damping force. The rate of dissipation of energy, or the rate of doing work against the damping force, is

$$F\dot{y} = R\dot{y}^2$$

From the law of conservation of energy, this must be equal to the rate of loss of mechanical energy. This statement can readily be proved analytically by the equation of motion, Eq. 5.25. Thus

$$m\ddot{y} + R\dot{y} + ky = 0$$

Multiplying each term by $\dot{y}$ gives

$$m\ddot{y}\dot{y} + R\dot{y}^2 + ky\dot{y} = 0$$

or

$$\frac{m}{2} \frac{d}{dt}(\dot{y}^2) + R\dot{y}^2 + \frac{k}{2} \frac{d}{dt}(y^2) = 0$$

The successive terms in this equation are the time rate of change of kinetic energy, the time rate of doing work by the damping force, and the time rate of change of potential energy. If the kinetic energy is denoted by T, its rate of change by $\dot{T}$, and similarly for the potential energy V, then

$$\dot{T} + \dot{V} = -R\dot{y}^2 \tag{5.34}$$

or the rate of change of the mechanical energy is equal to the rate of doing work against the damping force. If the damping force is zero, then

$$\dot{T} + \dot{V} = 0$$

or

$$T + V = A$$

where A is a constant. This equation shows that the law of conservation of mechanical energy applies to an idealized spring oscillating without damping.

Let us calculate the amount of energy lost in the half-cycle from $t = 0.181$ to $t = 0.501$ sec when the mass on the spring moves from its equilibrium position to its first maximum and back to the equilibrium position. Calling the points of zero displacement corresponding to the times given above A and B, we have for the loss of kinetic energy during the half-cycle

$$T_A - T_B = \tfrac{1}{2}m\dot{y}_A{}^2 - \tfrac{1}{2}m\dot{y}_B{}^2$$

From Eq. 5.31 for the velocity $\dot{y}$, it follows that $\dot{y}^2$ is given by

$$\dot{y}^2 = K^2 e^{-4t}[4\cos^2(9.80t - \phi) + 9.80^2 \sin^2(9.80t - \phi)$$
$$+ 39.2 \sin(9.80t - \phi)\cos(9.80t - \phi)]$$

Now at times t_A and t_B, the value of the cosine terms are zero and of the sine terms unity. Substituting numerical values:

$$\tfrac{1}{2}m\dot{y}_A{}^2 - \tfrac{1}{2}m\dot{y}_B{}^2 = \tfrac{2.5}{2} \times 5.10^2 \times 9.80^2(e^{-0.724} - e^{-2.004})$$
$$= 10.929 \text{ ergs}$$

This loss of energy is equal to the work done against the damping force. From Eq. 5.34 we have

$$d(T + V) = -R\dot{y}^2 \, dt$$

At the points A and B the potential energy is zero since the displacement is zero. Hence by integration

$$T_A - T_B = -\int_B^A R\dot{y}^2 \, dt$$

Now this integral which represents the work done against the damping force can be evaluated from mathematical tables, and it will be found to be equal to the loss in kinetic energy in the interval.

It is interesting to note that the maximum kinetic energy does not occur at the times corresponding to zero displacement but slightly before these times. You are asked to evaluate these times in a problem at the end of this chapter.

In a system undergoing damped oscillations it is common to introduce a figure of merit Q of the system, which is defined as

$$Q = \frac{m\omega_0}{R}$$

where ω_0 is the natural angular frequency of the system without damping, i.e., $\omega_0 = \sqrt{k/m}$. Hence, Q may be written as

$$Q = \sqrt{\frac{k}{m}\frac{m^2}{R^2}} = \sqrt{\frac{km}{R^2}}$$

The angular frequency ω' of the damped oscillations is given as

$$\omega' = \sqrt{\frac{k}{m} - \frac{R^2}{4m^2}} = \omega_0\sqrt{1 - \frac{R^2 m}{4m^2 k}} = \omega_0\sqrt{1 - \left(\frac{1}{2Q}\right)^2}$$

Thus, if Q is large compared to unity, so that the term $1/4Q^2$ may be neglected, then $\omega_0 = \omega'$. Making this substitution for ω_0 and for R in terms Q in Eq. 5.30, it follows that

$$y = e^{-Rt/2m} \cos{(\omega_0 t - \phi)} = e^{-\omega_0 t/2Q} \cos{(\omega_0 t - \phi)}$$

The energy of the oscillating system is proportional to $\ddot{y}^2$, that is, the energy E is proportional to $e^{-\omega_0 t/Q}$ or $E = Ae^{-\omega_0 t/Q}$ where A is a constant. If the damping factor R is small and Q large, so that the loss in energy per cycle is small compared to the total energy E, then the decrease in energy per cycle is

$$-P(dE/dt) = (P\omega_0/Q)Ae^{-\omega_0 t/Q} = P\omega_0 E/Q$$

where P is the period of the oscillations, approximately equal to the period of the undamped oscillations. Hence, to this approximation $\omega_0 = 2\pi/P$ or $P\omega_0 = 2\pi$, and the result may be written as

$$\frac{\text{Decrease in Energy per Cycle}}{\text{Total Energy}} = \frac{2\pi}{Q}$$

or

$$Q = 2\pi\frac{\text{(Total Energy)}}{\text{Decrease in Energy per Cycle}}$$

This result is valid only when Q is large compared to unity. For example, if $R = 2.5$ dyne sec/cm for the data used in Fig. 5.7c where $\omega_0 = 10$ rad/sec and $\omega' = \sqrt{100 - 0.0625} \approx \omega_0$ with the value of $Q = m\omega_0/R = 100$, the above theory is applicable. For values of R much greater than 2.5 the theory does not apply, but a more exact analysis could be made.

5.13   Forced Oscillations of a Damped Oscillator

In every oscillating system there is dissipation of mechanical energy and, consequently, if the oscillations are to be maintained indefinitely, energy must be supplied to the system. The slowly falling weight in a grandfather's clock and the spring in a watch or clock are means of supplying energy to the system to maintain the oscillations. In these the energy is supplied by some mechanism which produces a periodic driving force.

Suppose that a spring is subjected to a periodic force $F_0 \cos \omega t$, or $F_0 \sin \omega t$, where F_0 is the maximum value of the applied force and $f = \omega/2\pi$ is its frequency. Rather than using the sinusoidal functions, it is somewhat simpler to employ an exponential form, namely $F_0 e^{j\omega t} = F_0 \cos \omega t + jF_0 \sin \omega t$. Any physical quantity such as displacement must be real, and when solving equations involving complex quantities it is necessary that the reals be equated to the reals and the imaginaries to the imaginaries. For a mass m hung on the lower end of a spring, whose force constant is k and whose damping constant is R, the equation of motion is

$$m\ddot{y} + R\dot{y} + ky = F_0 e^{j\omega t} \tag{5.35}$$

Once a steady state is set up, the system will oscillate with the same frequency as the applied force. This is a necessary physical condition of the oscillations. Let us therefore try as a solution of Eq. 5.35

$$\hat{y} = Ae^{j(\omega t - \theta)} \tag{5.36}$$

where θ is the angle by which the displacement $\hat{y}$ lags behind the impressed force and A is the amplitude of the oscillations. By differentiation of Eq. 5.36 we have

$$\dot{y} = j\omega Ae^{j(\omega t - \theta)}$$
$$\ddot{y} = -\omega^2 Ae^{j(\omega t - \theta)}$$

Substituting these values in Eq. 5.35 and cancelling the common factor $e^{j\omega t}$, gives

$$-m\omega^2 A + jR\omega A + kA = F_0 e^{j\theta} = F_0 \cos \theta + jF_0 \sin \theta \tag{5.37}$$

Equating the real and imaginary quantities:

$$A(k - m\omega^2) = F_0 \cos \theta \quad \text{and} \quad R\omega A = F_0 \sin \theta \tag{5.38}$$

Hence

$$\tan \theta = \frac{\omega R}{k - m\omega^2} \tag{5.39}$$

and squaring and adding Eqs. 5.38 gives for the amplitude A of the oscillations

$$A = \frac{F_0}{\sqrt{(k - m\omega^2)^2 + (\omega R)^2}} \qquad (5.40)$$

Hence the solution, taking only the real part of Eq. 5.36, may be written as

$$y = \frac{F_0 \cos (\omega t - \theta)}{\sqrt{(k - m\omega^2)^2 + (\omega R)^2}} \qquad (5.41)$$

That this is a solution of the equation of motion, Eq. 5.35, may be verified by substitution. However, it cannot be a complete solution since it contains no arbitrary constants. The equation of motion, Eq. 5.35, is a linear, second-order, inhomogeneous differential equation with constant coefficients. The inhomogeneity arises from the term containing F_0, which does not contain the dependent variable y or any of its derivatives. The related homogeneous equation is Eq. 5.25, which does not contain the applied force term. In the theory of differential equations, the related homogeneous equation is called the *auxiliary* equation and its solution the *complementary* function. The complementary function in this problem is Eq. 5.27. A solution of the inhomogeneous equation is called the particular integral and is given in Eq. 5.41. The complete solution of an inhomogeneous equation is the sum of the complementary function and the particular integral. The proof of this follows from the linearity of the differential equation.

If y_1 is the complementary function or a solution of

$$m\ddot{y} + R\dot{y} + ky = 0$$

and y_2 is the particular integral or a solution of Eq. 5.35, then the complete solution of Eq. 5.35 is the sum of these or

$$y = y_1 + y_2$$

For this to be a solution

$$m\ddot{y}_1 + R\dot{y}_1 + ky_1 + m\ddot{y}_2 + R\dot{y}_2 + ky_2 - F_0 \cos \omega t = 0$$

and this follows from the fact that

$$m\ddot{y}_1 + R\dot{y}_1 + ky_1 = 0$$

and

$$m\ddot{y}_2 + R\dot{y}_2 + ky_2 - F_0 \cos \omega t = 0$$

For this problem, in which the independent variable is the time t, the complementary function is called the *transient solution* inasmuch as its value decreases with time owing to the damping factor. The particular integral is called the *steady-state solution*. Thus the complete solution of our problem is the sum of the transient solution y_t and the steady-state solution y_{ss}. The initial or boundary conditions must be applied to the complete solution:

$$y = y_t + y_{ss}$$

The choice of the solution for the transient depends on the particular value of the damping force constant R. An example will probably make this point clearer.

Example of a forced oscillation.   Let us use the data for the spring of the previous problem, namely, $k = 2500$ dynes/cm, $m = 25$ gm, and a damping constant R of 200 dynes sec/cm together with an applied force F_0 of 12,500 dynes having a frequency corresponding to $\omega = 8$ radians/sec. The problem is to find the displacement y at any time t if, at time $t = 0$, $\dot{y} = 0$ and $y = -5$ cm.

First we must examine the transient solution and determine whether this motion is overdamped, critically damped, or oscillatory. This depends on the relative sizes of k/m and $R^2/4m^2$. From the data we find that

$$\frac{k}{m} = 100 \text{ sec}^{-2} \qquad \frac{R^2}{4m^2} = 16 \text{ sec}^{-2}$$

Since k/m is greater than $R^2/4m^2$, the motion is oscillatory and the solution is given by Eq. 5.30. Thus

$$y_t = Ke^{-Rt/2m} \cos(\omega' t - \phi)$$

where

$$\omega' = \sqrt{\frac{k}{m} - \frac{R^2}{4m^2}} = \sqrt{84} = 9.17 \text{ sec}^{-1} \quad \text{(approximately)}$$

The steady-state solution is given by Eq. 5.41, and the phase angle θ is given by

$$\tan \theta = \frac{\omega R}{k - m\omega^2} = 1.778$$

and

$$\theta = 60° 39' = 1.058 \text{ radians}$$

The complete solution using the numerical values is

$$y = Ke^{-4t} \cos(9.17t - \phi) + 6.80 \cos(8t - 1.058)$$

The initial condition $y = -5$ cm at $t = 0$ gives

$$-5 = K \cos \phi + 6.80 \cos(1.058)$$

or

$$K \cos \phi = -5 - 3.33 = -8.33 \text{ cm}$$

Also

$$\dot{y} = -4Ke^{-4t} \cos(9.17t - \phi) - 9.17Ke^{-4t} \sin(9.17t - \phi)$$

$$- 8 \times 6.80 \sin(8t - 1.058)$$

at $t = 0$, $\dot{y} = 0$. Hence

$$0 = -4K \cos \phi + 9.17K \sin \phi + (54.4 \times 0.872)$$

or

$$K \sin \phi = \frac{(-47.44 - 33.32)}{9.17} = -8.81 \text{ cm}$$

Hence

$$\tan \phi = 1.058$$

and

$$\phi = 0.810 \text{ radian} = 46° \, 36'$$

or

$$\phi = 3.952 \text{ radians} = 226° \, 36'$$

and

$$K = \pm \sqrt{(8.81)^2 + (8.33)^2} = \pm 12.12$$

Now $K \sin \phi$ is a negative quantity. Hence, if we choose the angle ϕ as 46° 24′, then we must take the negative value of K. This is consistent with $K \cos \phi$ being negative. Hence the displacement of the mass m from its equilibrium position at any time t is

$$y = -12.12e^{-4t} \cos (9.17t - 0.810) + 6.80 \cos (8t - 1.058)$$

where all the angles are expressed in radians. We see that the first term of this equation representing the transient decreases rapidly with time, for after 2.5 sec the factor e^{-4t} is equal to 0.00005 of its value at zero time. After a few seconds the contribution from the transient term is effectively zero, and then only the steady-state term remains, which has the same frequency as the applied force. In the steady state the displacement y lags behind the applied force by 1.058 radians or 60° 39′. A plot of this motion is shown in Fig. 5.8. From this it can be seen that the effect of the transient is negligible after about 1 sec.

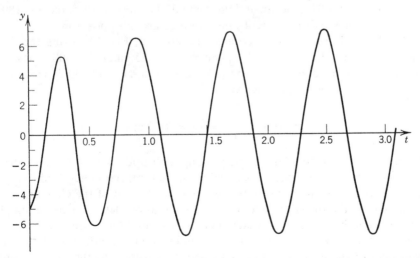

Fig. 5.8 The displacement y of a spring undergoing forced oscillations plotted against the time t.

5.14   Phase Difference between Displacement and Applied Force

As we have seen in Eq. 5.36, the displacement y of the mass m in the steady state of motion lags behind the applied force by an angle θ which is given in Eq. 5.39 as

$$\tan \theta = \frac{\omega R}{k - m\omega^2} \qquad (5.39)$$

The quantity $\sqrt{k/m}$ is the angular velocity ω_0 of the system vibrating without damping. This we shall call the *natural angular velocity* of the undamped simple harmonic motion of the system, and the corresponding frequency f_0 is

$$f_0 = \frac{\omega_0}{2\pi} = \frac{1}{2\pi}\sqrt{\frac{k}{m}} \qquad (5.42)$$

With this notation the phase angle θ may be written

$$\tan \theta = \frac{\omega R}{m(\omega_0{}^2 - \omega^2)} = \frac{fR}{2\pi m(f_0{}^2 - f^2)} \qquad (5.43)$$

where f is the frequency of the applied force.

As the applied frequency f is increased from zero to the natural frequency f_0 of the system, the phase angle θ increases from zero to 90°. When f is larger than f_0, then $\tan \theta$ is negative and θ increases beyond 90° and approaches 180° as f becomes very large compared to f_0. Thus, when the applied frequency is very large, the displacement lags behind the impressed force by 180°. In other words the displacement and the force are in opposition.

The change in phase with frequency is shown in Fig. 5.9, using the data for the spring problem given earlier, namely, $m = 25$ gm, $k = 2500$ dynes/cm. Three values of the damping constant R are used: 0, 50, and 500 dyne sec/cm. The natural frequency f_0 of the system is

$$f_0 = \frac{1}{2\pi}\sqrt{\frac{k}{m}} = \frac{10}{2\pi} = 1.59 \text{ cps}$$

When R is equal to zero, the angle θ changes abruptly from 0 to 180° where the applied frequency f becomes equal to the natural frequency f_0 of the system. The proof of this is given as a problem at the end of the chapter.

The lag of the displacement relative to the applied force can be simply demonstrated by swinging a meter stick, as a pendulum, about an axis near one end. If the axis is oscillated horizontally, then we notice that the displacement of the lower end is in an opposite direction to the applied force

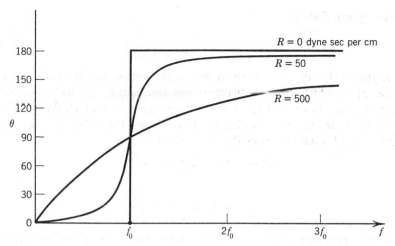

Fig. 5.9 Lag of the displacement θ behind the applied force plotted against the frequency f for different values of the resistance force constant.

when the frequency f of the applied force is greater than the natural frequency f_0 of the stick. Initially the displacement is in the same direction as the applied force, and it is only later when the steady state is set up that the displacement is in the opposite direction to the applied force.

5.15 Displacement Resonance in Forced Oscillations

The amplitude of the steady-state oscillations varies with the frequency of the impressed force. There is some frequency called the **resonance frequency** at which the amplitude becomes a maximum. This resonance frequency can be recognized in many vibrating systems. Automobiles, especially the older types, frequently have some part which vibrates violently at a particular speed. Below or above this speed the vibrations are small and usually not noticeable. In some cases, such as the marching in step of soldiers crossing a long bridge, the increase in amplitude of the vibrating system may produce disastrous results, whereas in a child's swing or the tuning of a radio a large amplitude may be what is desired. If two tuning forks of the same frequency are placed close together and one is set in vibration, the other will commence to vibrate. There are important examples of resonance in sound and in light. In fact, resonance can occur in any system capable of vibrating unless the damping force is too large.

The amplitude A of the oscillations at any applied frequency $f = \omega/2\pi$

is given by Eq. 5.40 as

$$A = \frac{F_0}{\sqrt{(k - m\omega^2)^2 + (\omega R)^2}} \qquad (5.40)$$

The angular velocity, $_r\omega_y$, at which the displacement has its maximum amplitude, or at which there is displacement resonance, can be found either by plotting A against ω and determining the value of ω at which the maximum amplitude occurs or, analytically, by determining the value of ω at which $\partial A/\partial\omega = 0$. This we find gives $\partial A/\partial\omega = 0$ when

$$R^2 - 2km + 2m^2{}_r\omega_y{}^2 = 0$$

Thus

$$_r\omega_y = 2\pi_r f_y = \sqrt{\frac{k}{m} - \frac{R^2}{2m^2}} \qquad (5.44)$$

From this equation we see that there can be no resonance if $R^2/2m^2$ is greater than k/m; i.e., if the damping of a system is large, then the amplitude A continually decreases as ω increases.

Figure 5.10 represents the variation of amplitude with ω for the data given in the earlier problem, namely, $m = 25$ gm, $k = 2500$ dynes/cm, and $F_0 = 12,500$ dynes. The smaller the value of R, the sharper the resonance or the tuning of the circuit to the applied force. Equation 5.44 shows that the resonance frequency varies with the value of the damping force constant R. It can be seen from Eq. 5.40 that the amplitude becomes infinite when the applied frequency becomes equal to the resonance frequency and the system has zero damping. When the damping constant R is not zero and $\omega = _r\omega_y$, then there is a maximum amplitude of $_rA_y$. The amplitude at which resonance takes place is

$$A_{\max} = _rA_y = \frac{F_0}{R\sqrt{[(k/m) - (R^2/4m^2)]}} \qquad (5.44a)$$

When the damping is small such that the Q of the circuit is large compared to unity, i.e., $m\omega_0 \gg R$ or $\omega_0{}^2 = k/m \gg R^2/m^2$, the resonant angular frequency $_r\omega_y$ is almost equal to ω_0. For this condition (Eq. 5.44a) the maximum amplitude at resonance is

$$A_{\max} = _rA_y \doteq \frac{F_0}{R\sqrt{k/m}} = \frac{F_0}{R\omega_0} \qquad (5.45)$$

Consider now the frequency ω of the driving force being nearly equal to the natural frequency ω_0 such that

$$\Delta\omega = \omega - \omega_0$$

where $\Delta\omega$ is very small compared to ω_0 and

$$(\omega_0{}^2 - \omega^2) = (\omega_0 + \omega)(\omega_0 - \omega) \approx -2\omega_0\Delta\omega$$

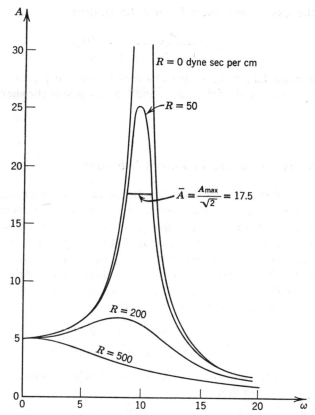

Fig. 5.10 Variation of the amplitude A plotted against the angular velocity ω of the forced oscillations—displacement resonance. The half width of the resonance curve for $R = 50$, $A_{max} = 25$ is $\Delta\omega = R/2m = 1$ unit along the axis of ω; and the corresponding amplitude $\bar{A}$ is $A_{max}/\sqrt{2} = 17.5$.

where the first term $(\omega_0 + \omega)$ has been set equal to $2\omega_0$. From Eq. 5.40 the amplitude A at frequency ω, reasonably close to ω_0, can be found in terms of A_{max} at frequency $_r\omega_y$. Thus, setting $k/m = \omega_0^2$, it follows that

$$A = \frac{F_0}{m[(\omega_0^2 - \omega^2)^2 + (\omega R/m)^2]^{\frac{1}{2}}} = \frac{F_0}{m[4\omega_0^2(\Delta\omega)^2 + (\omega R/m)^2]^{\frac{1}{2}}}$$

If in the second term under the radical, ω is set equal to ω_0, then from Eq. 5.45:

$$A = \frac{A_{max}R\omega_0}{2m\omega_0[(\Delta\omega)^2 + (R/2m)^2]^{\frac{1}{2}}}$$

Consider the special case where $R/2m = (\Delta\omega)$, then

$$\bar{A} = \frac{A_{max}R}{2m\sqrt{2}\,(R/2m)} = \frac{A_{max}}{\sqrt{2}}$$

Thus $R/2m$ is the half width of the resonance curve for points $1/\sqrt{2}$ or 0.71 of the maximum or peak of the curve, and is a measure of the sharpness of the peak.

5.16   Velocity Resonance in Forced Vibrations

The velocity of the moving particle attached to the spring executing forced vibrations varies from zero to a maximum in a manner somewhat similar to the displacement. However, the frequency of the applied force that produces the maximum velocity is not the same as that which produces the maximum displacement. We shall now determine the frequency necessary to produce the maximum value of the velocity, or to produce velocity resonance.

The velocity $\dot{y}$ of the moving particle in the steady state at any instant of time is given by differentiating Eq. 5.41. Thus

$$\dot{y} = \frac{-F_0\omega \sin(\omega t - \theta)}{\sqrt{(k - m\omega^2)^2 + (\omega R)^2}}$$

The amplitude of the velocity $\dot{A}$ is the maximum value of $\dot{y}$ obtained by setting the sine term equal to unity.

$$\dot{A} = \frac{F_0\omega}{\sqrt{(k - m\omega^2)^2 + (\omega R)^2}} \tag{5.46}$$

To obtain the frequency at which the velocity amplitude $\dot{A}$ is a maximum, one finds $\partial\dot{A}/\partial\omega$ and sets the result equal to zero. Carrying this out, we find that $\partial\dot{A}/\partial\omega = 0$ when $k - m\omega^2 = 0$. Thus the angular velocity, $_r\omega_v$, required for velocity resonance is given by

$$_r\omega_v = \sqrt{\frac{k}{m}} \tag{5.47}$$

This the value of the angular velocity of the natural oscillations of the free system without damping, as given in Eq. 5.5. The maximum velocity $\dot{A}$ is obtained from Eq. 5.46:

$$\dot{A}_{max} = \frac{F_0}{R} \tag{5.48}$$

It is interesting to note that the angular velocity at which velocity resonance occurs does not depend on the damping force whereas it does for displacement

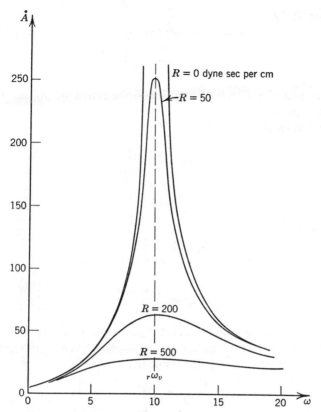

Fig. 5.11 Variation of the velocity amplitude $\dot{A}$ plotted against the angular velocity ω of the forced oscillations—velocity resonance.

resonance. Of course, the value of the velocity at velocity resonance does depend on the damping force as given by Eq. 5.48. Figure 5.11 shows the variation of the maximum velocity with the frequency of the applied force based on the data of the previous problem.

5.17   Phase Relations between Displacement and Velocity

In the steady state of the applied force oscillations the displacement and velocity differ in phase by 90° as may be seen from the appropriate equations. For the steady state the displacement is given by

$$y = A \cos (\omega t - \theta)$$

and the velocity by

$$\dot{y} = -A\omega \sin(\omega t - \theta)$$

$$= A\omega \cos\left(\omega t - \theta + \frac{\pi}{2}\right)$$

Thus at any time the velocity leads the displacement by 90° as it does for simple harmonic motion.

For convenience let $\theta - \pi/2 = \gamma$. Then

$$\tan \theta = \frac{-1}{\tan \gamma}$$

and

$$\dot{y} = A\omega \cos(\omega t - \gamma)$$

Since by Eq. 5.39 we have

$$\tan \theta = \frac{\omega R}{k - m\omega^2}$$

then

$$\tan \gamma = \frac{m\omega^2 - k}{\omega R} \tag{5.49}$$

Since the applied force at any instant of time t is given by $F = F_0 \cos \omega t$, it follows that the velocity lags behind the applied force by the angle γ.

The angular velocity at velocity resonance is

$$_r\omega_v{}^2 = \frac{k}{m}$$

so that when the applied force has this angular velocity

$$\tan \gamma = 0 \qquad \text{and} \qquad \gamma = 0$$

That is, at velocity resonance the velocity and applied force are in phase whereas the displacement lags behind the applied force by 90° Velocity resonance plays an important role in the theory of alternating-current circuits since the current in such a circuit is analogous to the velocity in a mechanical system.

5.18 Energy Relationships in Forced Oscillations

In the steady state, energy is being supplied to the oscillatory system by the applied force $F_0 \cos \omega t$ and is being dissipated through the agency of the damping force. From the general law of conservation of energy it must follow that in the steady state of motion the rate at which energy is being supplied to the system is equal to the rate at which it is being dissipated. We shall now prove this analytically.

The instantaneous value of the time rate of doing work $\dot{W}$ or the rate at which energy is being supplied to the system is the product of the force and velocity at the same instant of time or

$$\dot{W} = \dot{y}F_0 \cos \omega t$$

$$= A\omega F_0 \cos (\omega t - \gamma) \cos \omega t$$

$$= A\omega F_0(\cos^2 \omega t \cos \gamma + \tfrac{1}{2} \sin 2\omega t \sin \gamma) \qquad (5.50)$$

The average value of $\dot{W}$ over a complete cycle or several complete cycles may easily be obtained from the above expression. As was proved earlier in section 5.3, the average value of $\cos^2 \omega t$ over a complete cycle or period is one-half and the average value of $\sin 2\omega t$ is zero. Hence the average rate at which energy is supplied to the system is

$$\overline{W} = \frac{A\omega F_0 \cos \gamma}{2} \qquad (5.51)$$

This expression may be rewritten in different terms by substituting the value for A from Eq. 5.40 and noting that Eq. 5.49 gives for $\cos \gamma$ the value

$$\cos \gamma = \frac{\omega R}{(m\omega^2 - k)^2 + (\omega R)^2}$$

Hence

$$\overline{W} = \frac{F_0^2 \cos^2 \gamma}{2R} \qquad (5.52)$$

where γ is the angle by which the velocity lags the applied force. The variation of $\overline{W}$ with the angular velocity or frequency of the applied force is shown in Fig. 5.12, using the data of the earlier problem. Since at velocity resonance the angle γ is zero, it follows that at this resonance frequency the power supplied to the system is a maximum. This is to be expected, since at this frequency the velocity and kinetic energy have their maximum values.

The instantaneous rate of dissipation of energy D is the instantaneous rate of doing work against the damping force. The rate of doing work against the damping force is the produce of the damping force $R\dot{y}$ and the velocity $\dot{y}$ at the same instant. Thus

$$D = R\dot{y}^2$$

$$= RA^2\omega^2 \cos^2 (\omega t - \gamma) \qquad (5.53)$$

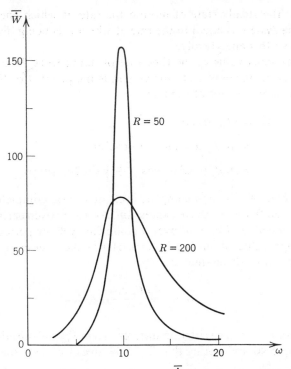

Fig. 5.12 Average rate of dissipation of energy $\bar{W}$ plotted against the angular velocity ω of the forced oscillations.

The average rate of dissipation of energy over a complete cycle is

$$\bar{D} = \frac{RA^2\omega^2}{2} \tag{5.54}$$

which by substituting for the amplitude A gives

$$\bar{D} = \frac{F_0{}^2 \cos^2 \gamma}{2R}$$

As predicted from the law of conservation of energy, this is equal to the power supplied to the system by the applied force.

5.19 Forced Vibrations in Other Fields of Physics

Since almost every branch of physics deals with motion of some kind, it is not surprising that some of the ideas developed for mechanics are of value in other branches of physics. It is neither desirable nor possible to present a long list of such examples in this text nor to give the complete theory in any one example. Frequently some of these systems are considered as analogous to a mechanical system. However, though analogies are very useful, one has to be careful. Often analogies are drawn on the basis of the similarity of equations and not on fundamental concepts. For instance, it is common in alternating-current theory to consider the inductance L as analogous to the mass m in the mechanical case, the capacitance C to the elastance $1/k$ or the inverse of the force constant of the spring, the current i to the velocity $\dot{y}$, the charge q to the displacement y, the resistance R to the damping force constant R, or the electromotive force E to the force F. These analogies are based on the similarity of the equations of motion, for the fundamental concepts in electricity are not the same as those in mechanics nor are they the only possible set of analogies.

The following examples may be omitted without any loss to the subject of mechanics.

An alternating-current circuit.   Let us consider a circuit composed of a resistance R, an inductance L, and a capacitance C placed in series with an alternating electromotive force of $E_0 \cos \omega t$ (Fig. 5.13) An alternating current is set up in the circuit, and, if at any instant of time the current is i, then by Kirchhoff's laws the equation for this circuit is given by

$$Ri + L\frac{di}{dt} + \frac{1}{C}\int i\, dt = E_0 \cos \omega t \tag{5.55}$$

Since the current i may be considered as the time rate of change of the charge at any point in the circuit, then

$$i = \frac{dq}{dt}$$

and Eq. 5.55 may be written in terms of the charge q as

$$L\frac{d^2q}{dt^2} + R\frac{dq}{dt} + \frac{q}{C} = E_0 \cos \omega t \tag{5.56}$$

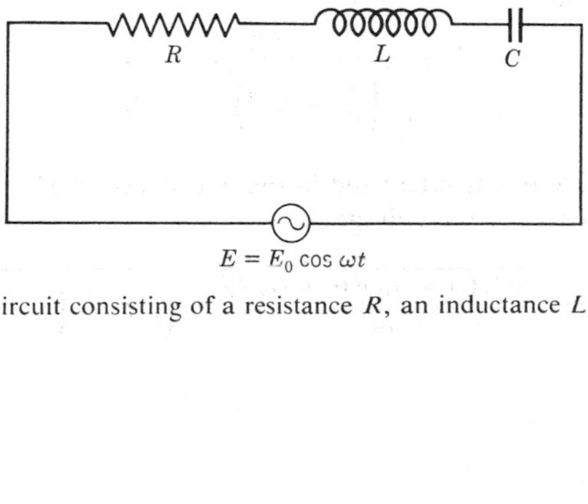

$$E = E_0 \cos \omega t$$

Fig. 5.13   An a-c circuit consisting of a resistance R, an inductance L, and a capacitance C placed in series.

This is an inhomogeneous linear differential equation of the second order similar to Eq. 5.35 for the forced oscillations of a spring. The analogous quantities mentioned above are immediately obtained by comparing Eq. 5.35 with Eq. 5.56.

Since it is the current I with which one is usually concerned in an electric circuit, then we must compare this with the velocity $\dot{y}$ of the oscillating mass. The resonance angular velocity $_r\omega_v$ given by Eq. 5.47 becomes for this electric circuit

$$_r\omega_v = \sqrt{\frac{1}{LC}}$$

or

$$_rf_v = \frac{1}{2\pi}\sqrt{\frac{1}{LC}} \tag{5.57}$$

where $_rf_v$ is the resonance frequency at which the current reaches its maximum value. The value of this maximum current I_0 given by the analogous Eq. 5.48 is

$$I_0 = \frac{E_0}{R} \tag{5.58}$$

In alternating-current theory the so-called root-mean-square values of current and voltage are used. If these are denoted by I and E respectively, then by the definition of the root-mean-square values

$$I = \frac{I_0}{\sqrt{2}} = 0.707 I_0$$

and

$$E = 0.707 E_0$$

so that Eq. 5.58 may be written

$$I = \frac{E}{R}$$

When the frequency of the applied voltage is the resonance frequency given above, then the phase angle γ, Eq. 5.49, is zero; i.e., the applies voltage and current are in phase at every instant of time.

When the applied voltage E has an angular velocity ω, then from Eq. 5.46 the current I is given by

$$I = \frac{E\omega}{\sqrt{\left(\dfrac{1}{C} - L\omega^2\right)^2 + (\omega R)^2}} \tag{5.59}$$

In the electrical case one is interested in the impedance Z of the circuit, which is the ratio of the current to the voltage, or

$$Z = \frac{E}{I} = \frac{\sqrt{[(1/C) - L\omega^2]^2 + (\omega R)^2}}{\omega} = \sqrt{R^2 + \left(L\omega - \frac{1}{\omega C}\right)^2} \tag{5.60}$$

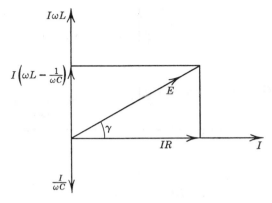

Fig. 5.14 A diagram showing the magnitude and direction of the voltages in an a-c circuit.

The angle γ by which the current lags behind the applied voltage is given by Eq. 5.49:

$$\tan \gamma = \frac{m\omega^2 - k}{\omega R} = \frac{L\omega^2 - (1/C)}{\omega R} = \frac{L\omega - (1/\omega C)}{R} \qquad (5.61)$$

If one cares to follow this analogy further, one can show relatively easily that the voltage and current are in phase in the resistance, that the voltage leads the current by 90° in the inductance and lags by 90° behind the current in the capacitance. These results are usually represented as in Fig. 5.14. In this diagram the voltage drop in the resistance is IR in phase with the current, the voltage drop in the inductance is $IL\omega$, leading the current by 90°, and in the capacitance it is $I/\omega C$, lagging the current by 90°.

We could have analyzed the mechanical or electrical oscillation systems by making use of imaginaries (see material on complex numbers in the Mathematical Appendix). Then the applied force $F_0 \cos \omega t$ would be the real part of the quantity $F_0 e^{j\omega t}$, where j represents $\sqrt{-1}$. By Euler's theorem

$$e^{j\omega t} = \cos \omega t + j \sin \omega t$$

where $\cos \omega t$ is the real part and $j \sin \omega t$ is the imaginary part of the complex quantity $e^{j\omega t}$. A complex quantity such as $2 + 3j$ can be represented graphically by employing an axis of reals and an axis of imaginaries, as shown in Fig. 5.15. In this mode of representation, multiplying a real number by j corresponds to rotating the point through 90°, multiplying by j^2 or -1 to rotating the point through 180°, multiplying by j^3 to rotating the point through 270°, and by j^4 or $+1$ to rotating the point through 360° or 0°. Thus we may represent both the magnitude and the direction of a complex quantity in the complex plane.

For example, the voltage drop in the resistance R is IR in phase with the current, the voltage drop in the inductance is written $(j\omega L)I$ and leads the current by 90°, and the voltage drop in the capacitance is written as $-jI/\omega C$ or $I/j\omega C$. The applied

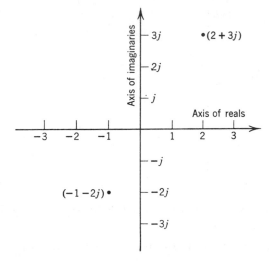

Fig. 5.15   Representation of complex quantities on a plane having axis of real and axis of imaginaries.

voltage E may then be expressed as

$$E = I\left(R + j\omega L - \frac{j}{\omega C}\right) \qquad (5.62)$$

The magnitude of the impedance Z of the circuit is given by Eq. 5.60, and the phase angle by Eq. 5.61. This mode of representation is very convenient for express-ing both the magnitudes and phase relations of current and voltage.

The derivation of Eq. 5.62 is readily obtained by assuming that the applied voltage in Eq. 5.56 is represented by the real part of $E_0 e^{j\omega t}$. As the steady-state solution assume that $q = Q_0 e^{j\omega t}$. Substituting for $q, \dot{q}, \ddot{q}$ in Eq. 5.56 gives

$$\dot{q} = I = \frac{E_0 \omega e^{j\omega t}}{j(L\omega^2 - 1/C) + \omega R} = \frac{E_0 e^{j\omega t}}{R + j[L\omega - (1/\omega C)]}$$

The damping of a galvanometer.   This is an example of the free oscillations of a galvanometer coil. In the ballistic galvanometer a quantity of electricity is dis-charged quickly through the coil of the galvanometer. A torque due to the electro-magnetic forces is exerted on the coil which causes a deflection. If θ is the angular deflection of the coil at some instant of time, then the equation of motion of the coil at this instant is

$$I\frac{d^2\theta}{dt^2} = -b\frac{d\theta}{dt} - c\theta \qquad (5.63)$$

In this expression $c\theta$ is the torque due to the twisting of the suspension of the coil through the angle θ and c is the torsion constant of the suspension, $b(d\theta/dt)$ is the torque due to the damping of the coil which has an angular velocity $d\theta/dt$ at the particular instant, $I\,d^2\theta/dt^2$ is the product of the moment of inertia of the coil about its axis of oscillation, and $d^2\theta/dt^2$ is its angular acceleration. This motion is the damped simple harmonic motion as is seen by comparing Eq. 5.63 with Eq. 5.25. Just as in the linear motion of the spring, so in the angular motion of the coil the motion may be overdamped, critically damped, or damped oscillatory. Which of these three types of motion results depends on whether b^2 is greater than, equal to, or less than $4cI$.

Let us suppose that the galvanometer is so constructed as to produce damped oscillatory motion. Then in the ballistic case the initial deflection is smaller than it would have been without any damping. The deflection without damping can be calculated from the initial observed deflection by means of the logarithmic decrement δ. The logarithmic decrement is the natural logarithm of the ratio of two maxima a period apart. In the linear case δ is given by Eq. 5.33 as

$$\delta = \frac{RP}{2m}$$

and this corresponds in the galvanometer problem to

$$\delta = \frac{bP}{2I}$$

In general, the quantities b and I cannot be calculated separately but, by observing successive maximum swings, the value of δ may be obtained. Suppose then that δ is known for the galvanometer. If θ_1 is the initial observed deflection and since it takes a quarter of a period $P/4$ to go from the central undeflected position to θ_1, then the deflection θ_0, which would have occurred had there been no damping, is given approximately by

$$\log_e \frac{\theta_0}{\theta_1} = \frac{b}{2I}\frac{P}{4} = \frac{\delta}{4}$$

or

$$\theta_0 = \theta_1 e^{\delta/4} \tag{5.64}$$

Dispersion theory in light and x-rays. In the dispersion theory of light and x-rays it is assumed that the electric field of the electromagnetic wave falling on a substance acts on the electrons of the atoms of the substance. The electrons are displaced from their normal positions, causing the electromagnetic wave to be refracted. This refraction is measured in terms of the refractive index. The refractive index varies with the frequency of the electromagnetic wave, and it is the object of the dispersion theory to provide a relationship between the refractive index and the frequency.

Let us assume the electron to be a point charge having a mass of m gm and a charge of e electrostatic units. If an electromagnetic wave having an electric field intensity given by $E_0 \cos \omega t$ acts on the electron, then a force $E_0 e \cos \omega t$ is exerted on the electron. The electron is displaced a distance y from its equilibrium position and, owing to this displacement, a restoring force ky is set up. Let us further assume that the motion of the electron is not damped. The problem then is one of forced oscillations without damping.

The equation of motion of the electron is similar to Eq. 5.35 and is given by

$$m\ddot{y} + ky = E_0 e \cos \omega t \tag{5.65}$$

In Eq. 5.65 the frequency of the electromagnetic wave producing the forced oscillations of the electron is $f = \omega/2\pi$.

The solution of Eq. 5.65 giving the displacement y at any time t is obtained for the steady state from Eq. 5.41 as

$$y = \frac{F_0 \cos(\omega t - \theta)}{k - m\omega^2} = \frac{F_0 \cos(\omega t - \theta)}{m(\omega_0^2 - \omega^2)}$$

$$= \frac{e E_0 \cos(2\pi f t - \theta)}{4\pi^2 m(f_0^2 - f^2)} \tag{5.66}$$

where $\omega_0 = 2\pi f_0 = \sqrt{k/m}$ and f_0 is the natural frequency of the electron within the atom.

In the present example we are interested only in the amplitude of the oscillations. The displacement y of the electron from its equilibrium position separates the positive from the negative charge of the atom in such a manner that the atom becomes an electric dipole. This electric dipole has an electric moment given by the magnitude of either the positive or the negative charge multiplied by the distance between the charges. At time t when the separation is y the dipole moment of the atom is ey.

Now from electromagnetic theory, by a proof which would take us too far afield, it follows that the dielectric constant K of a gas having a refractive index n and N electrons of natural frequency f_0 per unit volume is

$$K = n^2 = 1 + \frac{4\pi N e y}{E_0}$$

so that from Eq. 5.66, using the amplitude only, it follows that the square of the refractive index is

$$n^2 = 1 + \frac{N e^2}{\pi m (f_0^2 - f^2)}$$

The phenomenon of anomalous dispersion is which the refractive index becomes very large is explained by this theory by assuming that in this region the frequency f of the electromagnetic waves becomes almost equal to the frequency f_0 of the electron in the atom. For many substances it appears that the frequency f_0 is much larger than the frequency f of the visible light waves, so that in this region we may write

$$n^2 = 1 + \frac{N e^2}{\pi m f_0^2}$$

and, since the second term on the right is very small for gases, the refractive index n, by the binomial theorem, becomes

$$n = 1 + \frac{N e^2}{2\pi m f_0^2}$$

5.20   Coupled Systems and their Oscillations

In this discussion we shall be concerned with the motion of two vibrating systems in which the motion of one system can influence that of the other. Such systems could be mechanical (Fig. 5.16) or electric. The degree to which the motion in the one system influences that in the other is called the *coupling* of the systems. This coupling may be strong or weak but, since the analysis of weak coupling is the simpler, we shall generally be concerned with this type.

If the two systems are similar, e.g., identical pendulums or springs and masses, as shown in Fig. 5.16, there is resonance between the two elements of the system such that if object 1 is initially oscillated, with object 2 at rest, then the energy alternates between the two objects, and when one is vibrating with maximum amplitude the other is at rest, as shown in Fig. 5.17. Suppose now that both similar pendulums or systems are set in motion at the same time with equal amplitudes either in the same or opposite directions (Fig 5.18*a* and *b*), then there is no energy interchange between the two pendulums. The two modes of vibration shown in Fig. 5.18*a* and *b* are called the *normal modes of oscillation* of this system having two degrees of freedom. It is a general rule that an oscillatory coupled system having *n* degrees of freedom has *n* normal modes of oscillation.

For the analysis of coupled oscillators, let us consider the two similar simple pendulums (Fig. 5.19), each having a length *l* and a bob of mass *m*

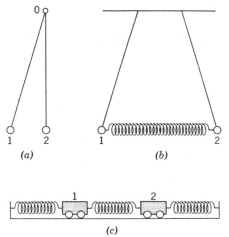

Fig. 5.16 Coupled systems. (*a*) End-on view of two pendulums coupled by twisting the support rod at 0. (*b*) Two pendulums coupled by means of a spring. (*c*) Two cars with coupling springs.

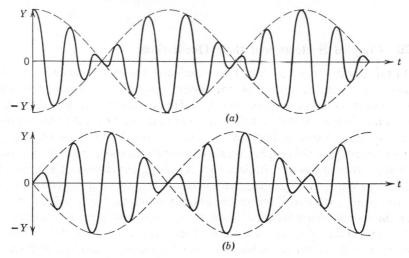

Fig. 5.17   (*a*) Motion of pendulum 1. (*b*) Motion of pendulum 2.

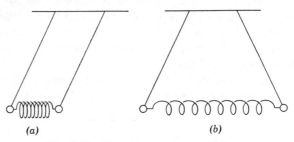

Fig. 5.18   Normal modes of oscillations.

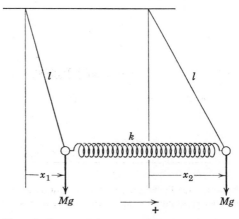

Fig. 5.19   Two similar pendulums coupled with a horizontal spring.

connected by a horizontal spring of negligible mass which is unstretched when the pendulums are in their positions of equilibrium, i.e., $x_1 = x_2 = 0$. The coupling between the pendulums is provided by the spring whose force constant or force per unit extension is k. At some instant of time, let the positive horizontal displacements of the bobs be x_1 and x_2 respectively, so that the spring is stretched by an amount $(x_2 - x_1)$ and the spring exerts a force of $k(x_2 - x_1)$. Assume that the motion is undamped and that the displacement x of a pendulum bob is so small that the restoring force due to the weight mg of the bob is mgx/l. The equations of motion for the two pendulums are:

$$m\ddot{x}_1 = -\frac{mgx_1}{l} + k(x_2 - x_1) \tag{5.67}$$

$$m\ddot{x}_2 = -\frac{mgx_2}{l} - k(x_2 - x_1) \tag{5.68}$$

We may assume a general solution of the form $x_1 = P_1 \cos \omega t + Q_1 \sin \omega t$ or $x_1 = Pe^{j\omega t}$, where $j = \sqrt{-1}$. For convenience we shall use the cosine and sine, and assume

$$x_1 = P_1 \cos \omega t + Q_1 \sin \omega t \tag{5.69}$$

$$x_2 = P_2 \cos \omega t + Q_2 \sin \omega t \tag{5.70}$$

Notice that we have assumed that the two masses each move with simple harmonic motion of the same angular frequency ω. By substitution from Eqs. 5.69 and 5.70 in Eqs. 5.67 and 5.68, it follows that

$$-mP_1\omega^2 \cos \omega t - mQ_1\omega^2 \sin \omega t = -\frac{mgP_1 \cos \omega t}{l} - \frac{mgQ_1 \sin \omega t}{l}$$

$$+ kP_2 \cos \omega t + kQ_2 \sin \omega t - kP_1 \cos \omega t - kQ_1 \sin \omega t \tag{5.71}$$

and

$$-mP_2\omega^2 \cos \omega t - mQ_2\omega^2 \sin \omega t = -\frac{mgP_2 \cos \omega t}{l} - \frac{mgQ_2 \sin \omega t}{l}$$

$$- kP_2 \cos \omega t - kQ_2 \sin \omega t + kP_1 \cos \omega t + kQ_1 \sin \omega t \tag{5.72}$$

For these equations to be valid at all times, the coefficients of the cosine and sine terms must algebraically add to zero. From Eq. 5.71 the coefficients for the cosine and sine terms give respectively

$$-mP_1\omega^2 = -\frac{mgP_1}{l} + kP_2 - kP_1 \tag{5.73}$$

$$-mQ_1\omega^2 = -\frac{mgQ_1}{l} + kQ_2 - kQ_1 \tag{5.74}$$

and, similarly, from Eq. 5.72, it follows that

$$-mP_2\omega^2 = -\frac{mgP_2}{l} - kP_2 + kP_1 \tag{5.75}$$

$$-mQ_2\omega^2 = -\frac{mgQ_2}{l} - kQ_2 + kQ_1 \tag{5.76}$$

From Eqs. 5.73 and 5.75 for P_1 and P_2, it follows that:

$$P_1\left(-m\omega^2 + \frac{mg}{l} + k\right) = kP_2 \tag{5.77}$$

$$kP_1 = P_2\left(-m\omega^2 + \frac{mg}{l} + k\right) \tag{5.78}$$

The two equations above give:

$$\left(-m\omega^2 + \frac{mg}{l} + k\right)^2 = k^2$$

or

$$\left(-m\omega^2 + \frac{mg}{l}\right)^2 + 2k\left(-m\omega^2 + \frac{mg}{l}\right) + k^2 = k^2$$

or

$$(\omega^2)^2 - \omega^2\left(\frac{2g}{l} + \frac{2k}{m}\right) + \left(\frac{g}{l}\right)^2 + \frac{2kg}{ml} = 0$$

Solving for ω^2 gives

$$\omega^2 = \frac{g}{l} + \frac{k}{m} \pm \sqrt{\left(\frac{g}{l} + \frac{k}{m}\right)^2 - \left(\frac{g}{l}\right)^2 - \frac{2kg}{ml}}$$

$$= \frac{g}{l} + \frac{k}{m} \pm \frac{k}{m}$$

The two values of ω^2 which may be designated ω_h^2 for the higher angular frequency and ω_l^2 for the lower value are

$$\omega_h^2 = \frac{g}{l} + \frac{2k}{m} \quad \text{or} \quad \omega_h = \pm\sqrt{\frac{g}{l} + \frac{2k}{m}} \tag{5.79}$$

$$\omega_l^2 = \frac{g}{l} \quad \text{or} \quad \omega_l = \pm\sqrt{\frac{g}{l}} \tag{5.80}$$

These values would have been obtained if Eqs. 5.74 and 5.76, having Q_1 and Q_2, had been used. Substituting the value for $\omega_h{}^2$ in Eq. 5.73 gives

$$-mP_1\left(\frac{g}{l} + \frac{2k}{m}\right) = -\frac{mgP_1}{l} + kP_2 - kP_1$$

or

$$P_1\left(-\frac{mg}{l} - 2k + \frac{mg}{l} + k\right) = kP_2$$

or

$$P_2 = -P_1$$

From Eq. 5.74 for $\omega^2 = \omega_h{}^2$ it follows that $Q_2 = -Q_1$. Similarly we may show for $\omega^2 = \omega_l{}^2$ that $Q_1 = Q_2$ and $P_1 = P_2$.

Summarizing these results:

For $\omega = \pm\omega_h = \pm\sqrt{g/l + 2k/m}$; $P_1 = -P_2$ and $Q_1 = -Q_2$.

For $\omega = \pm\omega_l = \pm\sqrt{g/l}$; $P_1 = P_2$ and $Q_1 = Q_2$.

The general solution of Eqs. 5.69 and 5.70 can be written as

$$x_1 = A \cos \omega_h t + B \sin \omega_h t + C \cos \omega_l t + D \sin \omega_l t \qquad (5.81)$$

$$x_2 = -A \cos \omega_h t - B \sin \omega_h t + C \cos \omega_l t + D \sin \omega_l t \qquad (5.82)$$

The two second-order differential equations, Eqs. 5.67 and 5.68 require four constants of integration, and these are given above as A, B, C, D. Notice that for $\omega = \omega_h$; $P_1 = -P_2$, $Q_1 = -Q_2$ so that the terms in x_2 are the negative of those in x_1 for the coefficients A and B, whereas for $\omega = \omega_l$; $P_1 = P_2$, $Q_1 = Q_2$, and the terms C and D accompanying the ω_l have the same signs for x_1 and x_2.

5.21   Normal Coordinates

It is possible to set the two pendulums in motion in such a manner that only one frequency in involved. To investigate this condition we must combine Eqs. 5.81 and 5.82 so that only one frequency is present in the result. Adding and subtracting these equations gives

$$x_1 + x_2 = Y_1 = 2C \cos \omega_l t + 2D \sin \omega_l t \qquad (5.83)$$

$$x_1 - x_2 = Y_2 = 2A \cos \omega_h t + 2B \sin \omega_h t \qquad (5.84)$$

The vibrations given by the above equations are called normal vibrations inasmuch as the variable $Y_1 = x_1 + x_2$ involves only the frequency ω_l while $Y_2 = x_1 - x_2$ involves the frequency ω_h. In the general motion of the pendulums both the motions of x_1 and x_2 or the frequencies ω_l and ω_h are present.

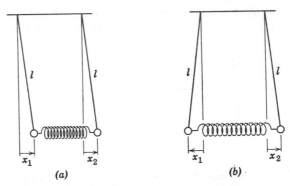

Fig. 5.20   Normal vibrations. (a) ω_l present, $\omega_h = 0$. (b) ω_h present, $\omega_l = 0$.

Suppose that the pendulums are started off with Y_2 zero and the Y_1 motion present (Fig. 5.20a). For similar pendulums, Y_2 will remain zero at all times. With Y_2 zero, then from Eq. 5.84 it follows that $x_1 = x_2$ and there is no vibration with angular frequency ω_h but there is a vibration with angular frequency ω_l associated with Y_1. From Eq. 5.80, $\omega_l = \sqrt{g/l}$, so that the two pendulums oscillate with the same frequencies with which they would were they free. In this case, the coupling spring has no effect or exerts no forces so that the pendulums are effectively free or separated.

This is not the situation had the initial motion been such as to make $Y_1 = 0$, Fig. 5.20b, or $x_1 = -x_2$, and produces only the motion of frequency $\omega_h = \sqrt{g/l + 2k/m}$. For this case the spring is exerting a restoring force. From Eqs. 5.83 and 5.84 it is seen that the normal coordinates Y_1 and Y_2 are independent of each other in that the system can oscillate with only one frequency while the other is suppressed. (For systems having more than two degrees of freedom only one frequency is present.)

It will now be shown that the equations of motion take the form of a set of linear differential equations with constant coefficients when they are expressed in terms of the normal coordinates. From Eqs. 5.83 and 5.84 it is seen that

$$x_1 = \frac{Y_1 + Y_2}{2} \quad \text{and} \quad x_2 = \frac{Y_1 - Y_2}{2} \tag{5.85}$$

Substituting these values in Eqs. 5.67 and 5.68 we have,

$$\frac{m}{2}(\ddot{Y}_1 + \ddot{Y}_2) = -\frac{mg}{2l}(Y_1 + Y_2) - kY_2 \tag{5.86}$$

$$\frac{m}{2}(\ddot{Y}_1 - \ddot{Y}_2) = -\frac{mg}{2l}(Y_1 - Y_2) + kY_2 \tag{5.87}$$

Adding and subtracting these equations gives,

$$m\ddot{Y}_1 = -\frac{mg}{l}Y_1 \quad \text{or} \quad \ddot{Y}_1 + \frac{g}{l}Y_1 = 0 \tag{5.88}$$

$$m\ddot{Y}_2 = -\frac{mg}{l}Y_2 - 2kY_2 \quad \text{or} \quad \ddot{Y}_2 + \left(\frac{g}{l} + \frac{2k}{m}\right)Y_2 = 0 \tag{5.89}$$

Eqs. 5.88 and 5.89 are linear differential equations with constant coefficients, and they represent physically simple harmonic motions in the normal coordinates.

5.22 Energy of Coupled Systems and the Transfer of Energy within the System

The energy of the coupled pendulum system shown in Fig. 5.19 is partly potential and partly kinetic. The potential energy is due to the extension of the spring and the raising of the pendulum bobs against gravity. The potential energy of the spring is

$$V_s = \int_0^{x_2-x_1} kx\,dx = \frac{k}{2}(x_2 - x_1)^2$$

and that of the pendulum bobs is $\int F_x\,dx$, which to the accuracy we are using for small displacements is

$$V_g = \int_0^{x_1} \frac{mgx}{l}\,dx + \int_0^{x_2} \frac{mgx}{l}\,dx = \frac{mg}{2l}x_1^2 + \frac{mg}{2l}x_2^2$$

Thus the total potential energy V is

$$V = \frac{k}{2}(x_2 - x_1)^2 + \frac{mg}{2l}(x_1^2 + x_2^2)$$

which, by substitution from Eqs. 5.83 and 5.84, becomes

$$V = \frac{k}{2}Y_2^2 + \frac{mg}{2l}\left[\frac{(Y_1 + Y_2)^2 + (Y_1 - Y_2)^2}{4}\right]$$

$$= \frac{mg}{4l}Y_1^2 + \left(\frac{mg}{4l} + \frac{k}{2}\right)Y_2^2 \tag{5.90}$$

The kinetic energy T of the pendulum is

$$T = \frac{m}{2}(\dot{x}_1^2 + \dot{x}_2^2) = \frac{m}{4}(\dot{Y}_1^2 + \dot{Y}_2^2) \tag{5.91}$$

Notice that the potential energy V contains terms in $x_1 x_2$, or cross-coupling terms, whereas when V is expressed in the normal coordinates, there are no cross terms of the $Y_1 Y_2$ type but only terms in $Y_1{}^2$ and $Y_2{}^2$. Thus another property of the normal coordinates, proved here only for a special case, is that the energy of an undamped system can be expressed as the sum of the squares of the normal coordinates and their first derivatives multiplied by the appropriate coefficients.

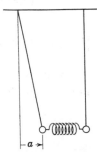

To understand the mechanism of the transfer of energy from one pendulum to the other, let us consider pendulum 1 at rest and displaced a distance a to the right while pendulum 2 is at rest at its equilibrium position (Fig. 5.21). In mathematical terms the initial conditions are: At $t = 0$,

Fig. 5.21 Initially, one of the pendulums has displacement a and the other has zero displacement.

$$x_1 = a \qquad \dot{x}_1 = 0 \qquad x_2 = 0 \qquad \dot{x}_2 = 0$$

These conditions must now be imposed on the expressions for x_1 and x_2, Eqs. 5.81 and 5.82, to evaluate the coefficients A, B, C, and D. From the initial conditions it follows that

$$a = A + C \qquad 0 = B\omega_h + D\omega_l$$
$$0 = -A + C \qquad 0 = -B\omega_h + D\omega_l$$

Thus

$$A = C = \frac{a}{2} \qquad B = D = 0$$

With the initial conditions the displacements x_1 and x_2 at any time t are

$$x_1 = \frac{a}{2}(\cos \omega_h t + \cos \omega_l t) \tag{5.92}$$

$$x_2 = \frac{a}{2}(\cos \omega_l t - \cos \omega_h t) \tag{5.93}$$

By trigonometric transformations Eqs. 5.92 and 5.93 may be written

$$x_1 = a \cos \left(\frac{\omega_h + \omega_l}{2}\right) t \cos \left(\frac{\omega_h - \omega_l}{2}\right) t \tag{5.94}$$

$$x_2 = a \sin \left(\frac{\omega_h + \omega_l}{2}\right) t \sin \left(\frac{\omega_h - \omega_l}{2}\right) t \tag{5.95}$$

If ω_h, the higher angular frequency, is very little different from ω_l, the lower angular frequency, then from Eqs. 5.79 and 5.80 it follows that the value of

k or the coupling between the pendulums is very weak. With this condition of very weak coupling, let

$$\frac{\omega_h + \omega_l}{2} = \omega'$$

then Eqs. 5.94 and 5.95 may be written as

$$x_1 = a \cos\left(\frac{\omega_h - \omega_l}{2}\right) t \cos \omega' t \qquad (5.96)$$

$$x_2 = a \sin\left(\frac{\omega_h - \omega_l}{2}\right) t \sin \omega' t \qquad (5.97)$$

Since ω' is much larger than $(\omega_h - \omega_l)/2$, we may regard Eqs. 5.96 and 5.97 as implying that x_1 and x_2 are oscillating with an angular frequency ω' and a slowly varying amplitude given respectively by the cosine and the sine of the difference of the frequencies, as shown in Fig. 5.17. From Eqs. 5.96 and 5.97 it is seen that, if x_2 or the sine terms are increasing, then x_1 or the cosine terms are decreasing, and vice versa. Thus x_2 has a maximum value when x_1 is zero and x_1 has its maximum value when x_2 is zero, so that the displacements and energy transfer from the one pendulum to the other with a period P are given by

$$\left(\frac{\omega_h - \omega_l}{2}\right) P = 2\pi \qquad \text{or} \qquad P = \frac{4\pi}{\omega_h - \omega_l}$$

Experiments on coupled systems, which illustrate the theory given above, may be found in the literature.*

PROBLEMS

1. The end of one of the prongs of a tuning fork which executes simple harmonic motion of frequency 1000 per second has an amplitude of 1.5 mm. If damping is neglected, find (a) the maximum velocity and acceleration of the

* F. Miller, Jr., "A Laboratory Experiment with Coupled Linear Oscillators," *Am. J. Phys.*, **20**, 23–25, 1952.
   C. R. Kannewurf and Harald C. Jensen, "Coupled Oscillations," *Am. J. Phys.*, **25**, 442–445, 1957.
   R. B. Runk, J. L. Stull, and O. L. Anderson, "A Laboratory Linear Analog for Lattice Dynamics," *Am. J. Phys.* **31**, 915–921, 1963. This is an experiment with a frictionless linear air track.

end of the prong; (b) the velocity and acceleration of the end of the prong when it has a displacement of 1.25 mm.

2. A block is placed on the top of a piston which is executing simple harmonic motion in a vertical direction. The piston has an amplitude of 0.05 m. Find (a) at what position in the motion of the piston the block is likely to leave the piston; (b) the maximum frequency of the piston at which the block will remain continuously in contact with the piston at all times during the motion.

3. Suppose that a hole is bored along a diameter of the earth and a particle is dropped into it at the surface of the earth. Assuming the earth is of constant density ρ and radius R, show that the particle will execute simple harmonic motion with a period of $2\pi \sqrt{R/g}$ where g is the acceleration due to gravity at the surface of the earth. When the particle is at a depth of 1/4 of the radius of the earth, show the acceleration is $0.75\,g$ and the velocity is $0.661\ \sqrt{Rg}$. (Neglect the rotation of the earth.)

4. Suppose a hole is bored through the earth which does not pass through the center of the earth and a particle slides through the hole without any friction. If the particle starts from rest at the surface of the earth show that the particle executes simple harmonic motion with the same period as in Problem 3. (Neglect the rotation of the earth.)

5. An object of mass m is moving clockwise in a circle of radius R with constant speed v_0. If the center of the circle is at the origin $(0, 0)$ of rectangular coordinates and at $t = 0$ the object is at $(0, R)$, then find the coordinates, velocities, and accelerations of the object as functions of time and show $\ddot{y} + \omega^2 y = 0$ where $\omega = v_0/R$.

6. A particle whose mass is 0.1 kg which is free to move in a plane is attracted towards a fixed point O with a force proportional to its distance from the fixed point. When its distance from the fixed point is 0.5 m, the force of attraction is 0.2 newton. (a) Find the period of the motion. (b) Suppose that the particle is started with a displacement of 0.5 m from O in the X direction and a velocity of 0.5 m/sec in the perpendicular Y direction. Find the equation for and plot the path of the particle.

7. A chalk mark is made on the lower edge of a tire of radius R. The tire is rolling with speed v_0 without slipping along a horizontal road. During a time t, a line from the axle to the chalk mark turns through an angle θ such that $\theta = \omega t = v_0 t/R$. Find the x and y coordinates of the chalk mark at any time t or angle θ, and also the velocity and acceleration of the chalk mark. (The path of the chalk mark is a cycloid.)

8. Mud is thrown from the tire in Problem 7 as it moves along a horizontal road. Show that the maximum height to which the mud can be thrown is $R + v_0^2/2g + gR^2/2v_0^2$. Use the results obtained in Problem 7, and note that the

height h to which the mud goes is $y + \dot{y}^2/2g$. Show the mud attaining this maximum height comes off the wheel at a point on the rim gR^2/v_0^2 higher than the center.

9. Electrons in an oscilloscope are deflected by two mutually perpendicular electric fields in such a manner that the displacement at any time t is given by

$$x = E_1 \sin (\omega t + \delta) \qquad y = E_2 \sin \omega t$$

(a) If $E_1 = E_2 = 10$ cm and the phase angle δ is $90°$, show that the path of the electrons is a circle whose radius is 10 cm. (b) If $\delta = 0°$, show that the path is a straight line making an angle with the X axis whose tangent is E_2/E_1. (c) If $\delta = 30°$, find the equation of the path and show that it is an ellipse.

10. A particle of mass m moving along the X axis is attracted towards the origin with a force of mkx. At a time t of 2 sec the particle is at the origin, and at time t of 3 sec the velocity of the particle is 6 cm/sec. If the period of the simple harmonic motion is 6 sec, then show that the force constant k is $\pi^2/9$ and the amplitude of the motion is $-36/\pi$.

11. A uniform iron bar of mass M and length L is placed in a horizontal position across the top of two similar cylindrical rollers which are rotating in opposite directions as shown in Fig. P11. The centers of the two rollers are in a horizontal

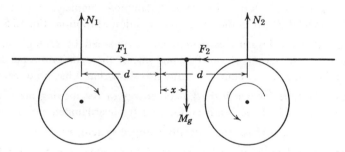

Fig. P11

line a distance of $2d$ apart, which is less than the length of the bar L. If μ is the coefficient of friction between the iron bar and the cylinders show that if the bar is at a displacement of x from its central position, the net horizontal force on the bar is $F = -Mg\mu x/d$. Show the bar will execute simple harmonic motion with a period of $P = 2\pi\sqrt{d/\mu g}$.

12. Lissajous' figures are produced by a particle being subjected to two simple harmonic motions at right angles, along the X and Y axes respectively. Suppose that the frequency along the Y axis is twice that along the X axis, that the two simple harmonic motions have equal amplitudes A and a phase difference

of 270°, and that they are started with $x = A$ and $y = 0$. Show that the equation of the path is

$$y^2 = \frac{4x^2}{A^2} (A^2 - x^2)$$

and plot the path for an amplitude A of 4 cm.

13. A uniform rod of length L and mass M slides with its ends in a frictionless verticle circle of radius R. If $L = \sqrt{3}\,R$, show that the rod undergoes simple harmonic oscillations with a period of $2\pi\sqrt{R/g}$. (Note that the rod subtends an angle of 120° at the center of the circle, and oscillates about the center of the circle c such that $I_c = ML^2/6$.)

14. If a pendulum clock has a period of exactly 1 sec when its amplitude is 2°, find the number of seconds gained or lost per day when the amplitude is 10°, assuming that the clock behaves as a simple pendulum. Find the maximum tension in the member holding the bob if the bob weighs 2 lb.

15. Show by integration of Eq. 5.18, $ml\ddot{\theta} = -mg \sin\theta$, that the kinetic energy at any angular displacement θ is equal to the loss in potential energy at this displacement or that there is conservation of mechanical energy for the simple pendulum.

16. Show that the solution for critical damping, namely, Eq. 5.28 or $y = (A + Bt)e^{-Rt/2m}$, is a solution of the equation of motion, Eq. 5.25.

17. Show that the logarithmic decrement δ, obtained by taking two successive maximum displacements on the same side of the equilibrium position for the data in Fig. 5.7c, is about 1.28, and also show that δ has no dimensions.

18. Show the time for which the kinetic energy of the moving mass of 25 gm in Section 5.7, Fig. 5.7a, is a maximum for the overdamped motion, is 0.076 sec.

19. Repeat Problem 18 for the critically damped system, Fig. 5.7b.

20. Repeat Problem 18 for the damped oscillatory system, Fig. 5.7c. Find the energy lost between the displacements of -5.0 cm and zero, and between zero and 2.63 cm. Give a physical explanation for the kinetic energy's not having its maximum value at zero displacement.

21. Show that the work done against the damping force during the half-cycle from the times 0.181 to 0.501 sec in Fig. 5.7c is equal to the change in the kinetic energy during this interval, namely, about 11,000 ergs.

22. Calculate the loss in energy for the mass executing damped oscillations as shown in Fig. 5.7c in the half-vibration between the times 0.501 and 0.822 sec.

23. Find the equation for the displacement with time for the forced oscillation, using the data employed in Fig. 5.8 except for the damping force constant which is now $R = 1000$ dyne sec/cm. Plot a curve of the displacement against time.

24. An electric charge sent through a ballistic galvanometer produces an initial deflection of 10 cm to the right of the zero mark of a scale. Subsequent deflections on alternate sides of the zero mark are $-9.71, 9.41, -9.14, 8.87$ cm. From these data find (a) the logarithmic decrement; (b) the initial deflection which would have occurred if there had been no damping.

25. An electric circuit is composed of a resistor having a resistance of 40 ohms, a capacitor having a capacitance of 100 μf, and an inductance of 0.2 h connected in series to an alternating voltage of 120 sin (400t) v. (a) Find the maximum value of the current in the circuit. (b) Find the root-mean-square current in the circuit. (c) If the capacitance can be varied, find what value it must have if the current is to be a maximum. Find the value of this current.

26. Using Eqs. 5.79 and 5.80 for the higher and lower frequencies show that when g/l is very much greater than k/m, then $\omega_h - \omega_l = \sqrt{(k^2 l)/(m^2 g)}$.

27. Solve Eqs. 5.67 and 5.68 by assuming the solutions $x_1 = Ae^{j\omega t}$, $x_2 = Be^{j\omega t}$. Find ω_h^2 and ω_l^2 and show that, for $\omega = \pm\omega_h$, $A = -B$, whereas for $\omega = \pm\omega_l$, $A = B$. Show that x_1 and x_2 may be written as

$$x_1 = Ae^{j\omega_h t} + Be^{-j\omega_h t} + Ce^{j\omega_l t} + De^{-j\omega_l t}$$

$$x_2 = -Ae^{j\omega_h t} - Be^{-j\omega_h t} + Ce^{j\omega_l t} + De^{-j\omega_l t}$$

and that these equations are equivalent to Eqs. 5.81 and 5.82.

28. An object having a mass of 0.05 kg is hung on the end of a vertical spring which has a force constant of 5 nt/m. The motion is started with an initial displacement of 0.05 m and with a damping force constant of 0.5 nt sec/m. Show that the displacement y at any time t is given by $y = 0.0577e^{-5t} \times \cos (8.66t - 0.523)$. Plot the curve of y against t, and also plot the curves of the potential and kinetic energies against time.

29. In Fig. 5.16c the masses called m_1 and m_2 move along a frictionless horizontal plane, are connected with springs of negligible mass, and have force constants, from the left-hand side, of k_1, k, k_2. If m_1 has a positive displacement of x_1 and m_2 a positive displacement of x_2 from their positions of equilibrium, show that the equations of motion are: $m_1\ddot{x}_1 = -k_1 x_1 - k(x_1 - x_2)$; $m_2\ddot{x}_2 = -k_2 x_2 + k(x_1 - x_2)$. For convenience, let $y_1 = x_1\sqrt{m_1}$, $y_2 = x_2\sqrt{m_2}$, $k_1 + k = m_1\omega_1^2$, $k_2 + k = m_2\omega_2^2$, and $k = c\sqrt{m_1 m_2}$, so that the equations of motion are:

$$\ddot{y}_1 + \omega_1^2 y_1 - cy_2 = 0 \qquad \ddot{y}_2 + \omega_2^2 y_2 - cy_1 = 0$$

Solve these equations by setting $y_1 = A \sin \omega t$, $y_2 = B \sin \omega t$, and show that:

$$\omega_h^2 = \frac{\omega_1^2 + \omega_2^2}{2} + \tfrac{1}{2}\sqrt{(\omega_1^2 - \omega_2^2)^2 + 4c^2}$$

$$\omega_l^2 = \frac{\omega_1^2 + \omega_2^2}{2} - \tfrac{1}{2}\sqrt{(\omega_1^2 - \omega_2^2)^2 + 4c^2}$$

Show that A/B is positive for $\omega = \omega_l$ or for the in-phase component, while A/B is negative for $\omega = \omega_h$ or for the out-of-phase component. Also show that $\omega_h^2 + \omega_l^2 = \omega_1^2 + \omega_2^2$, and that ω_h and ω_l are the angular frequencies of the normal modes of vibration.

30. Suppose the two coupled pendulums (Eqs. 5.81, 5.82) are started with $x_1 = 0$, $\dot{x}_1 = 0$, $x_2 = a$, $\dot{x}_2 = 0$. Find the displacements x_1 and x_2 as functions of time. Repeat with the initial conditions of $x_1 = -a$, $\dot{x}_1 = 0$, $x_2 = a$, $\dot{x}_2 = 0$ and find the displacements at any time. Qualitatively discuss the motions.

TRANSLATIONAL AND ROTATIONAL MOTION OF RIGID BODIES ABOUT FIXED AXIS

6.1 Internal Forces in Rigid Bodies

Up to the present we have been concerned with the motion of particles or of bodies that can be treated as particles, where no rotational motion is involved. In this chapter we shall extend the concepts introduced earlier to bodies of finite size that can possess both rotational motion and translational motion. In dealing with these bodies we shall assume them to be rigid, i.e., they do not change size or shape under the action of forces. Although this assumption may not be strictly accurate, it is sufficiently so for our present purposes.

The rigidity of a body is due to the nature of the internal attractive forces between the molecules composing the body. We shall assume that the internal force between any two molecules acts along the line joining them, and that Newton's third law is applicable. Thus for any two molecules or particles i and j in a body, it is assumed that the internal force $\mathbf{F}_{ij}$ exerted on i by j is equal and opposite to the force $\mathbf{F}_{ji}$ exerted on j by i. In symbols

$$\mathbf{F}_{ij} = -\mathbf{F}_{ji}$$

Therefore the sum of the internal forces within any body is zero.

6.2   Translational Motion and Center of Mass

In this section we shall show that for linear or translational motion the total external force acting on the particles composing a body is equal to the summation of the time rate of change of momentum of the particles. Also the total external force is equal to the total mass of the particles multiplied by the acceleration of their center of mass.

To prove this, let us consider a body made up of a number of particles whose masses are $m_1, m_2, \ldots, m_i$, etc., respectively, and whose positions relative to some fixed origin O in an inertial system are given by the vector distances $\mathbf{r}_1, \mathbf{r}_2, \ldots, \mathbf{r}_i$, etc., as shown in Fig. 6.1. Let the external force acting on the particle of mass m_i be $\mathbf{F}_i$ and the corresponding internal force be $\mathbf{F}_i'$. By Newton's second law

$$m_i \ddot{\mathbf{r}}_i = \mathbf{F}_i + \mathbf{F}_i'$$

Taking the summation for all the n particles of the body, we have

$$\sum_1^n m_i \ddot{\mathbf{r}}_i = \sum_1^n \mathbf{F}_i + \sum_1^n \mathbf{F}_i'$$

where the symbol Σ denotes a summation. Now the internal forces consist of equal and opposite pairs, and for the body as a whole the sum of these forces, $\sum_1^n \mathbf{F}_i'$, is zero. Thus

$$\sum_1^n m_i \ddot{\mathbf{r}}_i = \sum_1^n \frac{d^2}{dt^2}(m_i \mathbf{r}_i) = \sum_1^n \mathbf{F}_i = \mathbf{F} \qquad (6.1)$$

where $\mathbf{F}$ is the resultant external force acting on the whole of the particles.

An important development of this theory is brought about by the introduction of the concept of *center of mass*. The center of mass (C.M.) of a body

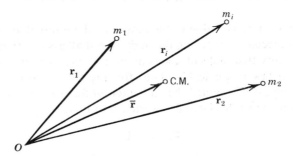

Fig. 6.1   Center of mass of a number of particles.

or of a system of particles is that point in the body or system of particles at which, for purposes of calculation, all the mass may be assumed to be concentrated. Relative to the origin O, the vector position $\bar{\mathbf{r}}$ of the center of mass of the n particles in Fig. 6.1 is defined as

$$\bar{\mathbf{r}} = \frac{\sum\limits_{1}^{n} m_i \mathbf{r}_i}{\sum\limits_{1}^{n} m_i} = \frac{\sum\limits_{1}^{n} m_i \mathbf{r}_i}{M} \qquad (6.2)$$

where M is the total mass, $\sum\limits_{1}^{n} m_i$, of all the particles. By differentiation of Eq. 6.2, we have

$$M\ddot{\bar{\mathbf{r}}} = \sum\limits_{1}^{n} m_i \ddot{\mathbf{r}}_i$$

and from Eq. 6.1 it follows that

$$M\ddot{\bar{\mathbf{r}}} = \mathbf{F} \qquad (6.3)$$

This is an important result showing that *for a group of particles or a rigid body the translational motion of the center of mass behaves as though all the mass of the particles were concentrated at the center of mass and the resultant external force also acted at that point.* In other words, the motion of the center of mass of a body is the same as that of a particle having the total mass of the body and the same resultant external force acting on it. This result applies either to a system of particles in motion, as the molecules of a gas, or to a rigid body. In all cases the resultant internal force must be zero if Eq. 6.3 is to be applied.

If a hammer or other body is thrown at some angle with the vertical, then its center of mass will follow a parabolic path just as would a particle with the same initial motion. This applies whether the hammer has an initial rotational motion or not. The resultant external force is the weight of the hammer, and the acceleration of the center of mass is the acceleration due to gravity. On a larger scale, if we assume that there is no resultant external force acting on the solar system as a whole, its center of mass will either be at rest or move with constant speed in a straight line.

For many purposes of calculation it is convenient to express the position of the center of mass in terms of its three Cartesian coordinate components. Thus, from the vector position $\bar{\mathbf{r}}$ of the center of mass given by Eq. 6.2, three equations may be written giving the X, Y, Z coordinates, $\bar{x}, \bar{y}, \bar{z}$, respectively, of the center of mass. These are

$$\bar{x} = \frac{\sum\limits_{1}^{n} m_i x_i}{M} \qquad \bar{y} = \frac{\sum\limits_{1}^{n} m_i y_i}{M} \qquad \bar{z} = \frac{\sum\limits_{1}^{n} m_i z_i}{M} \qquad (6.4)$$

where x_i, y_i, z_i are the coordinates of the particle m_i. The vector $\mathbf{r}_i$ giving the position of m_i may be written

$$\mathbf{r}_i = \mathbf{i}x_i + \mathbf{j}y_i + \mathbf{k}z_i$$

and the vector $\bar{\mathbf{r}}$ giving the position of the center of mass

$$\bar{\mathbf{r}} = \mathbf{i}\bar{x} + \mathbf{j}\bar{y} + \mathbf{k}\bar{z}$$

If ΣF_x, ΣF_y, ΣF_z are the components of the external force $\mathbf{F}$, then Eq. 6.3 may be written

$$\Sigma F_x = M\ddot{x} \qquad \Sigma F_y = M\ddot{y} \qquad \Sigma F_z = M\ddot{z} \qquad (6.5)$$

Thus the component of the external force in the X direction is equal to the product of the total mass and the component of the acceleration of the center of mass in the X direction.

Example of translational motion of the center of mass of a number of particles. Consider that four masses of 30 gm, 50 gm, 20 gm, and 100 gm are placed in the XY plane and have coordinates of $(4, -1)$, $(5, 2)$, $(-3, -4)$, and $(-2, 4)$ respectively. These are shown in Fig. 6.2 with the external forces acting on them. Let us first determine the coordinates of the center of mass. From Eq. 6.4

$$\bar{x} = \frac{\sum_1^4 m_i x_i}{M}$$

$$= \frac{(30 \times 4) + (50 \times 5) + (20 \times -3) + (100 \times -2)}{200}$$

$$= \tfrac{110}{200} = 0.55$$

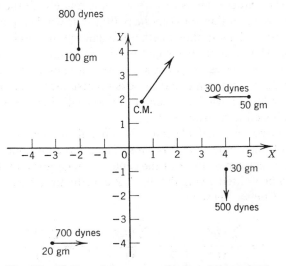

Fig. 6.2 Motion of the center of mass of four particles.

Similarly

$$\bar{y} = [(30 \times -1) + (50 \times 2) + (20 \times -4) + (100 \times 4)]/200$$
$$= \frac{390}{200} = 1.95$$

Thus the coordinates of the center of mass of the particles are (0.55, 1.95).

Suppose that the external forces act on the masses as shown in Fig. 6.2. The problem is to calculate the acceleration of the center of mass. In the X direction the resultant force is

$$\sum F_x = 700 - 300 = 400 \text{ dynes}$$

and in the Y direction

$$\sum F_y = 800 - 500 = 300 \text{ dynes}$$

The resultant force on the particles has a magnitude of

$$F = \sqrt{400^2 + 300^2} = 500 \text{ dynes}$$

and a direction that makes an angle θ with the X axis given by

$$\tan \theta = \frac{300}{400} = \frac{3}{4}$$

From Eq. 6.5 it follows that the instantaneous acceleration of the center of mass is

$$a = \frac{F}{M} = \frac{500}{200} = 2.5 \text{ cm/sec}^2$$

and it makes an angle with the X axis whose tangent is $\frac{3}{4}$.

6.3   The Center of Mass of a Portion of a Body

It is readily shown that, if the center of mass of the whole of a body and that of a portion of the body are known, then the center of mass of the remainder may be found. Rather than giving a formal proof, let us take the example given in Fig. 6.3.

In this a circular hole is cut out of a uniform rectangular piece of metal. If k is the mass per unit area of the metal, then:

Mass of rectangle $= 0.24k$

Mass of hole cut out $= \pi(0.1)^2 k$

Mass of remainder of metal $= (0.24 - 0.0314)k$

The center of mass of this remainder has the coordinates $\bar{x}$ and $\bar{y}$. Suppose the circular material is placed back in the hole, then the C.M. of the remainder and the circular material in the hole is at the C.M. of the original rectangle. This latter is at the geometrical center of the rectangle with coordinates $x = 0.3$ and $y = 0.2$. From Eqs. 6.4, or $M\bar{x} = M_1 x_1 + M_2 x_2$, it follows that:

$$(0.24k)0.30 = (0.24 - 0.0314)k\bar{x} + (0.0314k)0.45$$
$$(0.24k)0.20 = (0.24 - 0.0314)k\bar{y} + (0.0314k)0.25$$

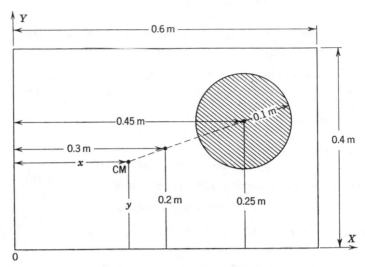

Fig. 6.3 Uniform rectangle with circular hole cut out.

From these it follows that the coordinates of the center of mass of the rectangle with the circular hole in it are

$$\bar{x} = 0.277 \text{ m} \quad \text{and} \quad \bar{y} = 0.193 \text{ m}$$

6.4 Calculation of the Centers of Mass of Various Symmetrical Bodies

In a uniformly distributed mass the summation in the expression for the center of mass is replaced by an integration. Thus the equivalent equations replacing Eq. 6.4 are

$$\bar{x} = \frac{\int x \, dm}{M} \qquad \bar{y} = \frac{\int y \, dm}{M} \qquad = \frac{\int z \, dm}{M} \tag{6.6}$$

where the integrations are carried out over the whole of the body of mass M.

Center of mass of a uniform right circular solid cone. Suppose that the cone has a vertical height h and a base with a radius a. For convenience we shall place the vertex of the cone at the origin of the coordinate system and the axis of symmetry of the cone along the Y axis, as in Fig. 6.4. Consider a horizontal thin slab of the cone whose thickness is dy, whose mean radius is r, and whose height above the origin is y. If the density or mass per unit volume of the cone is ρ, then the mass dm of the thin slab is

$$dm = \pi r^2 \rho \, dy$$

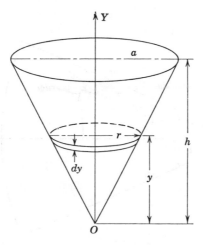

Fig. 6.4   Center of mass of a cone.

The radius r of the slab is proportional to its height y, and from the figure it can be seen that

$$\frac{r}{a} = \frac{y}{h} \quad \text{or} \quad r = \frac{ay}{h}$$

Thus

$$dm = \frac{\pi \rho a^2 y^2}{h^2} \, dy$$

and the total mass M of the cone is

$$M = \frac{\pi \rho a^2}{h^2} \int_0^h y^2 \, dy = \frac{\pi \rho a^2 h}{3}$$

The vertical distance of the center of mass from O is

$$\bar{y} = \int_0^h \frac{y \, dm}{M} = \frac{\pi \rho a^2}{M h^2} \int_0^h y^3 \, dy$$

Integrating and substituting the value for M, we have

$$\bar{y} = \frac{3\pi \, a^2 h^4}{4 h^2 \pi \rho a^2 h} = \frac{3h}{4}$$

or the center of mass is three-fourths of the altitude measured from the vertex.

Center of mass of a uniform plane sector of a circle of angle θ.   Suppose the X axis goes through the axis of symmetry of the section as shown in Fig. 6.5. Consider an elementary area of the sector at a distance r from the center O, and having a width $r \, d\phi$ and length dr. If the mass per unit is σ, then the mass of the element of

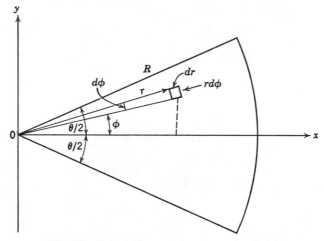

Fig. 6.5   Center of mass of a sector of a circle.

area is $\sigma r \, d\phi \, dr$. The x distance of this element of area from the Y axis is $r \cos \phi$ where ϕ is the angle which r makes with the X axis. From Fig. 6.5 it is seen that the limits of integration on r are O to R and on ϕ are $-\theta/2$ to $+\theta/2$. Hence

$$\bar{x} = \frac{\displaystyle\int_{-\theta/2}^{\theta/2}\int_0^R r^2 \cos \phi \, dr \, d\phi \, \sigma}{\displaystyle\int_0^R \int_{-\theta/2}^{\theta/2} r \, dr \, d\phi \, \sigma} = \frac{\dfrac{R^3}{3}\left[\sin\left(\dfrac{\theta}{2}\right) - \sin\left(\dfrac{-\theta}{2}\right)\right]}{\dfrac{R^2}{2}\left[\dfrac{\theta}{2} + \left(\dfrac{\theta}{2}\right)\right]}$$

$$= \frac{4R^3 \sin}{3R^2\theta}\left(\frac{\theta}{2}\right) = \frac{4R}{3\theta} \sin\left(\frac{\theta}{2}\right)$$

It can be readily appreciated that the center of mass can be calculated only for bodies having some degree of symmetry. For such bodies one must use judgment as to how to place the axes relative to an axis or plane of symmetry. With a little care, considerable time and effort may be saved.

6.5   Experimental Determination of the Center of Mass of an Unsymmetrical Body

In unsymmetrical bodies the center of mass cannot be calculated simply. It must be obtained experimentally. The principle of such an experiment may be illustrated for a plane object, as shown in Fig. 6.6. A smooth peg passing through a small hole as at O is used to suspend the object. A piece of string with a weight on the end is fastened to the peg at O to indicate the vertical direction. The line OA in the direction of the string is drawn on the object. Now, in principle, it is the center of weight or *center of gravity* (C.G.) which lies on the line OA. Each element of mass dm has a weight $g\,dm$, where g is the acceleration due to gravity. If we define the center of gravity of a body as the point in the body at which all the weight may be assumed to be concentrated, then the center of gravity of the plane in Fig. 6.6 must lie somewhere along the line OA. Similarly, another line $O'A'$ may be drawn on which the center of gravity must lie. The intersection of these lines gives the position of the center of gravity of the body.

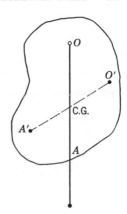

Fig. 6.6 Experimental determination of center of mass.

We may now ask whether the position of the center of gravity is also the position of the center of mass. The equations giving the position of the center of gravity relative to the X, Y, Z axes are similar to Eqs. 6.6 for the center of mass. The x coordinate of the center of gravity is defined as

$$\bar{x} = \frac{\int gx\,dm}{\int g\,dm} \tag{6.7}$$

If the acceleration due to gravity g is assumed to be the same in magnitude and direction at all points on the body, then the center of gravity coincides with the center of mass. This can be assumed to be correct for all ordinary objects. In dealing with the center of gravity we are finding the resultant of a series of forces. If g is not considered constant in direction and magnitude throughout the body, then the forces are neither parallel to each other nor proportional to the elements of mass, and almost impossible mathematical equations result.

6.6 Rotation of a Uniform Symmetrical Disk about a Fixed Axis through Its Center

In this section we shall prove in an elementary manner that for rotational motion about a fixed axis the total moment of the forces or torques about this fixed axis is equal to the time rate of change of the angular momentum of the body about the fixed axis. This rotational motion introduces a new concept called the **moment of inertia** or the **rotational inertia** of a body. In

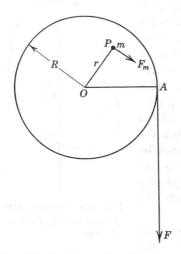

Fig. 6.7 Rotation of a disk about a fixed axis.

terms of this, the torque about a fixed axis is equal to the product of the moment of inertia and the angular acceleration of the body about the axis.

Consider a plane circular disk free to rotate about a fixed axis through its center O and perpendicular to its plane. A light cord is coiled around the periphery of the disk, and a constant vertical force F is applied as shown in Fig. 6.7. (Note that this is not equivalent to replacing the constant force F with a hanging weight when the system is moving.) As the cord is pulled downward by the force F, the disk rotates with an increasing angular velocity. Suppose that the angular velocity at some instant of time is ω radians/sec. Then its angular acceleration α is the time rate of change of this angular velocity:

$$\alpha = \frac{d\omega}{dt}$$

This angular acceleration depends on both the force F and the perpendicular distance of the force F from the axis of rotation. The product of the force F and this perpendicular distance is called the *moment of the force F*, or *the torque* τ about O. From Fig. 6.7 the torque τ for the force F about O is

$$\tau = F \times OA = FR$$

where R is the radius of the disk.

Consider now a particle of mass m in the disk at a distance OP or r from the axis of rotation. If at the instant considered the angular velocity of the disk is ω, then the tangential velocity of the mass m in a direction perpendicular to OP is $r\omega$. The instantaneous linear momentum of the mass m is $mr\omega$, and the *moment of this momentum* or the *angular momentum* about the axis of rotation is $mr^2\omega$. For all the particles of the disk, the total momentum of J of the disk about O is

$$J = \sum mr^2\omega = \omega\sum mr^2$$

where ω is the same for all the particles in the disk and the summation is taken over every particle in the disk. The quantity $\sum mr^2$ plays a very important role in rotational motion and is called the *moment of inertia* or the *rotational inertia I* of the body about the axis of rotation. Thus we may write

$$J = I\omega \tag{6.8}$$

To obtain the relationship between the torque τ and the angular acceleration α, consider the tangential force F_m acting on the particle m at P. From Newton's second law this is

$$F_m = mr\alpha = mr\frac{d\omega}{dt}$$

and the moment of this force about the axis of rotation is

$$F_m r = mr^2\alpha = mr^2\frac{d\omega}{dt}$$

For the whole disk the total moment of the forces acting on all the particles of the disk must be equal to the moment of the applied force τ. Thus

$$\tau = FR = \sum mr^2\alpha = \sum mr^2\frac{d\omega}{dt} = I\alpha = I\frac{d\omega}{dt}$$

From Eq. 6.8 it follows that the time rate of change of the angular momentum J is equal to the product of the moment of inertia I and the rate of change of the angular velocity. Thus we may write for the rotation of a body about a fixed axis

$$\tau = I\alpha = \dot{J} \tag{6.9}$$

Thus the torque about a fixed axis is equal to the product of the moment of inertia and the angular acceleration about the fixed axis.

6.7 Work and Rotational Kinetic Energy

We shall now determine the relationship between the work done on the disk by the force F and the increase in the kinetic energy of the disk.

If at some instant of time the angular velocity of the disk is ω, then the disk will turn through an infinitesimal angle $d\theta$ in an infinitesimal time dt where

$$\omega = \frac{d\theta}{dt} \quad \text{or} \quad d\theta = \omega\,dt$$

During this time dt a point on the periphery of the disk moves through a distance ds given by

$$ds = R\,d\theta$$

The work done by the force F in this time is $F\,ds$, Fig. 6.8. This work can be expressed as

$$F\,ds = FR\,d\theta = \tau\,d\theta$$

Thus the work done is equal to the product of the torque and the angle through which the disk turns. By substitution it follows that

$$\tau\,d\theta = I\frac{d\omega}{dt}\,\omega\,dt = I\omega\,d\omega$$

where $d\omega$ is the change in the angular velocity during the time dt. Suppose that the disk has an angular velocity ω_0 when the force is applied and that, after the disk has turned through an angle θ, the angular velocity has become ω. With these boundary conditions we have

$$\int_0^{\theta} \tau\,d\theta = \int_{\omega_0}^{\omega} I\omega\,d\omega$$

or

$$\tau\theta = \tfrac{1}{2}I\omega^2 - \tfrac{1}{2}I\omega_0^2 \qquad (6.10)$$

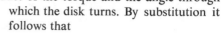

R

ds

$d\theta$

F

$ds = R\,d\theta$

Fig. 6.8 Kinetic energy of a rotating disk.

where $\tfrac{1}{2}I\omega^2$ is called the *angular or rotational kinetic energy* of a body having a moment of inertia I and an angular velocity ω. Thus the work done by the torque in turning the disk about a fixed axis through some angle is equal to the change in the angular or rotational kinetic energy of the disk. The quantity $\tfrac{1}{2}I\omega^2$ plays the same role in rotational motion about a fixed axis that $\tfrac{1}{2}mv^2$ does in linear motion. Equation 6.10 expresses the law of conservation of mechanical energy as applied to rotational motion.

To emphasize the ideas presented above, we shall take a numerical example. Suppose that the disk shown in Fig. 6.7 has a mass of 500 gm and a radius of 10 cm. A constant force of 20,000 dynes is applied at the end of the cord. If the disk has an initial angular velocity of 4 radians/sec when the force is applied, the problem is to find the angular velocity and rotational kinetic energy after 2 sec.

First we must calculate the value of the moment of inertia I of the circular disk about its axis of rotation. The definition given for the moment of inertia is

$$I = \sum mr^2$$

where the summation is taken over all the particles composing the body. For a continuously distributed mass such as a uniform solid body, for example, the disk, one replaces the summation $\sum mr^2$ used for a discrete series of masses by an integral. Thus for a uniform solid the moment of inertia is given by

$$I = \int r^2\,dm \qquad (6.11)$$

The calculation of I for the disk about an axis through its center perpendicular to its plane can be made by considering the disk as made up of a large number of concentric circular rings. One of these having a radius of r and a width of dr is shown in Fig. 6.9. If the mass of the disk of radius R is M, then the mass of the ring of radius r and width dr is

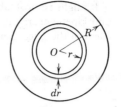

Fig. 6.9 Calculation of the moment of inertia of a circular disk.

$$dm = \frac{M}{\pi R^2}\, 2\pi r\, dr = \frac{2M}{R^2}\, r\, dr$$

Since, to a first approximation, all points on this ring lie at the same distance from the center, the moment of inertia of this ring about an axis perpendicular to its plane and through its center is

$$r^2\, dm = \frac{2M}{R^2}\, r^3\, dr$$

Thus I for the whole disk is

$$I = \int r^2\, dm = \int_0^R \frac{2M}{R^2}\, r^3\, dr = \frac{MR^2}{2} \tag{6.12}$$

In our problem $M = 500$ gm and $R = 10$ cm so that

$$I = \frac{MR^2}{2} = 2.5 \times 10^4 \text{ gm cm}^2$$

The torque τ exerted on the disk is

$$\tau = FR = 2 \times 10^5 \text{ dyne cm}$$

From $\tau = I\alpha$ the angular acceleration of the disk α is

$$\alpha = \frac{2 \times 10^5 \text{ dyne cm}}{2.5 \times 10^4 \text{ gm cm}^2} = 8 \text{ radians/sec}^2$$

If the angular velocity at zero time is ω_0 and at time t is ω, then, since $\alpha = d\omega/dt$, it follows that

$$\int_{\omega_0}^{\omega} d\omega = \int_0^t \alpha\, dt$$

or

$$\omega - \omega_0 = \alpha t$$

Since the initial angular velocity ω_0 in this problem is 4 radians/sec, then the angular velocity ω after 2 sec is

$$\omega = \omega_0 + \alpha t$$
$$= 4 + (8 \times 2) = 20 \text{ radians/sec}$$

The rotational kinetic energy after 2 sec is

$$\tfrac{1}{2}I\omega^2 = \tfrac{1}{2} \times 2.5 \times 10^4 \times 400 = 5 \times 10^6 \text{ ergs}$$

It might be profitable to show numerically that Eq. 6.10 dealing with both work and kinetic energy is valid. We must first find the angle θ through which the disk turns in the 2 sec. Since $\omega = d\theta/dt$, it follows that, if $\theta = 0$ at $t = 0$, then

$$\int_0^{\theta} d\theta = \int_0^t \omega \, dt = \int_0^t (\omega_0 + \alpha t) \, dt$$

or

$$\theta = \omega_0 t + \tfrac{1}{2}\alpha t^2$$

In our example

$$\theta = (4 \times 2) + \tfrac{1}{2}(8 \times 4) = 24 \text{ radians}$$

so that the work done by the force in the 2 sec is:

$$\tau\theta = (2 \times 10^5 \times 24) = 4.8 \times 10^6 \text{ ergs}$$

This quantity should be equal to the change in kinetic energy during the 2 sec or

$$\tau\theta = \tfrac{1}{2}I\omega^2 - \tfrac{1}{2}I\omega_0^2$$
$$= (5 \times 10^6) - (2 \times 10^5) = 4.8 \times 10^6 \text{ ergs}$$

which, of course, agrees with the value of $\tau\theta$ given above.

6.8   Two Theorems Relating to Moments of Inertia

Theorem a.   *The moment of inertia of a uniform plane body about an axis perpendicular to its plane is equal to the sum of its moments of inertia about any two perpendicular axes in the plane that intersect on the first axis.*

Let the plane object lie in the XY plane and let the perpendicular axis be the Z axis. From Fig. 6.10 the moment of inertia I_x about the X axis is given by

$$I_x = \int y^2 \, dm$$

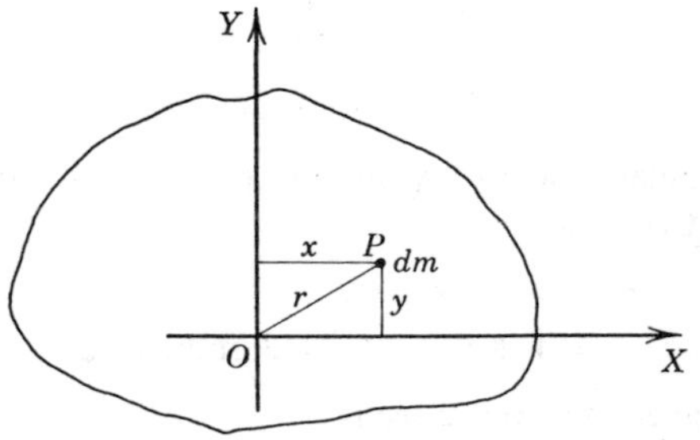

Fig. 6.10   Moments of inertia about perpendicular axes.

and similarly that about the Y axis is

$$I_y = \int x^2 \, dm$$

For any point P in the plane having coordinates x, y, the distance r of P from O is

$$r^2 = x^2 + y^2$$

The moment of inertia I of the plane about the Z axis is

$$I_z = \int r^2 \, dm = \int (x^2 + y^2) \, dm$$

Thus

$$I_z = I_x + I_y \tag{6.13}$$

As an example of the application of this theorem let us find the moment of inertia of a plane circular disk about any diameter using Eq. 6.12. If the disk lies in the XY plane with its center at the origin of the coordinates, then

$$\frac{MR^2}{2} = I = I_x + I_y$$

For the circular disk

$$I_x = I_y$$

Hence the moment of inertia of a plane circular disk about any diameter is

$$I_x = I_y = \frac{MR^2}{4}$$

Theorem b, or the Theorem of Parallel Axes.   *The moment of inertia I of a body about any axis is equal to the moment of inertia $I_{\text{C.M.}}$ of the body about a parallel axis through its center of mass plus the mass M of the body times the square of the perpendicular distance l between the axes:*

$$I = I_{\text{C.M.}} + Ml^2$$

For convenience let the axis about which the moment of inertia is to be taken be the Z axis which is perpendicular to the XY plane. Let the coordinates of the center of mass be $\bar{x}$, $\bar{y}$, $\bar{z}$. If a particle at P in Fig. 6.11 of mass dm has coordinates x, y, z relative to the X, Y, Z axes and x', y', z' relative to a set of parallel axes X', Y', Z' through the center of mass of the body, then

$$x = x' + \bar{x} \qquad \text{and} \qquad y = y' + \bar{y}$$

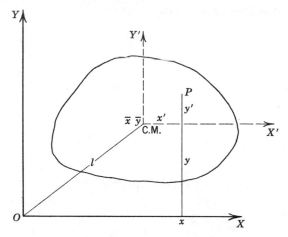

Fig. 6.11   Moment of inertia about parallel axes.

The moment of inertia I of the body about the OZ axis is

$$I = \int (x^2 + y^2)\,dm = \int [(x' + \bar{x})^2 + (y' + \bar{y})^2]\,dm$$

$$= \int (x'^2 + y'^2)\,dm + (\bar{x}^2 + \bar{y}^2)\int dm + 2\bar{x}\int x'\,dm + 2\bar{y}\int y'\,dm$$

Since the coordinates x', y' are chosen with respect to axes through the center of mass, then by definition of the center of mass

$$\int x'\,dm = 0 \qquad \text{and} \qquad \int y'\,dm = 0$$

Let the moment of inertia about the Z' axis through the center of mass be denoted by $I_{\text{C.M.}}$, and let the distance between the Z and Z' axes be l so that

$$I_{\text{C.M.}} = \int (x'^2 + y'^2)\,dm$$

and

$$l^2 = (\bar{x}^2 + \bar{y}^2)$$

Then

$$I = I_{\text{C.M.}} + Ml^2 \tag{6.14}$$

As an example of the application of this theorem, we can find the moment of inertia of a plane disk about an axis at its edge perpendicular to its plane. For the axes through the center of the disk and perpendicular to its plane

$$I_{\text{C.M.}} = \frac{MR^2}{2}$$

Since we have assumed the circular disk to be uniform, the center of mass must be at its geometrical center. Hence, for a parallel axis through the edge of the disk at a distance R from the center of mass, the moment of inertia I is

$$I = \frac{MR^2}{2} + MR^2 = \frac{3MR^2}{2}$$

6.9 Radius of Gyration

The radius of gyration k of a body of mass M and of moment of inertia I is defined as

$$I = Mk^2 \tag{6.15}$$

Thus we see that the moment of inertia I of a body of mass M and radius of gyration k is the same as that of a particle of mass M placed at a distance k from the axis of rotation. Notice that the radius of gyration applies to a particular axis. For the plane circular disk and for an axis perpendicular to its plane through its center of mass the radius of gyration is

$$k = \frac{R}{\sqrt{2}}$$

and for axis perpendicular to its plane through the edge of the disk the radius of gyration is

$$k = \sqrt{\tfrac{3}{2}}R$$

6.10 Calculations of Moments of Inertia

Moment of inertia of a narrow uniform rod about an axis through its center and perpendicular to its length.   Let the mass of the rod be M and its length be l. Then the mass per unit length of the rod σ is

$$\sigma = \frac{M}{l}$$

If the X axis lies along the length of the rod with its origin at the center of the rod, then the moment of inertia about O_1O_2 in Fig. 6.12 is

$$I_{\text{C.M.}} = \int_{-l/2}^{+l/2} x^2 \, dm = \int_{-l/2}^{+l/2} x^2 \sigma \, dx = \frac{\sigma l^3}{12} = \frac{Ml^2}{12}$$

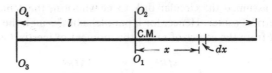

Fig. 6.12 Moment of inertia of a uniform rod.

The moment of inertia about an axis $O_3 O_4$ parallel to $O_1 O_2$ is

$$I = I_{\text{C.M.}} + \frac{Ml^2}{4} = \frac{Ml^2}{3}$$

Moment of inertia of a uniform plane rectangle about an axis through its geometrical center perpendicular to its plane.   Let the length of the rectangle be a, its width b, and its mass M, as in Fig. 6.13. Let us consider the rectangle as made up of narrow strips parallel to the Y axis. From the previous example the moment of inertia I_x about the X axis is

$$I_x = \frac{Mb^2}{12}$$

Similarly,

$$I_y = \frac{Ma^2}{12}$$

From theorem a given above, the moment of inertia I about an axis through the center O and perpendicular to the plane of the rectangle is given by

$$I = I_x + I_y$$
$$= \frac{M(a^2 + b^2)}{12}$$

Moment of inertia of a solid uniform sphere about any diameter.   Let the sphere

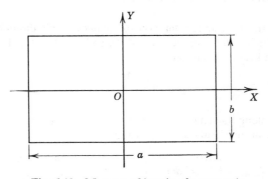

Fig. 6.13 Moment of inertia of a rectangle.

TABLE 6.1 Moments of Inertia (m is mass of body)

Body	Axis	Moment of Inertia
Uniform thin rod length l	Normal to length through center	$\dfrac{ml^2}{12}$
Uniform thin rod length l	Normal to length at one end	$\dfrac{ml^2}{3}$
Thin rectangular sheet, sides a and b	Through center parallel to side b	$\dfrac{ma^2}{12}$
Thin rectangular sheet, sides a and b	Through center perpendicular to sheet	$\dfrac{m(a^2 + b^2)}{12}$
Thin circular disk, radius r	Through center perpendicular to disk	$\dfrac{mr^2}{2}$
Thin circular ring, radii r_1, r_2	Through center perpendicular to plane of ring	$\dfrac{m(r_1^2 + r_2^2)}{2}$
Thin circular ring, radii r_1, r_2	Through center along any diameter	$\dfrac{m(r_1^2 + r_2^2)}{4}$
Uniform solid sphere, radius r	Any diameter	$\dfrac{2mr^2}{5}$
Thin spherical shell, radius r	Any diameter	$\dfrac{2mr^2}{3}$
Spherical shell, radii r_1, r_2	Any diameter	$\dfrac{2m(r_1^5 - r_2^5)}{5(r_1^3 - r_2^3)}$
Right cone, radius of base r	Along axis of cone	$\dfrac{3mr^2}{10}$
Uniform circular cylinder, length l, radius r	Through center perpendicular to axis	$m\left(\dfrac{r^2}{4} + \dfrac{l^2}{12}\right)$
Hollow circular cylinder, length l, radii r_1, r_2	Through center perpendicular to axis	$m\left(\dfrac{r_1^2 + r_2^2}{4} + \dfrac{l^2}{12}\right)$
Very thin circular cylinder, length l, radius r	Through center perpendicular to axis	$m\left(\dfrac{r^2}{2} + \dfrac{l^2}{12}\right)$

have a radius a and a mass M. To find the moment of inertia about a diameter, consider the sphere as made up of a series of disks, one of which is at a distance x from the center and has a thickness dx, Fig. 6.14. The area of the disk is $\pi(a^2 - x^2)$ and, if the density of the sphere is ρ, the mass of the disk is $\pi\rho(a^2 - x^2)\, dx$. From Eq. 6.12 the moment of inertia of the circular disk about the axis OX is

$$dI = \pi\rho(a^2 - x^2)\, dx\, \frac{(a^2 - x^2)}{2}$$

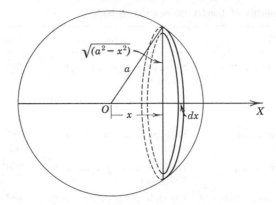

Fig. 6.14   Moment of inertia of a sphere about a diameter.

Hence the moment of inertia of the sphere is

$$I = \int_{-a}^{+a} \frac{\pi \rho}{2} (a^2 - x^2)^2 \, dx$$

$$= \frac{\pi \rho}{2} \left[a^4 x + \frac{x^5}{5} - \frac{2a^2 x^3}{3} \right]_{-a}^{+a} = \frac{8\rho \pi}{15} a^5$$

Since the mass of the sphere is $M = (4\pi a^3 \rho)/3$, then

$$I = \tfrac{2}{5} M a^2$$

From these examples it can be seen that it would not be possible to calculate the moment of inertia of an unsymmetrical body about any axis. For such bodies the moments of inertia must be obtained experimentally. Table 6.1 gives the moments of inertia of some common symmetrical objects.

6.11   The Kinetic Energy of a Rotating Body

We shall now find the kinetic energy of a body rotating with respect to a fixed point or axis and also with respect to its center of mass. Taking the fixed point O (Fig. 6.15) as the origin of the polar coordinates, the kinetic energy T of the body about this fixed point is

$$T = \sum_{1}^{n} \tfrac{1}{2} m_i \dot{r}_i{}^2 \tag{6.16}$$

This kinetic energy can be written in terms of the translational and rotational kinetic energies of the center of mass by using the relationship shown in

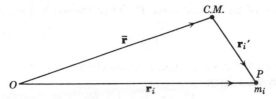

Fig. 6.15 Rotational motion of a plane object with reference to a fixed point O and to center of mass $C.M.$

Fig. 6.15, namely

$$\mathbf{r}_i = \bar{\mathbf{r}} + \mathbf{r}_i'$$

From this it follows by differentiation that

$$\dot{\mathbf{r}}_i = \dot{\bar{\mathbf{r}}} + \dot{\mathbf{r}}_i'$$

If the dot or scalar product of each side of this equation and itself is taken, then the kinetic energy T in Eq. 6.16 can be written

$$T = \sum_1^n \tfrac{1}{2}m_i(\dot{r}_i^2) = \sum_1^n \tfrac{1}{2}m_i(\dot{\mathbf{r}}_i \cdot \dot{\mathbf{r}}_i)$$

$$= \sum_1^n [\tfrac{1}{2}m_i(\dot{\bar{\mathbf{r}}} + \dot{\mathbf{r}}_i') \cdot (\dot{\bar{\mathbf{r}}} + \dot{\mathbf{r}}_i')]$$

$$= \tfrac{1}{2}M\dot{\bar{\mathbf{r}}}^2 + \sum_1^n \tfrac{1}{2}m_i\dot{\bar{r}}_i'^2 + \dot{\bar{\mathbf{r}}} \cdot \sum_1^n m_i\dot{\mathbf{r}}_i'$$

where M is the total mass $\sum_1^n m_i$ of the body and $\dot{\bar{r}}$ is a constant. Since $\mathbf{r}'$ is measured relative to the center of mass of the body, it follows that

$$\sum_1^n m_i\mathbf{r}_i' = 0 \qquad \text{and} \qquad \sum_1^n m_i\dot{\mathbf{r}}_i' = 0$$

Thus the kinetic energy of the body relative to the origin of coordinates at its center of mass is

$$T = \tfrac{1}{2}M\dot{\bar{r}}^2 + \sum_1^n \tfrac{1}{2}m_i\dot{\mathbf{r}}_i'^2 \tag{6.17}$$

The first term on the right-hand side of Eq. 6.17 is the translational kinetic energy of a particle having the mass M of the body moving with the linear velocity $\dot{\bar{r}}$ of the center of mass, and the second is the total kinetic energy of all the inidividual particles taken with respect to the center of mass.

This increase in the kinetic energy is equal to the work done by the applied forces. Suppose that the ith particle of the body undergoes a displacement $d\mathbf{r}_i$

under the action of an applied force $\mathbf{F}_i$. Then the work done by the applied force is

$$\mathbf{F}_i \cdot d\mathbf{r}_i$$

Since the applied force $\mathbf{F}_i$ produces an acceleration $\ddot{\mathbf{r}}_i$ on the particle, the work done may be written

$$\mathbf{F}_i \cdot d\mathbf{r}_i = m_i \ddot{\mathbf{r}}_i \cdot d\mathbf{r}_i$$

Now there is also an internal force acting on m_i, but, for the body as a whole, the work done by the internal forces is zero since the internal forces are in equilibrium. Hence the work done by the applied forces on the body in the displacement is $\sum_1^n \mathbf{F}_i \cdot d\mathbf{r}_i$. From the above, this work can be written

$$\sum_1^n \mathbf{F}_i \cdot d\mathbf{r}_i = \sum_1^n m_i \ddot{\mathbf{r}}_i \cdot d\mathbf{r}_i$$

$$= \sum_1^n \frac{1}{2} \frac{d}{dt} (m_i \dot{\mathbf{r}}_i \cdot \dot{\mathbf{r}}_i) \, dt$$

$$= \sum_1^n \tfrac{1}{2} d(m_i \dot{r}_i^2)$$

Suppose that the body undergoes a displacement from $\mathbf{r}_1$ to $\mathbf{r}_2$ and that in consequence the velocity changes from $_1\mathbf{v}_i$ to $_2\mathbf{v}_i$. Then by integration, we have for the work done

$$\sum_1^n \int_{r_1}^{r_2} \mathbf{F}_i \cdot d\mathbf{r}_i = \sum_1^n \tfrac{1}{2} m_i ({}_1v_i^2 - {}_2v_i^2) \tag{6.18}$$

Thus the total work done on the body in a given displacement is equal to the change in kinetic energy of the body during the displacement. To give a better understanding of these principles, we shall apply them first to a problem in which there is a fixed axis of rotation and second to the problem in which there is an instantaneous axis of rotation which maintains a fixed direction in space. An example of the latter is a cylinder rolling down an inclined plane, Fig. 6.16.

For the first problem let us suppose that the body is rotating about a *fixed axis* through a point O with an angular velocity ω perpendicular to the plane of the paper. For this situation the velocity at a distance r_i from O has a magnitude of ωr_i, and the kinetic energy T of the body about the fixed axis from Eq. 6.16 is

$$T = \sum_1^n \tfrac{1}{2} m_i \dot{r}_i^2 = \sum_1^n \tfrac{1}{2} m_i r_i^2 \omega^2$$

$$= \frac{1}{2} \left(\omega^2 \sum_1^n m_i r_i^2 \right) = \frac{I_0 \omega^2}{2} \tag{6.19}$$

since ω is the same for every particle in the body and $\sum_1^n m_i r_i^2$ is by definition the moment of inertia I_0 of the body about the axis through O about which the rotation is taking place.

Relative to the center of mass the kinetic energy is given by Eq. 6.17

$$T = \tfrac{1}{2}M\dot{\bar{r}}^2 + \sum_1^n \tfrac{1}{2}m_i\dot{r}_i'^2$$

The first term on the right is the translational kinetic energy of the center of mass, and the second is the total kinetic energy of all the particles in the body relative to the center of mass. This second term is equal to $(I_{C.M.}\,\omega^2)/2$. Hence the kinetic energy of a body rotating about a fixed axis may be written

$$T = \tfrac{1}{2}I_0\omega^2 = \tfrac{1}{2}M\dot{\bar{r}}^2 + \tfrac{1}{2}I_{C.M.}\,\omega^2 \qquad (6.20)$$

where I_0 is the moment of inertia of the body about the fixed axis through the fixed point O, and $I_{C.M.}$ is the moment of inertia of the body about its center of mass. This kinetic energy can be equated to the work done by the applied forces as is done in Eq. 6.18.

6.12 Example: A Cylinder Rolling Down a Rough Plane

Let us consider the motion of a uniform cylinder of mass M and radius R rolling down a rough, inclined plane whose length is l and whose angle of inclination is θ.

The problem is to find the linear and the angular acceleration of the cylinder, its energy at the bottom of the plane, and the time taken for it to roll down the plane. We shall assume that on the rough plane the force of friction between the cylinder and the plane is large enough to cause the cylinder to roll without sliding. This force of friction, shown in Fig. 6.16 as f, is exerted by the plane on the cylinder

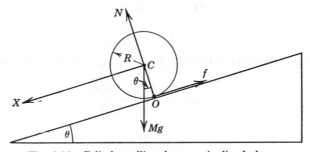

Fig. 6.16 Cylinder rolling down an inclined plane.

and causes the line of contact of cylinder and plane through O to be instantaneously at rest. Consider what would happen if this were not the case!

If the angular acceleration of the cylinder about the axis of rotation through O is α, then the linear acceleration $\ddot{x}$ of the center of mass down the plane is

$$\ddot{x} = \alpha R$$

If the cylinder is rolling and not sliding, the angular velocity and the acceleration about the axis through O must be respectively equal to the angular velocity and acceleration about the axis through the center of mass C. In order to satisfy yourself on this point, consider the displacement of C and O when the cylinder rolls a short distance down the plane.

To obtain the angular acceleration, we shall use the relationship $\tau_0 = I_0 \alpha$ for the instantaneous motion of the cylinder about the axis through O. Thus

$$MgR \sin \theta = I_0 \alpha \tag{6.21}$$

The moment of inertia about a horizontal axis through C is the same as that for a circular disk. From Eq. 6.12,

$$I_{\text{C.M.}} = \frac{MR^2}{2}$$

and by the theorem of parallel axes

$$I_0 = \frac{MR^2}{2} + MR^2 = \frac{3}{2} MR^2$$

Substituting in Eq. 6.21 gives the angular acceleration α as

$$\alpha = \frac{2g \sin \theta}{3R}$$

and the acceleration of the center of mass parallel to the plane as

$$\ddot{x} = \alpha R = \tfrac{2}{3} g \sin \theta \tag{6.22}$$

This result may also be obtained by considering the motion of the center of mass. From Eqs. 6.3 and 6.9 for the translational and rotational accelerations of the center of mass, we have

$$M\ddot{x} = Mg \sin \theta - f$$

and

$$fR = \dot{J}_{\text{C.M.}} = I_{\text{C.M.}} \alpha$$

(You should prove that in this problem $J_{\text{C.M.}} = I_{\text{C.M.}} \alpha$.) Substituting for α and $I_{\text{C.M.}}$ gives

$$I_{\text{C.M.}} \alpha = \frac{MR^2 \ddot{x}}{2R} = \frac{MR\ddot{x}}{2}$$

Hence

$$f = \frac{M\ddot{x}}{2}$$

Substituting this value for f in the above equation, we have

$$M\ddot{x} = Mg \sin \theta - \frac{M\ddot{x}}{2}$$

or

$$\ddot{x} = \tfrac{2}{3}g \sin \theta$$

which agrees with Eq. 6.22.

Consider now the kinetic energy of the cylinder as it rolls down the plane. Let us assume that the cylinder starts from rest at the top of the plane and, after a time t_f, is at the bottom of the plane where it has an angular velocity of ω_f and a linear velocity of the center of mass equal to v. Thus at $t = 0$, $x = 0$, $\dot{x} = 0$, and $\omega = 0$; at $t = t_f$, $x = l$, $\dot{x} = v$, and $\omega = \omega_f$. The velocity v at the bottom of the plane can be obtained from Eq. 6.22. Multiplying this equation by the identity $\dot{x}\,dt = dx$ gives

$$\tfrac{1}{2} d(\dot{x}^2) = \tfrac{2}{3}g \sin \theta \, dx$$

Integrating and applying the boundary conditions gives

$$\frac{v^2}{2} = \frac{2}{3}gl \sin \theta \tag{6.23}$$

Now this result can also be obtained from Eq. 6.18, which, applied to this problem, gives for the work done and the change in kinetic energy

$$Mgl \sin \theta = \tfrac{1}{2}I_0\omega_f{}^2 = \tfrac{1}{2}I_{\text{C.M.}}\omega_f{}^2 + \tfrac{1}{2}Mv^2$$

where we have taken the work done by the external force $\int \mathbf{F} \cdot d\mathbf{r}$ as equal to the loss in potential energy $Mgl \sin \theta$. Substituting for I_0 and $I_{\text{C.M.}}$ and setting $\omega_f{}^2 = v^2/R^2$, we have

$$Mgl \sin \theta = \frac{3MR^2v^2}{4R^2}$$

or

$$gl \sin \theta = \tfrac{3}{4}v^2$$

Thus from the expression for the energy we have

$$v^2 = \tfrac{4}{3}gl \sin \theta$$

which agrees with Eq. 6.23.

The time taken for the cylinder to roll down the plane is obtained by integrating Eq. 6.22 with respect to time. Integrating $\ddot{x} = \tfrac{2}{3}g \sin \theta$ gives

$$\dot{x} = \tfrac{2}{3}tg \sin \theta + C_1$$

where C_1 is zero from the boundary conditions. Integrating a second time gives

$$\bar{x} = \frac{t^2}{3}g \sin \theta + C_2$$

where the boundary conditions give $C_2 = 0$ and $l = (t_f{}^2/3)g \sin \theta$ or

$$t_f = \sqrt{\frac{3l}{g \sin \theta}}$$

In obtaining the value of the velocity v at the bottom of the plane from the energy equation we assumed that the work done by the external force was $Mgl \sin \theta$ and that no work was done by or against the frictional force f. The force of friction f brings the cylinder instantaneously to rest along the line of contact through O. There is no work done by this force of friction since there is no displacement of the cylinder in the direction of the force. If work were done, then the cylinder would be sliding and not rolling.

6.13   Example: Rotation of a Heavy Pulley

Suppose that a uniform pulley of mass M and radius R is free to turn about a frictionless axle through its center. Two masses M_1 and M_2 are suspended at the end of a weightless inextensible cord passing over the pulley, Fig. 6.17. The problem is to find the acceleration of this system.

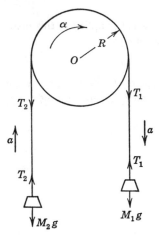

If mass M_1 is larger than mass M_2, then M_1 will descend with some acceleration and M_2 will ascend with an equal acceleration. Why must this be so? The pulley will rotate only if there is friction between the cord and the periphery of the pulley. If this is assumed to be so, the tensions in the cord on the two sides of the pulley must be different. They are denoted by T_1 and T_2, and from the direction of the accelerations it follows that T_1 must be greater than T_2. The upper arrow for T_1, Fig. 6.17, represents the direction of the force exerted by the cord on the pulley, and the lower arrow the direction of the force exerted by the cord on mass M_1.

If we apply Newton's second law to each of the masses M_1 and M_2 respectively, we have for the linear acceleration a

Fig. 6.17   Rotation of a heavy pulley.

$$M_1g - T_1 = M_1a$$

$$T_2 - M_2g = M_2a$$

Taking moments about the axis of rotation O

$$T_1R - T_2R = I_0\alpha = \frac{MR^2\alpha}{2}$$

where α is the angular acceleration of the pulley and I_0 the moment of inertia of the pulley about the axis through O. But $\alpha = a/R$. Hence

$$T_1 - T_2 = \frac{Ma}{2}$$

Combining the upper two equations, we have

$$(M_1 - M_2)g - (T_1 - T_2) = a(M_1 + M_2)$$

or

$$(M_1 - M_2)g = a\left(M_1 + M_2 + \frac{M}{2}\right)$$

or

$$a = \frac{(M_1 - M_2)g}{M_1 + M_2 + \dfrac{M}{2}} = \frac{(M_1 - M_2)g}{[M_1 + M_2 + (I_0/R^2)]}$$

Thus the effect of the mass of the pulley is to increase the inertia or effective mass of the system by I_0/R^2 or $M/2$.

6.14   Example: A Cord Unwinding from a Cylinder

A cord is wound around a cylinder and the free end of the cord is passed over a fixed smooth peg with a mass M' attached to this end, Fig. 6.18. Suppose that the cylinder has a mass M and a radius R. The cylinder has an instantaneous axis of rotation about O, the point of contact of the cord and cylinder. Let the constant angular acceleration about O be α. Then the axis of the cylinder has a linear acceleration $a = \alpha R$ relative to O. Suppose that mass M' moves upward with an acceleration a'. Then a' is the acceleration of M' relative to our assumed inertial system of which the peg is a part. Since the point O moves downward with this acceleration a', the acceleration of the center of mass of the cylinder relative to the peg, sometimes called the absolute acceleration, is $a + a'$. The problem is to find a, a', and also T, the tension in the cord, in terms of M, M', and R.

Consider first the angular acceleration and the torque about the center of mass of the cylinder. From Fig. 6.18 we see that

$$\tau_{C.M.} = TR = I_{C.M.}\alpha = I_{C.M.}\frac{a}{R}$$

For a uniform cylinder $I_{C.M.} = MR^2/2$ so that

$$T = \frac{Ma}{2}$$

From Newton's second law, $M\ddot{\mathbf{r}} = \Sigma\,\mathbf{F}$, for the linear acceleration of the center of mass of the cylinder

$$Mg - T = M(a + a')$$

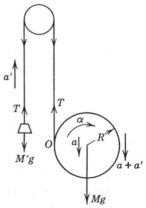

Fig. 6.18 Cord unwinding from a cylinder.

Similarly, for the mass M'

$$T - M'g = M'a'$$

These three equations are the dynamical equations of motion of the system. From them we obtain our required quantities. Substituting for T gives

$$g = \frac{3a}{2} + a'$$

and

$$\frac{Ma}{2} = M'a' + M'g \quad \text{or} \quad a' + g = \frac{M}{2M'}a$$

Thus

$$a = \frac{4gM'}{M + 3M'} \qquad a' = \frac{g(M - 3M')}{M + 3M'} \qquad T = \frac{2gMM'}{M + 3M'}$$

We could also have taken moments about O, giving

$$M(g - a')R = I_0\alpha = \frac{3R^2Ma}{2R}$$

Notice that the moment of the force about O is not MgR but $M(g - a')R$. Why? Thus

$$g - a' = \tfrac{3}{2}a$$

which can be obtained by adding the equations for the angular and linear accelerations of the center of mass. From these equations it follows that, if $3M'$ is larger than M, the acceleration of the hanging weight will be downward.

6.15   The Stability of a Car on a Horizontal Turn

Suppose that a car of mass M having a speed v is making a horizontal turn to the left, Fig. 6.19. The radius of this turn is R. The force Mv^2/R providing the centripetal acceleration is the force of friction F between the wheels and the ground. Thus

$$F = \frac{Mv^2}{R}$$

Since there is no vertical acceleration

$$P + Q = Mg$$

Also, since the angular acceleration α about the center of mass at C is zero, it follows that the torque about C is zero.

$$\frac{Pl}{2} - \frac{Ql}{2} + Fh = 0$$

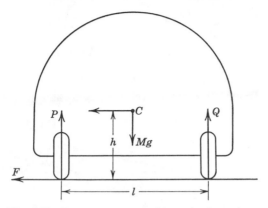

Fig. 6.19 Forces on a car making a horizontal turn.

Hence

$$Q - P = \frac{2hMv^2}{lR}$$

and by addition of these equations

$$2P = Mg - \frac{2hMv^2}{lR}$$

Now P is negative or there is no support on the inside wheel if

$$v^2 > \frac{lRg}{2h} \qquad \text{or} \qquad \frac{2hv^2}{lR} > g$$

Thus, if v is large enough, the wheel on the inside of the curve leaves the ground and the car tends to overturn outward.

The transverse force of friction F between the tires and the road can be eliminated by banking the road as shown in Fig. 6.20. It is a simple problem to find this appropriate angle of banking of the road θ for a car turning a corner of radius R with a speed v. In this example the centripetal force necessary for the turn is provided by the horizontal components of the forces P and Q acting on the wheels

$$\frac{Mv^2}{R} = (P + Q) \sin \theta$$

The vertical components must equal the weight

$$Mg = (P + Q) \cos \theta$$

Thus, if there is to be no horizontal force of friction between the road and the wheels, the road must be banked at an angle θ given by

$$\tan \theta = \frac{v^2}{Rg}$$

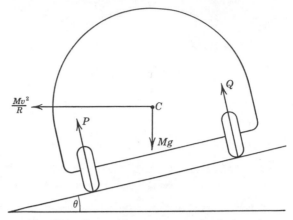

Fig. 6.20 Car making a turn on a banked road.

6.16 The Physical or Compound Pendulum

In chapter 5 we discussed the ideal simple pendulum which was assumed to consist of a point mass suspended by a weightless cord from a rigid support. We shall now drop some of these restrictions and investigate the motion of a rigid body oscillating about a fixed frictionless support. The rigid body might be a uniform rod such as a meter stick oscillating in a vertical plane about a fixed horizontal axis through a point near one end as in Fig. 6.21.

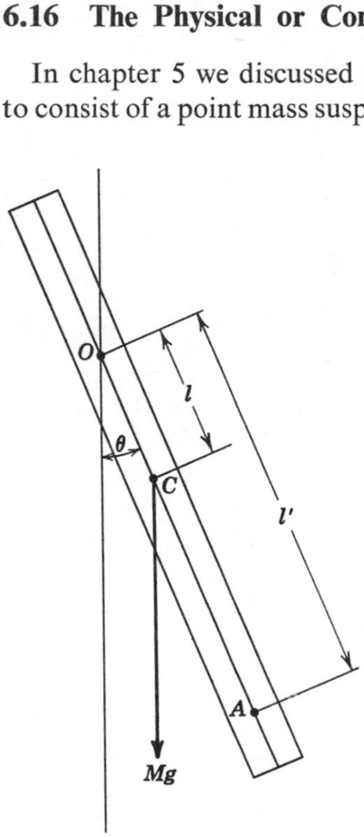

Let the rod have a mass M and be suspended at O from a frictionless support a distance l from the center of mass C. Suppose that at some instant of time the rod makes an angle θ with the vertical. At this time the torque τ_0 exerted by the rod about the axis through O is

$$\tau_0 = Mgl \sin \theta$$

This torque is related to the angular acceleration $\ddot{\theta}$ by Eq. 6.9, which in this case gives

$$Mgl \sin \theta = -I_0 \ddot{\theta} \qquad (6.24)$$

where I_0 is the moment of inertia of the rod about O. The negative sign is used since the torque tends to decrease θ whereas $\ddot{\theta}$ is positive in the direction of increasing θ. We are assuming that

Fig. 6.21 A physical pendulum showing the center of suspension O and the center of oscillation A.

there are no torques due to friction or air resistance, which is, of course, only approximately correct in practice.

We shall further assume that the maximum angle of oscillation is so small that $\sin \theta$ may be replaced by the angle θ. If this is not done, then the expression for the period becomes relatively complicated but may be evaluated by the same method as for the simple pendulum. With this assumption, Eq. 6.24 becomes

$$\ddot{\theta} = -\frac{Mgl\theta}{I_0}$$

This is the expression for angular simple harmonic motion, the period of which is

$$P_0 = 2\pi\sqrt{\frac{I_0}{Mgl}} \qquad (6.25)$$

It is well to check equations such as these for dimensions, i.e., to show that $\sqrt{I_0/Mgl}$ has the dimensions of time. The moment of inertia of the rod about O may be written by the theorem of parallel axes in terms of the moment of inertia about the center of mass as

$$I_0 = I_{C.M.} + Ml^2$$

If k is the radius of gyration of the rod about the center of mass, then

$$I_{C.M.} = Mk^2 \qquad \text{and} \qquad I_0 = M(k^2 + l^2)$$

Thus the period about O from Eq. 6.25 is

$$P_0 = 2\pi\sqrt{\frac{k^2 + l^2}{lg}} \qquad (6.26)$$

An ideal simple pendulum having this same period P_0 would have a length l' given by

$$P_0 = 2\pi\sqrt{\frac{l'}{g}}$$

This length l' is related to the length l between the point of support O and the center of mass C of the physical pendulum by

$$l' = \frac{k^2 + l^2}{l} = \frac{k^2}{l} + l \qquad (6.27)$$

We shall see that this length l', called the length of the equivalent simple pendulum, has physical importance. Suppose that the pendulum is now suspended from an axis through A a distance l' from O, the former axis of suspension. It is then readily shown that the period of oscillation about A is the same as that about O. The

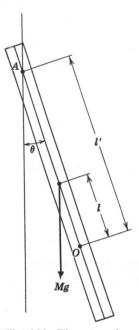

Fig. 6.22 The same physical pendulum with the positions of the center of suspension and oscillation reversed.

moment of inertia I_A of the rod about A in Fig. 6.22 is

$$I_A = Mk^2 + M(l' - l)^2$$

which by Eq. 6.27 becomes

$$I_A = Mk^2 + M\left(\frac{k^2}{l}\right)^2$$

$$= \frac{Mk^2}{l^2}(l^2 + k^2)$$

The period P_A of the oscillations about A is given from Eq. 6.25 as

$$P_A = 2\pi\sqrt{\frac{I_A}{Mg(l' - l)}} = 2\pi\sqrt{\frac{Mk^2(l^2 + k^2)l}{l^2 Mgk^2}}$$

$$= 2\pi\sqrt{\frac{(k^2 + l^2)}{lg}} = P_0$$

The two points O and A are called the *center of suspension* and the *center of oscillation* respectively. It follows that, if either point is the center of oscillation, the other is the center of suspension.

We shall next prove that there are two points of suspension at different distances from the center of mass and on the same side of the center of mass at which the periods are equal. From Eq. 6.27

$$l^2 - ll' + k^2 = 0$$

The solution of this quadratic equation is

$$l = \frac{l' \pm \sqrt{l'^2 - 4k^2}}{2}$$

so that the two lengths l_1, l_2 having the same period as that of the simple pendulum of length l' are

$$l_1 = \frac{l' + \sqrt{l'^2 - 4k^2}}{2} \qquad l_2 = \frac{l' - \sqrt{l'^2 - 4k^2}}{2} \tag{6.28}$$

The period of the compound pendulum of length l_1 is

$$P = 2\pi\sqrt{\frac{M(k^2 + l_1^2)}{Mgl_1}}$$

which may readily be shown by substitution to be equal to

$$P = 2\pi\sqrt{\frac{l'}{g}}$$

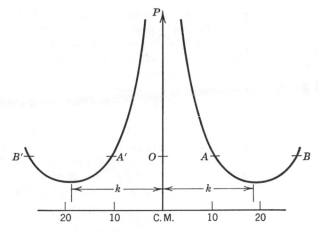

Fig. 6.23 The period P of a physical pendulum plotted against the distance from its center of mass.

As shown in Fig. 6.23, the period of a physical pendulum varies with the distance of its center of suspension from the center of mass. When the pendulum is suspended at its center of mass, the period in infinite. The figure shows that the same period is obtained when any of the four points A, B, A', B' is the point of suspension. The distance of the point of the suspension from the center of mass for which the period is a minimum may be calculated by setting dP/dl equal to zero in the expression

$$P = 2\pi\sqrt{\frac{k^2 + l^2}{lg}}$$

When the differentiation is carried out, it is found that $dP/dl = 0$ when $l = k$. The minimum period of oscillation is therefore

$$P_{\min} = 2\pi\sqrt{\frac{2k}{g}}$$

If this minimum period is experimentally determined, then the radius of gyration k of the physical pendulum can be found. This radius of gyration k may also be found from the product of any of the two distances l_1, l_2, for from Eq. 6.28 it follows that

$$l_1 l_2 = k^2 \tag{6.29}$$

About 1818, Kater,* an English army captain, constructed a special form of physical pendulum to make a precise determination of the acceleration due to

* V. F. Lenzen and R. P. Multhauf, "Development of Gravity Pendulums in the 19th Century." *Bulletin 240: Contributions from the Museum of History and Technology,* Paper 44, 1965. Smithsonian Institution, Washington, D.C.

Fig. 6.24 Kater's pendulum. (Courtesy of the Central Scientific Company, Chicago.)

gravity g. In this apparatus, Fig. 6.24, there are two fixed knife edges on opposite sides of the center of mass of the pendulum. It is the object of the experiment to arrange that the periods of oscillation about each of these two knife edges are almost equal. This is done by adjusting the position of two masses, one relatively large for coarse adjustment and the other relatively small for fine adjustment.

Let us assume that the period about one knife edge at a distance of L_1 from the center of mass is P_1, and that the period about the other at a distance of L_2 is P_2. The adjustment is made so that P_1 is nearly equal to P_2. From Eq. 6.26 the period P of a compound pendulum is given by

$$P = 2\pi\sqrt{\frac{k^2 + L^2}{Lg}} \tag{6.26}$$

or P_1 for the length L_1

$$P_1{}^2L_1g = 4\pi^2(k^2 + L_1{}^2)$$

and P_2 for the length L_2

$$P_2{}^2L_2g = 4\pi^2(k^2 + L_2{}^2)$$

By subtraction

$$g(P_1{}^2L_1 - P_2{}^2L_2) = 4\pi^2(L_1{}^2 - L_2{}^2)$$

or

$$\frac{4\pi^2}{g} = \frac{P_1{}^2L_1 - P_2{}^2L_2}{L_1{}^2 - L_2{}^2} = \frac{1}{2}\left(\frac{P_1{}^2 + P_2{}^2}{L_1 + L_2} + \frac{P_1{}^2 - P_2{}^2}{L_1 - L_2}\right) \tag{6.30}$$

If considerable time and care are taken, then P_1 may be made so nearly equal to P_2 that the second term is negligible. Assuming that P_1 may be taken as equal to P_2, which we shall set equal to P, then

$$\frac{4\pi^2}{g} = \frac{P^2}{L_1 + L_2}$$

or

$$g = \frac{4\pi^2}{P^2}(L_1 + L_2)$$

The distance between the knife edges, $L_1 + L_2$, can be measured very accurately with a comparator and a standard metal meter scale. If the period is accurately known, then a really accurate determination of the value of g can be made. If the period P_1 is not exactly equal to P_2, then the first term on the right in Eq. 6.30 can be evaluated accurately. The second term, which is very small, need not be evaluated so accurately. By balancing the pendulum on a knife edge at its center of mass, the distances L_1 and L_2 can be measured with sufficient accuracy.

Though hardly in keeping with the material in this book, we shall digress to give the method by which the period of oscillation of a Kater's pendulum is determined. In science, measurements are often made by comparing a standard quantity with an unknown quantity, using a suitable instrument to make the comparison. A mass is compared with a standard mass by means of a balance; a voltage is compared with a standard voltage or electromotive force by means of a potentiometer. In this experiment the standard timing device may be an auxiliary pendulum clock whose period, of perhaps 2 sec, is checked with the time signals from the Naval Station at Arlington. The standard pendulum and Kater's pendulum are then connected in an electric circuit so that, when both are in their central positions, the electric circuit is closed and a bell rings or a lamp lights. This is known as the method of coincidences.

Suppose that after a coincidence the standard pendulum makes n half-vibrations and the Kater's pendulum makes $(n - 1)$ half-vibrations before the next coincidence. The period P' of the Kater's pendulum is larger than the period P of the standard since

$$\frac{nP}{2} = \frac{(n - 1)P'}{2}$$

Half-periods are used since a coincidence takes place when the pendulums are in their central positions. Suppose that the period P of the standard pendulum is precisely 2 sec; then

$$\frac{P'}{2} = \frac{n}{n - 1} = \frac{1}{1 - \frac{1}{n}} = \left(1 - \frac{1}{n}\right)^{-1} = 1 + \frac{1}{n} + \frac{1}{n^2}\cdots$$

If, for example, there are 552 half-vibrations of the standard pendulum and 551 half-vibrations of the Kater's pendulum between successive coincidences, then $n = 552$ and

$$\frac{P'}{2} = \frac{552}{551} = 1 + \frac{1}{551}$$

$$= 1.001819$$

or

$$P' = 2.003638 \text{ sec}$$

6.17   Reactions on the Axis of Rotation

When a body is rotating about a fixed axis, there is present a force exerted by the rotating body on the axis. By Newton's third law there is a reacting

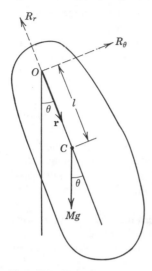

Fig. 6.25 Reactions on the axis of rotation of a physical pendulum.

force exerted by the axis on the body. The simplest but most common problems of this type are those in which both the forces and the body are symmetrical with respect to a plane through the center of mass perpendicular to the fixed axis. We shall assume here that the fixed axis is perpendicular to the plane of the paper and that the only external force acting on the system is that due to gravity, as shown in Fig. 6.25.

Let O be the fixed axis on which the body of mass M rotates. From the assumed symmetry of the body the reactions of the axis reduce to a single force in the plane of the paper whose components, along and perpendicular to the line OC joining the axis to the center of mass, are R_r and R_θ respectively. As the body vibrates about the axis O, the line OC changes direction and so do the forces R_r and R_θ. Let O be the origin of a radius vector $\mathbf{r}$ along OC. Then from Eqs. 1.14 and 1.15 the acceleration of a particle along $\mathbf{r}$ is $\ddot{r} - r\dot\theta^2$ and perpendicular to $\mathbf{r}$ in a direction of increasing θ is $r\ddot\theta + 2\dot r\dot\theta$. For the center of mass the vector $\mathbf{r}$ has a constant magnitude of l so that $\dot r$ and $\ddot r$ are zero. From Eq. 6.3, $M\ddot{\mathbf{r}} = \mathbf{F}$, we have for the equations of motion of the center of mass, along and perpendicular to OC respectively, at any instant of time,

$$Mg \cos \theta - R_r = -Ml\dot\theta^2 \qquad (6.31)$$

and

$$R_\theta = Mg \sin \theta = Ml\ddot\theta \qquad (6.32)$$

where θ is the instantaneous angle made by the line OC with the vertical. The final equation of motion comes from the relationship of the torque to the angular acceleration about the fixed axis. Applying Eq. 6.9 to this problem, we have

$$Mgl \sin \theta = -I_0 \ddot\theta \qquad (6.33)$$

where the negative sign is used since the torque is in the direction tending to decrease θ.

The moment of inertia I_0 about O may be written in terms of the radius of gyration k about the center of mass as

$$I_0 = I_{C.M.} + Ml^2 = M(k^2 + l^2)$$

Thus Eq. 6.33 becomes

$$\ddot\theta = - \frac{gl}{k^2 + l^2} \sin \theta \qquad (6.34)$$

To obtain the angular velocity $\dot\theta$, Eq. 6.34 is multiplied by the identity

$$d\theta = \frac{d\theta}{dt}\, dt = \dot\theta\, dt$$

giving

$$\tfrac{1}{2} d(\dot\theta^2) = -\frac{gl}{k^2 + l^2}\sin\theta\, d\theta = +\frac{gl}{k^2 + l^2}\, d(\cos\theta)$$

By integration

$$\tfrac{1}{2}\dot\theta^2 = \frac{gl}{k^2 + l^2}\cos\theta + C_1 \qquad (6.35)$$

where C_1 is the constant of integration determined from the initial conditions of motion. To complete the problem, let us assume that, at time $t = 0$, $\dot\theta = 0$ and $\theta = \pi/2$; i.e., the body is started from rest with the line OC horizontal. The constant of integration C_1 is thus zero. Substituting for $\dot\theta$ and $\ddot\theta$ in Eqs. 6.31 and 6.32 gives

$$R_r = Mg \cos\theta\left(1 + \frac{2l^2}{k^2 + l^2}\right)$$

and

$$R_\theta = Mg \sin\theta\left(1 - \frac{l^2}{k^2 + l^2}\right)$$

The magnitude of the resultant reaction R is

$$R = \sqrt{R_r{}^2 + R_\theta{}^2}$$

As an example of the reactions on an axis of rotation, let us consider the following problem. A uniform meter stick whose mass is 150 gm is free to turn on a horizontal axis 10 cm from one of its ends. It is started from rest when the rod is at an angle of 45° above the horizontal, Fig. 6.26. The problem is to find the resultant reaction of the axis of rotation.

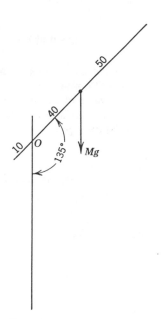

Fig. 6.26 A meter stick falling from an angle of 135° with the vertical.

We must first find the radius of gyration k. For a uniform rod of length L

$$I_{\text{C.M}} = Mk^2 = \frac{ML^2}{12}$$

or

$$k^2 = \frac{L^2}{12} = \frac{10^4}{12}\ \text{cm}^2$$

The constant of integration C_1 in Eq. 6.35 is evaluated from the initial conditions:

$t = 0$, $\dot{\theta} = 0$, $\theta = 135°$, so that

$$C_1 = - \frac{gl}{k^2 + l^2} \cos 135° = 0.707 \frac{gl}{k^2 + l^2}$$

From Eq. 6.35

$$\dot{\theta}^2 = \frac{2gl}{k^2 + l^2} (\cos \theta + 0.707)$$

Hence from Eqs. 6.31 and 6.32 the components of the force of reaction when the meter stick makes an angle θ with the vertical are

$$R_r = Mg \cos \theta + \frac{2Mgl^2}{k^2 + l^2} (\cos \theta + 0.707) = (347.5 \cos \theta + 139.6) \text{ gmw}$$

$$R_\theta = Mg \sin \theta \left(1 - \frac{l^2}{k^2 + l^2} \right) = 51.3 \sin \theta \text{ gmw}$$

When the meter stick is in the vertical position, $\theta = 0°$. Then R_θ is zero, and $R_r = 487.1$ gmw; i.e., the reaction by the support is entirely vertical.

This value of R_r when the rod is vertical can be obtained in a simpler manner, for in this position the support must exert a force equal to the sum of the weight and the centripetal force, or

$$R_r = Mg + Ml\dot{\theta}_0^2$$

where $\dot{\theta}_0$ is the angular velocity of the rod when it is vertical. The value of $\dot{\theta}_0$ can be obtained from the conservation of energy equation. If h is the vertical distance through which the center of mass falls, then

$$Mgh = \tfrac{1}{2} I_0 \dot{\theta}_0^2$$

From Fig. 6.26

$$h = 40 \times 0.707 + 40 = 68.3 \text{ cm}$$

and

$$I_0 = M(k^2 + l^2) = M \left(\frac{29,200}{12} \right) \text{ gm cm}^2$$

Hence

$$\dot{\theta}_0^2 = \frac{2 \times 68.3 \times 980 \times 12}{29,200} \text{ sec}^{-2}$$

giving

$$R_r = Mg \left[1 + \frac{(2 \times 68.3 \times 12 \times 40)}{29,200} \right] = 487.1 \text{ gmw}$$

6.18   Impulsive Forces

In Chapter 2 we showed that the change in momentum of a particle is equal to the time integral of the force acting on the particle or is equal to the impulse of the force. A similar theorem holds for the linear motion of the center of mass of a rigid body.

From Eq. 6.3 the linear acceleration of the center of mass $\ddot{\mathbf{r}}$ of a body of mass M acted on by a resultant force $\mathbf{F}$ is given by

$$M\ddot{\mathbf{r}} = \mathbf{F} \tag{6.3}$$

Suppose that the force $\mathbf{F}$ is very large and that the time ΔT for which it acts is very small. Then the impulse $\mathbf{H}$ of this force is defined as

$$\mathbf{H} = \int_0^{\Delta T} \mathbf{F}\,dt$$

During the time ΔT the force may vary considerably in magnitude, as when a bat strikes a ball. From Eq. 6.3 it follows that

$$M\,d(\dot{\mathbf{r}}) = \mathbf{F}\,dt$$

If $\bar{\mathbf{v}}_0$ is the initial velocity of the center of mass when the force begins to act and $\bar{\mathbf{v}}$ that after the impulse $\mathbf{H}$ is given, then

$$\int_{\mathbf{v}_0}^{\bar{\mathbf{v}}} M\,d(\dot{\mathbf{r}}) = \int_0^{\Delta T} \mathbf{F}\,dt = \mathbf{H}$$

or

$$M(\bar{\mathbf{v}} - \bar{\mathbf{v}}_0) = \int_0^{\Delta T} \mathbf{F}\,dt = \mathbf{H} \tag{6.36}$$

This is an equation between vector quantities. It states that *the change in momentum of the center of mass of a body in any direction is equal to the component of the impulse in that direction.*

Consider now an impulse on a body free to rotate about a fixed axis through a point O. Then from Eq. 6.9 a torque τ_0 produces an angular acceleration $\ddot{\theta}$ about the axis given by

$$\tau_0 = I_0\ddot{\theta}$$

where I_0 is the moment of inertia of the body about the axis through O. Suppose that a large torque acts for a short time. Then $\int \tau_0\,dt$ is called the impulse of the torque. From Eq. 6.9 it follows that

$$\tau_0\,dt = d(I_0\dot{\theta}) = I_0\,d(\dot{\theta})$$

If ω_0 is the angular velocity of the body about O just before the torque starts to act and ω that immediately after the torque has ceased acting, then

$$\int_0^{\Delta T} \tau_0\,dt = I_0(\omega - \omega_0) \tag{6.37}$$

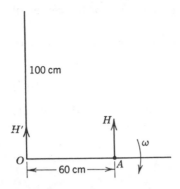

Fig. 6.27 A meter stick making an inelastic impact with rod A.

or *the impulse of a torque about a fixed axis is equal to the change in angular momentum about that axis.*

As an example of this analysis let us consider the following problem. A meter stick whose mass is 150 gm can rotate about an axis through O at one end, Fig. 6.27. Suppose that the stick starts from a vertical position and falls to the horizontal position where it strikes a fixed inelastic rod A at a distance of 60 cm from the end O. The problem is to find the impulse of the blow on the fixed inelastic rod and the impulse of the reaction at the support O.

The angular velocity ω of the rod when it is horizontal and just before it strikes the fixed rod at A is given by the law of conservation of energy

$$\frac{1}{2} I_0 \omega^2 = Mg \frac{l}{2}$$

where

$$I_0 = I_{\text{C.M.}} + \frac{Ml^2}{4} = \frac{Ml^2}{12} + \frac{Ml^2}{4} = \frac{Ml^2}{3}$$

Hence

$$\omega^2 = \frac{3g}{l} = 29.4 \text{ sec}^{-2}$$

If H is the impulse of the blow exerted on the meter stick when it strikes the rod A, then from Eq. 6.37 for the motion about the fixed axis

$$H \times 60 = I_0 \omega$$

since the meter stick comes to rest after the inelastic impact with the rod. Thus

$$H = \frac{Ml^2 \omega}{180} = \frac{15 \times 10^5 \sqrt{29.4}}{180} = 4.52 \times 10^4 \frac{\text{gm cm}}{\text{sec}}$$

If H' is the reaction of the blow on the rod at the support A, then from Eq. 6.36 for the motion of the center of mass

$$H + H' = \frac{Ml\omega}{2}$$

Thus

$$H' = (4.07 - 4.52) \times 10^4 = -4.50 \times 10^3 \frac{\text{gm cm}}{\text{sec}}$$

The blow exerted by the axis is therefore downward in the opposite direction to that shown in Fig. 6.27.

6.19   Center of Percussion

If a body that is capable of rotating freely about a fixed axis is given a blow at such a point that there is no impulsive blow on the axis, then the point is called the center of percussion of the body.

Let us consider that a uniform rod of mass M, suspended freely about an axis near one end O, is struck by a horizontal blow of impulse $\mathbf{H}$ at a distance $x + a$ from O, as shown in Fig. 6.28. In general the blow $\mathbf{H}$ causes an impulsive action $-\mathbf{H}'$ on the axis and a reaction $\mathbf{H}'$ on the rod. In this example $\mathbf{H}'$ is horizontal since H is horizontal. The problem is to find some distance x where $\dot{\mathbf{H}}'$ is zero.

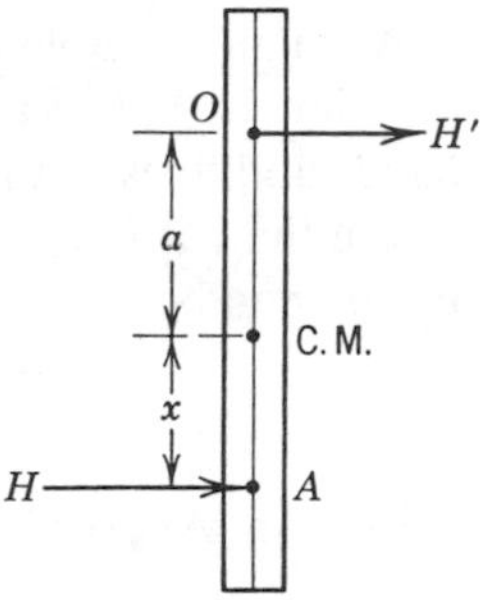

Let ω be the angular velocity of the center of mass immediately after the blow. Then from Eq. 6.36

$$H + H' = Ma\omega$$

From Eq. 6.37 for the moment of the impulse about O we have

$$H(x + a) = I_0\omega$$

where

$$I_0 = I_{C.M.} + Ma^2 = Mk^2 + Ma^2 = M(k^2 + a^2)$$

Hence

$$H = \frac{M(k^2 + a^2)\omega}{x + a}$$

and

$$H' = Ma\omega - \frac{M(k^2 + a^2)\omega}{x + a} \tag{6.38}$$

Fig. 6.28 The center of percussion of a rod.

If the blow is struck at the center of percussion, then by definition of the center of percussion H' is equal to zero. Thus the center of percussion lies at some distance x below the center of mass where

$$a - \frac{k^2 + a^2}{x + a} = 0$$

or

$$xa = k^2 \tag{6.39}$$

This is similar to Eq. 6.29 derived earlier for a compound pendulum. Thus, if A is the center of percussion where the impulsive blow H' on the rod is zero, then the distance OA is the length of the equivalent simple pendulum when the meter stick is regarded as a compound pendulum. In other words, the center of percussion with respect to a fixed axis coincides with the center of oscillation with respect to the same fixed axis. From Eq. 6.39 it follows that the blow H' at the axis is positive in the direction shown in Fig. 6.28 if the blow H is struck below the center of percussion A. Also, H' is negative or in the opposite direction to that shown in the figure if the blow H is struck above the center of percussion.

A rough example of the above theory, and one with which most of us are familiar, is that of a bat striking a baseball. Although the bat is held by the hands and does not move about any well-defined axis, there is a certain place where the bat must be held if no "sting" or impulsive blow is to be received by the hands.

6.20   An Approximate Theory of the Gyroscope

A gyroscope is any symmetrical body rotating on an axis, which axis is also free to turn about some point. A spinning top and a heavy wheel rotating on an axle supported at some point as in Fig. 6.29 are examples of gyroscopes. In these gyroscopes the axis of symmetry, about which the body rotates, precesses or rotates about another axis. Thus in Fig. 6.29a the wheel and axle precess with an angular velocity ω' about a vertical axis. We shall assume that the angular velocity of spin ω is very much larger than the angular velocity

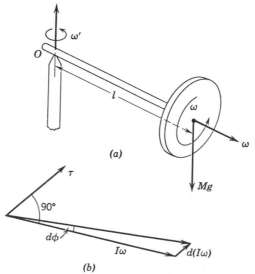

(a)

(b)

Fig. 6.29   A simple gyroscope and the components of angular momentum $\mathbf{J}$ and torque $\boldsymbol{\tau}$.

of precession ω' so that the total angular momentum of the system lies along the spin axis. Our approximation then consists in neglecting the angular momentum about the axis of precession.

If the wheel were not rotating it would fall vertically under gravity. When it is rotating rapidly the torque due to the weight of the wheel Mg causes precession about the vertical axis through O. The torque about O is Mgl and is represented by the vector τ in Fig. 6.29b. This torque τ is in a horizontal plane perpendicular to the plane formed by the axle and the direction of the weight Mg. If I is the moment of inertia of the wheel about the axle and ω is the angular velocity of the wheel, then the angular momentum of the wheel is $I\omega$ along the direction of the vector ω. In a short time dt, the torque τ produces a change in angular momentum given by Eq. 6.37.

$$\tau \, dt = d(I\omega)$$

and is in a direction parallel to the torque vector τ. This change in angular momentum is represented by the vector $d(I\omega)$ at right angles to $I\omega$ in Fig. 6.31b. The resultant of these two vectors is a vector which makes an angle $d\phi$ with the vector $I\omega$. In the time dt the vector $I\omega$ and hence the axle turns through an angle $d\phi$. This is the motion of precession. From the vector triangle in Fig. 6.31b it follows that

$$d(I\omega) = I\omega \, d\phi$$

Thus

$$Mgl \, dt = d(I\omega) = I\omega \, d\phi$$

or the angular velocity of precession ω' is given by

$$\omega' = \frac{d\phi}{dt} = \frac{Mgl}{I\omega} \tag{6.40}$$

From the discussion above it follows that the direction of the precession is such that the spin vector precesses towards the torque vector. If such a gyroscope as shown in Fig. 6.29a is watched carefully, it will be noticed that the axle does not precess exactly in a plane but rises above and falls below this plane. This motion, called nutation, cannot be analyzed by this simple theory, and a more complete theory is given in Section 7.8.

PROBLEMS

1. The linear density of a rod 1 m long decreases uniformly from 0.6 kg/m to 0.2 kg/m along its length. Find the mass of the rod and the distance of the center of mass from the end having the larger density.

2. A uniform circular disk whose radius is 40 cm has a rectangular area 12 cm by 16 cm cut out of it with a corner of the rectangle at the center of the disk. Find the distance of the center of mass of the remainder from the center of the disk along the line of the diagonal of the rectangle.

3. A right circular cylinder 20 cm in radius and 50 cm high has a conical portion of 30 cm height and radius of base 20 cm cut away so that the axis of the cone and cylinder coincide. Find the center of mass of the remainder measured from the base of the cone.

4. The mass of the moon is about $\frac{1}{80}$ the mass of the earth, and its distance from the center of the earth is about 60 times the radius of the earth. If the radius of the earth is 4000 miles, find the distance of the center of mass of the earth-moon system from the center of the earth.

5. Two particles A and B are placed along the X axis: A, whose mass is 50 gm, at the origin, and B, whose mass is 150 gm, 100 cm from the origin. The initial velocity of A is 10 cm/sec along the positive Y axis, and that of B is 20 cm/sec at an angle of 60° with the positive X axis. The particles act on each other with such a force that the particle B is at its starting point after 10 sec. Find (a) the center of mass of the system at zero time; (b) the x and y components of the velocity of the center of mass at zero time; (c) the x and y coordinates of the center of mass after the 10 sec; (d) the x and y coordinates of the 50-gm mass after the 10 sec.

6. Two particles initially at rest have masses m_1 and m_2 kg respectively and are at a distance of d meters apart. They attract each other with a constant force of F newtons. What is the motion of the center of mass of the two particles? Show that the two particles collide at the center of mass.

7. A circular disk having a radius of 4 ft and weighing 40 lbw has a small weight of 20 lbw attached to it at a distance of 2 ft from its center. If the disk is set in rotation about an axis through its center so that its minimum speed is 120 rpm, find its maximum speed.

8. A uniform plane circular disk whose mass is 500 gm and whose radius is 20 cm is free to turn in a vertical plane about a horizontal axis through its center. Suppose that the disk has a small light projection attached to its periphery and that originally the disk is at rest with the projection placed vertically above the center of the disk. A piece of wax whose mass is 200 gm is thrown horizontally with a velocity of 100 cm/sec so that it becomes attached to the projection. Find the resulting angular velocity of the disk and wax. Calculate the kinetic energy of the system just before and after the wax sticks to the projection.

9. Prove from fundamental principles the results given in the Table of Moments of Inertia (Table 6.1) in Section 6.10.

10. A so-called Borda's pendulum is made up of a spherical bob of radius a and a

long this wire of negligible weight. If the length of the wire from the point of suspension to the center of the bob is l, show that the period of the pendulum for small oscillations is

$$P = 2\pi\sqrt{\frac{l}{g}\left(1 + \frac{2}{5}\frac{a^2}{l^2}\right)}$$

If this pendulum is considered as a simple pendulum, show that the period calculated on this basis is in error by about eight parts in a hundred thousand when the radius of the spherical bob is one-fiftieth of the length of the wire.

11. A uniform rod of mass M and length $2h$ is suspended from the same horizontal level by two vertical cords of length l which are attached to its ends. Show that the period of this bifilar suspension for small oscillations about a vertical axis through the center of the rod is

$$P = 2\pi\sqrt{\frac{Il}{Mgh^2}}$$

where I is the moment of inertia of the rod about a vertical axis through its center of mass.

12. A steel sphere of radius r executes simple harmonic oscillations of small amplitude on a concave surface of radius R. Show that the period of the oscillations P is given by

$$P = 2\pi\sqrt{\frac{1.4(R - r)}{g}}$$

(*Hint:* Set up the equation of motion from the energy equation.)

13. A ring of inner radius r_1, outer radius r_2, and mass m has a rectangular cross section. This ring is suspended from a knife edge perpendicular to the plane of the disk. Show the period of the ring for small oscillations about the knife edge is $2\pi\sqrt{(r_2^2 + 3r_1^2)/2r_1g}$.

14. A Kater's pendulum has been carefully adjusted so that the period about one knife edge A is almost the same as that about the other knife edge B. The distance between the knife edges as measured by a comparator is 99.8503 cm. The periods are obtained by the method of coincidences, using an accurate 2-sec pendulum. It is found that for 500 half-oscillations about knife edge A there are 501 half-oscillations of the 2-sec pendulum, whereas after 10,000 half-oscillations about knife edge B there are 10,021 half-oscillations of the 2-sec pendulum. From an approximate measurement the center of mass is 53.9 cm from the knife edge B and 45.9 cm from A. From these data determine the value of the acceleration of gravity g.

15. The determination of the constant of gravitation G was made by an apparatus similar to that shown in Fig. 4.1. Two small spheres, each having a mass of

10 gm, were placed on the ends of a light rod 10 cm long. The light rod was supported horizontally from its center by a long, thin, vertical fiber. The period of torsional oscillation of this system was 769.2 sec. A fixed sphere whose mass is 10 kgm was placed near each suspended sphere so as to produce the maximum torsion. It was found that there was an angular deflection of the suspended rod of 0.02 radian when the distance between the centers of the large and small spheres was 10 cm. From these data calculate the constant of gravitation G.

16. A cylindrical rod 1 in. in radius, 6 in. long, and weighing 10 lb has two cords wound around it whose ends are attached to the ceiling, as shown in Fig. P16.

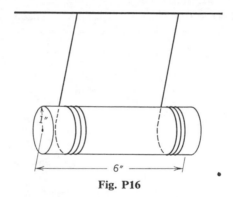

Fig. P16

The rod is held horizontally with the two cords vertical. When the rod is released, the cords unwind and the rod rotates. Find the translational acceleration of the rod and the tension in the cord as the rod falls vertically.

17. An 8-lb weight hangs from one end of a cord that passes over a pulley that is free to rotate about its horizontal axis and has a radius of 6 in. and a weight of 4 lbw. The other end of the cord is wound around a cylinder whose radius is 4 in. and whose weight is 40 lbw. When released, the 8-lb weight accelerates upward and the cylinder downward as the string unwinds. Find the upward acceleration of the hanging weight and the angular acceleration of the cylinder.

18. A fixed pulley 3 in. in radius and weighing 4 lbw is free to rotate about a frictionless horizontal axis through its center. A cord passes over the pulley, the left side of the cord being wrapped around a cylinder A, 4 in. in radius and weighing 6 lbw. The right side of the cord is wrapped around a cylinder B, having a radius of 6 in. and a weight of 8 lbw. Originally the cylinders are at rest with the cords vertical. When released the cords unwrap, causing the cylinders to rotate. Find the angular acceleration of the pulley and the linear accelerations of the centers of mass of the cylinders A and B relative to the fixed axis through the pulley.

19. A yard stick held vertically with one end resting on a table is allowed to fall over. Find the velocity of the far end when it hits the table, assuming that the end on the table does not slip. Find the reaction of the table on the end of the yard stick just before it strikes the horizontal table if the yard stick weighs $\frac{1}{4}$ lb.

20. A yard stick weighing $\frac{1}{4}$ lb lies on a frictionless horizontal table and is free to rotate about a vertical axis through one end. It is struck a blow whose impulse is 0.5 lbw sec at a point 2 ft from the pivoted end and at right angles to the length of the rod. Find the kinetic energy of the rod and the impulse on the axis.

21. A yard stick weighing $\frac{1}{4}$ lb lies on a horizontal frictionless table and is subject to no constraints. It is struck a blow whose impulse is 0.8 lbw sec at the 2.5 ft mark in a direction at right angles to the length of the rod. Find (a) the point about which the rod begins to turn and (b) the kinetic energy received by the rod.

22. A meter stick whose mass is 200 gm rests on a frictionless horizontal table and is free to turn about a smooth pivot at the 10-cm mark. Almost in contact with the rod at the 40-cm mark is an inelastic piece of putty having a mass of 50 gm. A horizontal blow whose impulse is 10^4 dyne sec is given to the meter stick at the 80-cm mark in a direction perpendicular to the rod. Find the resulting angular velocity of the rod and the impulsive actions at the pivot and on the putty.

23. A uniform rod of mass 800 gm, having one end attached to a smooth hinge, is allowed to fall vertically from a horizontal position. Show that the horizontal force on the hinge is a maximum when the rod makes an angle of 45° with the vertical and that the vertical force on the hinge at this angle is 1100 gmw.

24. A plane circular disk having a weight of 6 lbw is free to turn about a horizontal axis through a point O at its circumference. The axis is perpendicular to the plane of the disk. The motion starts when the diameter through O is horizontal. Prove that when this diameter makes an angle θ with the vertical the components of the force on the axis along, and perpendicular to, this diameter are respectively 14 cos θ lbw and 2 sin θ lbw.

25. A pendulum is made up of a solid sphere of radius R attached to a rod of length l. The system is suspended on a horizontal axis through the free end of the rod. The masses of the sphere and rod are equal. Show that there is no force on the axis if the pendulum is struck a horizontal blow at a distance of

$$(\tfrac{14}{5}R^2 + \tfrac{8}{3}l^2 + 4Rl)/(3l + 2R)$$

26. A uniform rod AB weighing 24 lbw and having a length of 8 ft hangs vertically from a smooth horizontal axis at A. The rod is struck normally at a point 6 ft below A by a blow whose impulse is 120 lbw sec. Find the instantaneous angular velocity of the rod and the impulse received by the axis.

27. A body of mass M can slide without friction along a horizontal rod. One end of a cord of length l is attached to the mass M and the other end to a bob of mass m so as to form a pendulum. Show that the period of the pendulum for small oscillations is $T = 2\pi\sqrt{(Ml)/[M + m)g]}$. If the system is started by holding the mass M and having the cord with mass m horizontal and taut and then letting go, show that the maximum angular velocity of the cord is $\sqrt{[2g(M + m)]/lM)}$. (Consider the resultant horizontal force on the center of mass of the system.)

28. A uniform solid cylinder rolls down a plane whose angle of inclination is θ. Show that, if there is to be rolling and no sliding, the coefficient of friction between the cylinder and the plane must be equal to or greater than $\frac{1}{3}\tan\theta$.

29. A sphere whose radius is 10 cm and whose mass is 200 gm starts from rest from the top of an inclined plane whose length is 196 cm and whose angle of inclination is 30°. (a) Show that there is slipping if the coefficient of friction μ between the sphere and the plane is less than $\frac{2}{7}\tan 30°$.

   Now suppose that the coefficient of friction is 0.12 so that there is both rolling and slipping. (b) Show that the velocity of the point of contact of the sphere on the plane after a time t is gt (sin 30° $-$ 3.5μ cos 30°). (c) Show that the time taken for the sphere to reach the bottom of the plane is 1.005 sec. (d) Find the linear velocity of the center of mass of the sphere at the bottom of the plane. (e) Find the angular velocity of the sphere at the bottom of the plane. (f) Find the translational and rotational kinetic energy of the sphere at the bottom of the plane. (g) Find the energy expended against friction by the sphere in moving down the plane. (h) Find the loss in potential energy and show that this is equal to the sum of the energies in f and g, or there is conservation of energy.

30. A gyroscope is constructed from a plane circular disk 0.1 m in radius and 1 kg in mass capable of rotating about an axle 0.5 kg in mass and 1 m in length, as illustrated in Fig. 6.29. The axle is supported at its far end on a vertical pivot and the disk set in rotation with an angular velocity of 1500 rad/sec. If the axle is originally horizontal, show the angular velocity of precession is 1.63 rad/sec.

ROTATIONAL MOTION ABOUT A FIXED POINT; LAGRANGE AND HAMILTON EQUATIONS

7.1 A More General Analysis of the Motion of Rigid Bodies

In this section we shall show the importance of the center of mass in dynamics. We have already proved that for linear motion the center of mass of a body moves as though all the mass of the body were concentrated at that point and the resultant external force acted there. We shall now show that there is the same relationship between the rotational acceleration and the applied torques about an axis through the center of mass as there is about a fixed axis. Finally we shall show that for a body having both translational and rotational motion the total kinetic energy may be expressed either as the sum of the transitional kinetic energy of the center of mass and the rotational kinetic energy about the center of mass, or as the rotational kinetic energy about the instantaneous axis of rotation.

Let us consider a particle of mass m acted on by a force F as shown in Fig. 7.1. Relative to an origin O fixed in an inertial system, the force F exerts a torque τ which has a magnitude of

$$F \times OA = Fr \sin \theta$$

and a direction that would tend to produce counterclockwise rotation about O. This torque τ can be expressed as the vector or *cross product* of the vectors $\mathbf{r}$ and $\mathbf{F}$. This product is written as

$$\tau = \mathbf{r} \times \mathbf{F} \qquad (7.1)$$

By definition the vector product of two vectors is a third vector whose magnitude is the product of the magnitudes of the two vectors and the sine of the angle between their positive directions, and whose direction is given by the direction in which a right-handed screw would progress when rotated from the first to the

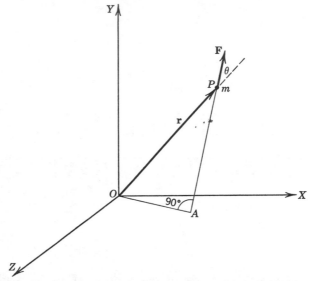

Fig. 7.1   Rotational motion of a body with reference to a fixed point.

second vector. If the vectors all lie in the plane of this paper, then rotating a right-handed screw from the positive direction of **r** into **F**, i.e., in a counterclockwise direction, would cause the screw to progress perpendicularly out of the plane of the paper. Notice that the positive direction of each vector must be used when considering the rotation. Applying this definition to the vectors **r** and **F** in Fig. 7.1, we see that their vector product has a magnitude of $rF \sin \theta$ and is directed perpendicularly out of the plane of the paper corresponding to the counterclockwise rotation of **r** to **F**.

Before proceeding further with the dynamical theory, let us consider some of the properties of this vector product. The vector product of a vector and itself such as **r** × **r** is zero since the angle between the vectors is zero. Thus for the unit vectors **i**, **j**, **k** along the X, Y, Z axes respectively, it follows that

$$\mathbf{i} \times \mathbf{i} = 0 \quad \mathbf{j} \times \mathbf{j} = 0 \quad \mathbf{k} \times \mathbf{k} = 0$$

Also, by applying the rule to a system of right-handed axes X, Y, Z, it can be seen that, if the unit vector **i** is rotated into the unit vector **j**, there results a unit vector **k** along the Z axis. Thus

$$\mathbf{i} \times \mathbf{j} = \mathbf{k} \quad \mathbf{j} \times \mathbf{k} = \mathbf{i} \quad \mathbf{k} \times \mathbf{i} = \mathbf{j}$$

and

$$\mathbf{j} \times \mathbf{i} = -\mathbf{k} \quad \mathbf{k} \times \mathbf{j} = -\mathbf{i} \quad \mathbf{i} \times \mathbf{k} = -\mathbf{j}$$

The results can readily be recalled by noticing that, if the vectors are taken in the cyclical order **i**, **j**, **k**, the cross product of two of them is the positive third unit vector, whereas if taken in a reverse direction the cross product is negative.

By writing out the vectors **r** and **F** in component form, their cross product can be written as a determinant. Suppose that the coordinates of a point P are x, y, z relative to the respective fixed axes X, Y, Z with the origin at O. The vector **r** can be expressed

$$\mathbf{r} = \mathbf{i}x + \mathbf{j}y + \mathbf{k}z$$

Also, if the components of the force **F** acting at P are F_x, F_y, F_z respectively, then

$$\mathbf{F} = \mathbf{i}F_x + \mathbf{j}F_y + \mathbf{k}F_z$$

The vector product is

$$\mathbf{r} \times \mathbf{F} = (\mathbf{i}x + \mathbf{j}y + \mathbf{k}z) \times (\mathbf{i}F_x + \mathbf{j}F_y + \mathbf{k}F_z)$$
$$= \mathbf{k}xF_y - \mathbf{j}xF_z - \mathbf{k}yF_x + \mathbf{i}yF_z + \mathbf{j}zF_x - \mathbf{i}zF_y$$
$$= \mathbf{i}(yF_z - zF_y) + \mathbf{j}(zF_x - xF_z) + \mathbf{k}(xF_y - yF_x)$$

which may be written in the form of a determinant, thus:

$$\boldsymbol{\tau} = \mathbf{r} \times \mathbf{F} = \begin{vmatrix} \mathbf{i} & \mathbf{j} & \mathbf{k} \\ x & y & z \\ F_x & F_y & F_z \end{vmatrix} \tag{7.2}$$

Let us now return to the problem of the particle of mass m in Fig. 7.1 acted on by the force **F** which produces a torque $\mathbf{r} \times \mathbf{F}$ about the fixed point O. Suppose that the instantaneous angular velocity of the mass m about the fixed point O is $\boldsymbol{\omega}$. Before proceeding with the main problem we should give some justification for considering an angular velocity $\boldsymbol{\omega}$ as a vector quantity.

We have already seen in Section 1.5 that finite rotations do not qualify as vectors. However, this does not apply to infinitesimal rotations where the second order of these rotations can be neglected. Suppose $\mathbf{r}_1$ is a unit vector along the direction of the axis of rotation and $\Delta\theta_1$ the infinitesimal angle of rotation about the axis. Similarly, let $\mathbf{r}_2$ be another unit vector and $\Delta\theta_2$ the infinitesimal rotation about the axis whose direction is along $\mathbf{r}_2$, where the two axes of rotation pass through the common point O (Fig. 7.2a). Then, it is found that, neglecting second order terms of the infinitesimal angles of rotation, the vector addition of $\mathbf{r}_1\Delta\theta_1$ and $\mathbf{r}_2\Delta\theta_2$ produces the resultant angle of rotation $\Delta\theta$, along the axis whose direction is given by unit vector $\mathbf{r}'$, such that

$$\mathbf{r}'\Delta\theta = \mathbf{r}_1\Delta\theta_1 + \mathbf{r}_2\Delta\theta_2 = \mathbf{r}_2\Delta\theta_2 + \mathbf{r}_1\Delta\theta_1$$

Suppose these infinitesimal angles of rotation take place in the infinitesimal time Δt; then the corresponding angular velocities are given in the limit as Δt approaches zero. Thus

$$\lim_{\Delta t \to 0} \frac{\Delta\theta}{\Delta t}\mathbf{r}' = \lim_{\Delta t \to 0} \frac{\Delta\theta_1}{\Delta t}\mathbf{r}_1 + \lim_{\Delta t \to 0} \frac{\Delta\theta_2}{\Delta t}\mathbf{r}_2$$

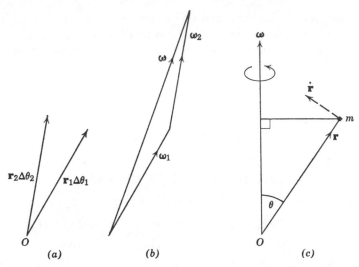

Fig. 7.2 (a) Two infinitesimal angles of rotation $\Delta\theta_1$ and $\Delta\theta_2$. (b) Addition of the two corresponding angular velocities ω_1 and ω_2. (c) The linear velocity $\dot{\mathbf{r}}$ of m about O due to angular velocity $\boldsymbol{\omega}$ is $\boldsymbol{\omega} \times \mathbf{r}$ directed into the page.

or

$$\boldsymbol{\omega} = \boldsymbol{\omega}_1 + \boldsymbol{\omega}_2$$

where the angular velocity vectors $\boldsymbol{\omega}$ are along the instantaneous axes of rotation and have a magnitude equal to the instantaneous rate of rotation in radians per second (Fig. 7.2b).

Let us now consider the particle of mass m which is at a vector distance $\mathbf{r}$ from the fixed point O and rotating with the angular velocity $\boldsymbol{\omega}$ about the fixed point O.* Owing to this angular velocity $\boldsymbol{\omega}$ the particle of mass m has an

* "The Teaching of Angular Momentum and Rigid Body Motion," J. I. Shonle, Resource Letter CM-1, A.I.P. or *Am. J. Phys.*, **33**, 879, 1965. A correction to the above article is given by Professor Shonle in *Am. J. Phys.*, **34**, 273, 1966. We have encountered two kinds of

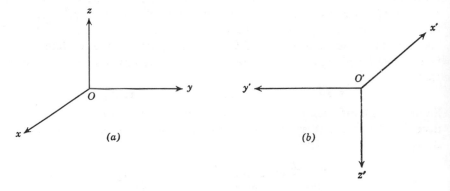

instantaneous linear velocity $\dot{\mathbf{r}}$ which, from Fig. 7.2c, may be seen to be

$$\dot{\mathbf{r}} = \boldsymbol{\omega} \times \mathbf{r} \qquad (7.2a)$$

The magnitude of $\dot{\mathbf{r}}$ is $\omega r \sin \theta$ and is in a direction at right angles to the plane containing the vectors $\boldsymbol{\omega}$ and $\mathbf{r}$. In this example we are considering the motion to be entirely one of rotation. The linear momentum of the mass m is $m\dot{\mathbf{r}}$, and the angular momentum or moment of momentum $\mathbf{J}$ of m about the axis through the fixed point O is defined as

$$\mathbf{J} = \mathbf{r} \times m\dot{\mathbf{r}} \qquad (7.3)$$

We shall now show that the torque $(\mathbf{r} \times \mathbf{F})$ exerted on the mass m in Fig. 7.1 is equal to the time rate of change of angular momentum of the mass m. By Newton's second law, $\mathbf{F} = m\ddot{\mathbf{r}}$. Hence the torque $\boldsymbol{\tau}$ about O is

$$\boldsymbol{\tau} = (\mathbf{r} \times \mathbf{F}) = (\mathbf{r} \times m\ddot{\mathbf{r}})$$

Now by carrying out the differentiation in the usual manner, it follows that

$$(\mathbf{r} \times m\ddot{\mathbf{r}}) = \frac{d}{dt}(\mathbf{r} \times m\dot{\mathbf{r}})$$

since $(\dot{\mathbf{r}} \times m\dot{\mathbf{r}}) = 0$ or the cross product of a vector and itself is zero. Hence from the above equations

$$\boldsymbol{\tau} = \frac{d\mathbf{J}}{dt} = \dot{\mathbf{j}} \qquad (7.4)$$

or the torque about any fixed point O as origin is equal to the time rate of change of the angular momentum of the particle about O.

Let us now consider the rotation of a rigid body relative to some fixed point which is produced by a torque. Let the mass m in Fig. 7.1 be one of the particles of the body, as for instance the ith, having a mass m_i at a distance $\mathbf{r}_i$ from the fixed point O. In other words, we are considering the particle m to be a typical particle m_i of the body. The particle m_i has an external force $\mathbf{F}_i$ and also an internal force acting on it. Both of these exert torques about the fixed point O. The sum of the moments of all the external forces acting on all the particles

---

vector quantities namely **polar** ones such as displacement $\mathbf{r}$, force $\mathbf{F}$ etc., and those associated with cross products such as torque $\boldsymbol{\tau}$, angular momentum $\mathbf{J}$, etc., called **axial** or **pseudovectors**. So long as one stays in one system of axes, right- or left-handed, there is no essential distinction between these two kinds of vectors. However, under an inversion from right-handed (a) to left-handed (b) system of axes, or vice versa (see the sketch), a polar vector changes sign but an axial vector does not. Thus for $\mathbf{C} = \mathbf{A} \times \mathbf{B}$, if the components such as A_x and B_y change sign, then their product $A_x B_y$ does not. Presumably a right-handed system of axes is generally used because there are more right-handed people in the world than left-handed ones. The inversion of axes is shown in the figure. As an exercise, show that $\mathbf{i} \times \mathbf{j} = \mathbf{k}$ etc., using the right-handed, R.H. rule while $\mathbf{i}' \times \mathbf{j}' = -\mathbf{k}$ (R.H.) $= \mathbf{k}'$ (L.H.) etc. for other components.

of the body is the total moment τ_0 of the applied forces, since we have as-sumed that the sum of all the internal forces acting on the particles com-posing a rigid body is zero. If the sum of the internal forces were not zero, then the body would spontaneously accelerate. In other words, the sum of the moments of all the internal forces acting on the particles about the fixed point O is zero. Hence the sum of all the moments of the external forces about O is

$$\tau_0 = \sum_1^n \mathbf{r}_i \times \mathbf{F}_i$$

Now

$$\mathbf{F}_i = m_i \ddot{\mathbf{r}}_i$$

Hence

$$\tau_0 = \sum_1^n (\mathbf{r}_i \times m_i \ddot{\mathbf{r}}_i)$$

$$= \sum_1^n \frac{d}{dt} (\mathbf{r}_i \times m_i \dot{\mathbf{r}}_i)$$

since

$$\dot{\mathbf{r}}_i \times \dot{\mathbf{r}}_i = 0$$

The total angular momentum $\mathbf{J}_0$ of the body about the fixed point O is defined as

$$\mathbf{J}_0 = \sum_1^n (\mathbf{r}_i \times m_i \dot{\mathbf{r}}_i)$$

Hence

$$\tau_0 = \frac{d\mathbf{J}_0}{dt} = \dot{\mathbf{J}}_0 \tag{7.5}$$

Thus the resultant external torque acting on a body about a fixed point is equal to the time rate of increase of angular momentum of the body about the fixed point. This principle applies equally well to either a group of par-ticles or a rigid body so long as the internal forces act along the lines joining them. If there are no external torques acting on a group of particles or on a rigid body, then from Eq. 7.5 it follows that the total angular momentum of the particles or the rigid body about any fixed point as origin is a constant quantity. Thus the angular momentum is conserved.

It will now be shown that a similar expression can be obtained for rotation about the center of mass of a body. Let P be the position of a particle of mass m_i in a body that is rotating about an axis through the point O fixed in an inertial system. In Fig. 7.3 the position of the center of mass of the body is at the point marked C.M., and the position of a particle m_i at P is given by the vector $\mathbf{r}_i$ relative to the fixed point O and by the vector $\mathbf{r}_i'$ relative to the moving center of mass of the body. If $\bar{\mathbf{r}}$ is the vector distance of the center of

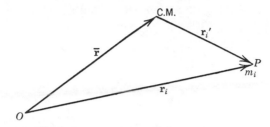

Fig. 7.3 Rotation of a body relative to its center of mass.

mass from the fixed point O, then at any instant of time

$$\mathbf{r}_i = \bar{\mathbf{r}} + \mathbf{r}_i'$$

The total angular momentum $\mathbf{J}_0$ of the body about the fixed point O is by definition

$$\mathbf{J}_0 = \sum_1^n (\mathbf{r}_i \times m_i \dot{\mathbf{r}}_i)$$

$$= \sum_1^n [(\bar{\mathbf{r}} + \mathbf{r}_i') \times m_i(\dot{\bar{\mathbf{r}}} + \dot{\mathbf{r}}_i')]$$

$$= (\bar{\mathbf{r}} \times M\dot{\bar{\mathbf{r}}}) + \bar{\mathbf{r}} \times \left(\sum_1^n m_i \dot{\mathbf{r}}_i'\right) + \left(\sum_1^n m_i \mathbf{r}_i'\right) \times \dot{\bar{\mathbf{r}}} + \sum_1^n (\mathbf{r}_i' \times m_i \dot{\mathbf{r}}_i')$$

where M is the total mass $\sum_1^n m_i$ of the body and $\bar{\mathbf{r}}$ is a constant. Since $\mathbf{r}_i'$ is measured relative to the center of mass of the body, it follows that $\sum_1^n m_i \mathbf{r}_i'$ must be zero for the whole body, and it also follows that

$$\sum_1^n m_i \dot{\mathbf{r}}_i' = \frac{d}{dt} \sum_1^n m_i \mathbf{r}_i' = 0$$

Thus the two middle terms in the expansion are zero and the total angular momentum $\mathbf{J}_0$ is

$$\mathbf{J}_0 = \bar{\mathbf{r}} \times M\dot{\bar{\mathbf{r}}} + \sum_1^n (\mathbf{r}_i' \times m_i \dot{\mathbf{r}}_i')$$

Now the total angular momentum of the body about its center of mass $\mathbf{J}_{\text{C.M.}}$ is by definition

$$\mathbf{J}_{\text{C.M.}} = \sum_1^n (\mathbf{r}_i' \times m_i \dot{\mathbf{r}}_i')$$

Hence

$$\mathbf{J}_0 = \bar{\mathbf{r}} \times M\dot{\bar{\mathbf{r}}} + \mathbf{J}_{\text{C.M.}} \qquad (7.6)$$

Let us now consider the total torque $\boldsymbol{\tau}_0$ exerted about O which is given by Eq. 7.1 as

$$\boldsymbol{\tau}_0 = \sum_1^n (\mathbf{r}_i \times \mathbf{F}_i)$$

$$= \bar{\mathbf{r}} \times \sum_1^n \mathbf{F}_i + \sum_1^n (\mathbf{r}_i' \times \mathbf{F}_i)$$

Now $\sum_1^n (\mathbf{r}_i' \times \mathbf{F}_i)$ is equal to the torque $\boldsymbol{\tau}_{\text{C.M.}}$ exerted by the external forces about the center of mass of the body so that

$$\boldsymbol{\tau}_0 = \bar{\mathbf{r}} \times \sum_1^n \mathbf{F}_i + \boldsymbol{\tau}_{\text{C.M.}} \tag{7.7}$$

Since the torque $\boldsymbol{\tau}_0$ and the total angular momentum $\mathbf{J}_0$ are taken with reference to the fixed origin O, we can apply Eq. 7.5, namely $\boldsymbol{\tau}_0 = \dot{\mathbf{J}}_0$. Substituting the values of $\mathbf{J}_0$ and $\boldsymbol{\tau}_0$ from Eqs. 7.6 and 7.7, we have

$$\bar{\mathbf{r}} \times \sum_1^n \mathbf{F}_i + \boldsymbol{\tau}_{\text{C.M.}} = \frac{d}{dt}(\bar{\mathbf{r}} \times M\dot{\bar{\mathbf{r}}}) + \dot{\mathbf{J}}_{\text{C.M.}}$$

$$= \bar{\mathbf{r}} \times M\ddot{\bar{\mathbf{r}}} + \dot{\mathbf{J}}_{\text{C.M.}}$$

since the vector product $\dot{\bar{\mathbf{r}}} \times \dot{\bar{\mathbf{r}}}$ of a vector and itself is zero. From Eq. 6.3 for the translational motion of the center of mass of the body it follows that

$$\sum_1^n \mathbf{F}_i = M\ddot{\bar{\mathbf{r}}} \quad \text{and} \quad \bar{\mathbf{r}} \times \sum_1^n \mathbf{F}_i = \bar{\mathbf{r}} \times M\ddot{\bar{\mathbf{r}}}$$

Hence by subtraction we have

$$\boldsymbol{\tau}_{\text{C.M.}} = \dot{\mathbf{J}}_{\text{C.M.}} \tag{7.8}$$

or *the torque about the center of mass of a body is equal to the time rate of change of the angular momentum about the center of mass.* This relationship between the torque and the rate of change of angular momentum holds whether these quantities are computed with reference to a point fixed in an inertial system or to the moving center of mass of the body.

We now have two important relationships that hold with respect to the center of mass of a body. The first of these, given in Eq. 6.3,

$$\mathbf{F} = M\ddot{\bar{\mathbf{r}}}$$

shows that the linear or translational acceleration of the center of mass of a body is the same as though all the mass of the body were concentrated at the center of mass and the resultant external force also acted at that point. The second relationship, given in Eq. 7.8, shows that the body turns about the center mass as though the center of mass were fixed and the same system of forces continued to act on the body as had been already acting. These two important principles regarding the translational and rotational motion of the center of mass can frequently be applied in solving problems, inasmuch as the translational and rotational motions about the center of mass of a body are independent of each other. We must now leave this subject and investigate the accelerations involved in a moving system of axes.

7.2   Velocity and Acceleration Relative to Moving Axes

In the discussion of Newton's laws of motion it was stated that these laws should be referred to a primary inertial system of axes such as those attached to the fixed stars. Since the earth is rotating on its axis and revolving about the sun, it follows that any system of axes attached to the earth cannot theoretically constitute an inertial system.

We shall now investigate what changes should be made in Newton's laws when they are they are referred to a system of moving axes such as the one attached to the earth. This investigation will be important in rigorously analyzing the motions of a spinning top or of a gyroscope.

Suppose that O_0 is the origin of a right-handed system of axes X_0, Y_0, Z_0 which is rigidly attached to a primary inertial system. Furthermore, let O be the origin of a system of axes X, Y, Z which is moving with respect to X_0, Y_0, Z_0, Fig. 7.4. The motion of the axes X, Y, Z may involve both that of translation and that of rotation with respect to the fixed axes X_0, Y_0, Z_0.

Suppose that a point P which is moving with respect to both sets of axes has an instantaneous position denoted by the vector $\mathbf{r}_0$ with respect to the fixed axes and by the vector $\mathbf{r}$ with respect to the moving axes. If $\mathbf{R}$ is the vector distance between the origins O_0O at the instant of time being considered, then from Fig. 7.4 it may be seen that

$$\mathbf{r}_0 = \mathbf{R} + \mathbf{r}$$

Suppose that the coordinates of P with respect to the fixed set of axes are x_0, y_0, z_0. Then

$$\mathbf{r}_0 = \mathbf{i}_0 x_0 + \mathbf{j}_0 y_0 + \mathbf{k}_0 z_0 \tag{7.9}$$

where $\mathbf{i}_0$, $\mathbf{j}_0$, $\mathbf{k}_0$ are the unit vectors along the X_0, Y_0, Z_0 axes respectively. These unit vectors are constant both in magnitude and direction. Similarly, if the coordinates of P with respect to the moving set of axes are x, y, z respectively, then

$$\mathbf{r} = \mathbf{i}x + \mathbf{j}y + \mathbf{k}z \tag{7.10}$$

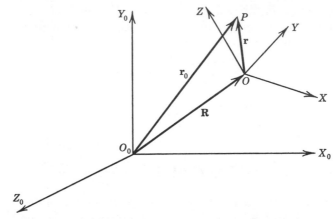

Fig. 7.4 A particle at P referred to fixed and moving axes.

where $\mathbf{i}$, $\mathbf{j}$, $\mathbf{k}$ are the unit vectors along the X, Y, Z axes respectively. These unit vectors have a constant magnitude, namely unity, but have the varying direction of the rotating axes.

The velocity of the point P relative to the fixed axes is $\dot{\mathbf{r}}_0$, and this is given by

$$\dot{\mathbf{r}}_0 = \dot{\mathbf{R}} + \dot{\mathbf{r}}$$

The velocity $\dot{\mathbf{r}}$ is

$$\dot{\mathbf{r}} = \frac{d}{dt}(\mathbf{i}x + \mathbf{j}y + \mathbf{k}z)$$

$$= \mathbf{i}\dot{x} + \mathbf{j}\dot{y} + \mathbf{k}\dot{z} + x\frac{d\mathbf{i}}{dt} + y\frac{d\mathbf{j}}{dt} + z\frac{d\mathbf{k}}{dt} \qquad (7.11)$$

In this expression

$$\mathbf{i}\dot{x} + \mathbf{j}\dot{y} + \mathbf{k}\dot{z} = \mathbf{v} \qquad (7.12)$$

is the apparent velocity of the point P relative to the moving axes, X, Y, Z. Thus, if the point P is at rest in the moving system, then its velocity $\mathbf{v}$, relative to the axes X, Y, Z would be zero. However, even in this case the point P has a velocity relative to the fixed axes X_0, Y_0, Z_0 due to the rotation of the axes X, Y, Z. This is given by the last three terms in Eq. 7.11.

To understand these terms, we must investigate the meaning of such terms as $d\mathbf{i}/dt$, the time rate of change of the unit vector along the X axis of the moving axes. Since the unit vector $\mathbf{i}$ has the constant magnitude of unity, any change in $\mathbf{i}$ must be perpendicular to $\mathbf{i}$, i.e., along the $\mathbf{j}$ and $\mathbf{k}$ axes. For pure rotation the rate of change with time of a vector $\mathbf{r}$ is given by Eq. 7.2a as $\dot{\mathbf{r}} = \boldsymbol{\omega} \times \mathbf{r}$ so that $d\mathbf{i}/dt = \dot{\mathbf{i}} = \boldsymbol{\omega} \times \mathbf{i}$. Hence

$$\frac{d\mathbf{i}}{dt} = (\mathbf{i}\omega_x + \mathbf{j}\omega_y + \mathbf{k}\omega_z) \times \mathbf{i} = \mathbf{j}\omega_z - \mathbf{k}\omega_y \qquad (7.13)$$

Similarly it may be proved that

$$\frac{dj}{dt} = \mathbf{k}\omega_x - \mathbf{i}\omega_z$$

and

$$\frac{dk}{dt} = \mathbf{i}\omega_y - \mathbf{j}\omega_x$$

These latter two equations may be obtained from Eq. 7.13 by cyclically rotating each of the items in the equation.

If the resultant angular velocity of rotation of the X, Y, Z axes relative to the X_0, Y_0, Z_0 axes is $\boldsymbol{\omega}$ with the components ω_x, ω_y, ω_z along the X, Y, Z axes respectively as given above, then

$$\boldsymbol{\omega} = \mathbf{i}\omega_x + \mathbf{j}\omega_y + \mathbf{k}\omega_z \tag{7.14}$$

As an exercise in manipulating vectors, it can be readily proved from the above relationships that

$$\frac{di}{dt}x + \frac{dj}{dt}y + \frac{dk}{dt}z = \boldsymbol{\omega} \times \mathbf{r} \tag{7.15}$$

Hence we may now write Eq. 7.11 as

$$\dot{\mathbf{r}} = \boldsymbol{\omega} \times \mathbf{r} + \mathbf{v} \tag{7.16}$$

and

$$\dot{\mathbf{r}}_0 = \dot{\mathbf{R}} + \boldsymbol{\omega} \times \mathbf{r} + \mathbf{v} \tag{7.17}$$

In words, these terms signify that:

$\dot{\mathbf{R}}$ is the velocity of the origin O of the moving axes relative to the origin O_0 of the fixed axes.

$\dot{\mathbf{r}}_0$ is the true velocity of the moving point P as measured relative to the fixed axes X_0, Y_0, Z_0.

$\dot{\mathbf{r}}$ is the velocity of the point P as would be measured by an observer who is located at the origin of the moving axes X, Y, Z and who is not rotating with the axes.

$\boldsymbol{\omega} \times \mathbf{r}$ is the velocity of the point P which is not moving relative to the axes X, Y, Z but whose motion is due to the rotation of these axes.

$\mathbf{v}$ is the velocity of the point P relative to the axes X, Y, Z or as recorded by an observer who is attached to, and rotating with, the moving axes X, Y, Z.

Thus the velocity of the point P which is given by the one term $\dot{\mathbf{r}}_0$ relative to the origin of the fixed axes is also given by three terms relative to the moving axes. These three terms are the motion of translation $\dot{\mathbf{R}}$ of the origin of the fixed set of axes relative to the origin of the moving set, $\boldsymbol{\omega} \times \mathbf{r}$ due to the rotation of the axes, and $\mathbf{v}$ due to the motion of the point P relative to the axes X, Y, Z which are now considered as stationary.

We next turn to the linear and angular accelerations referred to the fixed and moving axes. From Eq. 7.17, by differentiation, the acceleration of the point P is given by

$$\ddot{\mathbf{r}}_0 = \ddot{\mathbf{R}} + \dot{\boldsymbol{\omega}} \times \mathbf{r} + \boldsymbol{\omega} \times \dot{\mathbf{r}} + \dot{\mathbf{v}} \qquad (7.18)$$

Substituting for $\dot{\mathbf{r}}$ from Eq. 7.16 gives

$$\ddot{\mathbf{r}}_0 = \ddot{\mathbf{R}} + \dot{\boldsymbol{\omega}} \times \mathbf{r} + \boldsymbol{\omega} \times (\boldsymbol{\omega} \times \mathbf{r}) + \boldsymbol{\omega} \times \mathbf{v} + \dot{\mathbf{v}} \qquad (7.19)$$

Differentiating Eq. 7.12, we obtain for $\dot{\mathbf{v}}$:

$$\dot{\mathbf{v}} = \mathbf{i}\ddot{x} + \mathbf{j}\ddot{y} + \mathbf{k}\ddot{z} + \dot{x}\dot{\mathbf{i}} + \dot{y}\dot{\mathbf{j}} + \dot{z}\dot{\mathbf{k}} \qquad (7.20)$$

in which $\dot{\mathbf{i}}$ is used in place of $d\mathbf{i}/dt$, etc. Substituting from Eq. 7.13 for $\dot{\mathbf{i}}, \dot{\mathbf{j}}, \dot{\mathbf{k}}$ and rearranging gives

$$\dot{\mathbf{i}}\dot{x} + \dot{\mathbf{j}}\dot{y} + \dot{\mathbf{k}}\dot{z} = \mathbf{i}(\omega_y \dot{z} - \omega_z \dot{y}) + \mathbf{j}(\omega_z \dot{x} - \omega_x \dot{z}) + \mathbf{k}(\omega_x \dot{y} - \omega_y \dot{x})$$

$$= \begin{vmatrix} \mathbf{i} & \mathbf{j} & \mathbf{k} \\ \omega_x & \omega_y & \omega_z \\ \dot{x} & \dot{y} & \dot{z} \end{vmatrix} = \boldsymbol{\omega} \times \mathbf{v} \qquad (7.21)$$

The first three terms in Eq. 7.20 represent the apparent linear acceleration $\mathbf{a}$ of the point P relative to the moving axes X, Y, Z or

$$\mathbf{a} = \mathbf{i}\ddot{x} + \mathbf{j}\ddot{y} + \mathbf{k}\ddot{z}$$

Thus Eq. 7.20 may be written

$$\dot{\mathbf{v}} = \mathbf{a} + \boldsymbol{\omega} \times \mathbf{v}$$

It is left as an exercise to prove that

$$\dot{\boldsymbol{\omega}} = \mathbf{i}\dot{\omega}_x + \mathbf{j}\dot{\omega}_y + \mathbf{k}\dot{\omega}_z$$

Thus the term $\dot{\boldsymbol{\omega}}$ may be considered to be the angular acceleration of the moving axes relative to the fixed axes X_0, Y_0, Z_0. Collecting the terms, we then have for the acceleration of the point P from Eq. 7.18

$$\mathbf{a}_0 = \ddot{\mathbf{r}}_0 = \ddot{\mathbf{R}} + \mathbf{a} + 2\boldsymbol{\omega} \times \mathbf{v} + \dot{\boldsymbol{\omega}} \times \mathbf{r} + \boldsymbol{\omega} \times (\boldsymbol{\omega} \times \mathbf{r}) \qquad (7.22)$$

As a summary we shall give the physical meaning of the terms in Eq. 7.22.

$\mathbf{a}_0$ or $\ddot{\mathbf{r}}_0$: The acceleration of the point P relative to the fixed axes X_0, Y_0, Z_0 in the inertial system.

$\ddot{\mathbf{R}}$: The acceleration of the origin O of the moving axes X, Y, Z relative to the origin O_0 of the fixed axes X_0, Y_0, Z_0.

$\mathbf{a}$: The acceleration of the point P relative to the moving axes X, Y, Z.

$2\boldsymbol{\omega} \times \mathbf{v}$: The so-called *Coriolis acceleration*. It is the acceleration received by a point P moving relative to the axes which are rotating with the angular velocity $\boldsymbol{\omega}$.

$\dot{\boldsymbol{\omega}} \times \mathbf{r}$: The linear acceleration of a point P, at a distance $\mathbf{r}$ from the origin of the moving axes, due to the angular acceleration $\dot{\boldsymbol{\omega}}$ of these axes.

$\boldsymbol{\omega} \times (\boldsymbol{\omega} \times \mathbf{r})$: The centripetal acceleration of the point P due to the rotation of the axes X, Y, Z. It is readily shown that this term is equal to $\omega^2 p$ where p is the perpendicular distance from the point P, or the head of the vector $\mathbf{r}$, to the angular velocity vector $\boldsymbol{\omega}$.

7.3   Motion of a Particle Falling toward the Earth

We have repeatedly emphasized that the earth is not an inertial system, though up to the present we have assumed that Newton's laws of motion are valid for a coordinate system attached to the earth. We shall now investigate how far we have been in error in making this assumption.

Let the center of the earth be the origin of a rotating set of axes rigidly attached to the earth. Relative to the origin O_0 of a set of axes X_0, Y_0, Z_0 in an inertial system, the acceleration of the center of the earth O is negligibly small. Thus we may set $\ddot{\mathbf{R}}$ in Eq. 7.22 equal to zero. The angular velocity $\boldsymbol{\omega}$ of the earth is easily calculated to be about 7.3×10^{-5} radian/sec. Since this angular velocity $\boldsymbol{\omega}$ is constant, it follows that its angular acceleration $\dot{\boldsymbol{\omega}}$ is zero.

Thus from Eq. 7.22 the acceleration of a point P relative to the axes attached to the rotating earth is

$$\mathbf{a} = \mathbf{a}_0 - 2\boldsymbol{\omega} \times \mathbf{v} - \boldsymbol{\omega} \times (\boldsymbol{\omega} \times \mathbf{r}) \qquad (7.23)$$

Consider a particle falling toward the earth under gravity with the acceleration $\mathbf{g}$. Now when $\mathbf{g}$ is measured, as for example with a pendulum having a negligible velocity relative to the earth, its value is the resultant of the earth's gravitational attraction $\mathbf{a}_0$ and the centrifugal acceleration $\boldsymbol{\omega} \times (\boldsymbol{\omega} \times \mathbf{r})$. This effect was investigated in problem 19 of Chapter 1. Thus the value of the acceleration due to gravity is

$$\mathbf{g} = \mathbf{a}_0 - \boldsymbol{\omega} \times (\boldsymbol{\omega} \times \mathbf{r}) \qquad (7.24)$$

From Eqs. 7.23 and 7.24 it follows that the acceleration $\mathbf{a}$ of a point moving with a velocity $\mathbf{v}$ on the earth is

$$\mathbf{a} = \mathbf{g} - 2\boldsymbol{\omega} \times \mathbf{v} \qquad (7.25)$$

This acceleration is relative to the rotating axes X, Y, Z attached to the earth. Since the position of the point P on the earth, given by the vector $\mathbf{r}$, is not contained in Eq. 7.25, it follows that the origin of the rotating axes may be placed anywhere on the earth and not only at the center of the earth. Let us take the origin of the rotating axes at some point O on the earth at latitude θ,

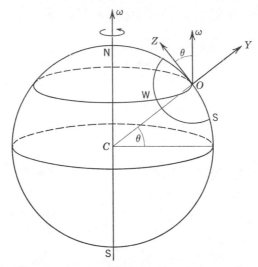

Fig. 7.5 Motion relative to the moving earth.

Fig. 7.5. Suppose that the Y axis is drawn vertically upward and the Z axis on the horizon circle through O pointing north. Then for a right-handed system of axes, the X axis is directed perpendicularly out of the plane of the paper towards the west on the horizon circle.

Now suppose that a particle falls from rest from a point on the Y axis near the surface of the earth. Its velocity v after a time t is given approximately by

$$v = gt$$

and is directed toward the center of the earth approximately along the $-Y$ axis. For this particle the vector $-2\boldsymbol{\omega} \times \mathbf{v}$ is directed into the plane of the paper, that is, along the $-X$ axis or in an easterly direction on the horizon circle through O. From Fig. 7.5 it can be seen that the magnitude of this vector $-2\boldsymbol{\omega} \times \mathbf{v}$ is

$$2v\omega \cos \theta = 2gt\omega \cos \theta \qquad (7.26)$$

This then is the easterly acceleration of a particle dropped from rest above the surface of the earth. It signifies that a particle does not fall vertically down toward the earth or along a plumb line but falls east of this line. The direction of this easterly acceleration, given in Eq. 7.26, is, by our convention, along the $-X$ axis. Hence we may write

$$-\ddot{x} = 2gtw \cos \theta$$

Since at time zero both the velocity $\dot{x}$ and the displacement x are zero, it follows that

$$-x = \frac{\omega g t^3 \cos \theta}{3}$$

This $-x$ is the easterly displacement of a particle starting from rest and falling for a time t under the constant gravitational acceleration g. It is the easterly displacement from the foot of a plumb line hung from the point where the particle is initially started. In this example it should be noticed that angular momentum is conserved.

7.4   Other Examples of Coriolis Forces

Horizontal deflection of moving air.   If you have ever observed a weather map, you have probably noticed that the winds are not from the high- to the low-pressure regions but rather along the lines of equal pressure. This result is given in some meteorological books as Buys Ballots' law, which states that, if a person stands with his back to the wind, the region of lowest barometric pressure is on his left hand and slightly towards his front. An explanation of this law is given in terms of the *Coriolis* acceleration.

Consider a westerly wind, that is, a horizontal mass of air moving from west to east with a velocity **v** (Fig. 7.6). In this horizon circle the effective angular rotation, from Fig. 7.5, is $\omega \sin \theta$, directed vertically upward out of the plane of the paper. The Coriolis acceleration $-2\boldsymbol{\omega} \times \mathbf{v}$ has a magnitude of $2\omega v \sin \theta$ and a direction toward the south in the northern hemisphere.

Thus a mass of air moving toward the east is deflected toward the south. It is easy to show from what has been given above that a low-pressure area gives rise to cyclones whose winds move in a counterclockwise direction, Fig. 7.7. In the southern hemisphere the winds would be in the opposite direction. A low-pressure area would give rise to winds in a clockwise direction.

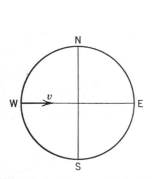

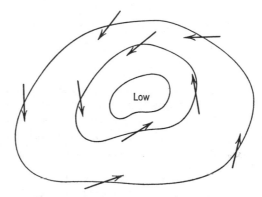

Fig. 7.6   Motion on a horizontal circle.

Fig. 7.7   The winds and curves of equal pressure in a cyclone.

Swirling of water going down a drain. Although there is a Coriolis force acting on the water in a sink or bathtub when the stopper is removed, this force is very small indeed and does not account for the swirling of the water around the exit hole. That a Coriolis force does act on the water has been shown by a careful experiment* and the magnitude of the acceleration of water flowing at the rate of 0.25 cm per sec is about 3×10^{-5} cm per sec^2 as may be readily shown.

Coriolis (1792–1843) is largely remembered for the explanation of the acceleration which bears his name and this was given in a paper of a dozen pages.† He also published a book on machines in which he gave the now accepted definition of work as equal to the product of force and distance.

7.5   Motion of Rigid Bodies Relative to a Fixed Point

The analogy between the equations for rotational and translational motion holds only when the rotation is about a fixed axis. For the translational motion of the center of mass of a rigid body, the resultant force is equal to the product of the mass and the linear acceleration of the center of mass. Similarly, for the rotational motion of a rigid body about a *fixed axis*, the resultant torque is equal to the product of the moment of inertia about the axis and the angular acceleration of the body about the axis. In these cases, mass and moment of inertia are both constants. The accelerations are then respectively proportional to the resultant force or the resultant torque.

Now, if the rotational motion is given with reference to a fixed point rather than with reference to a fixed axis, no simple relationship exists between the torque and the angular acceleration. This may seem incorrect since the torque referred to a fixed point is equal to the time rate of change of the angular momentum of the body about the point as given in Eq. 7.4. Although this is a simple relationship, it yields no simple result, for the moment of inertia is a tensor and, in general, is not a constant quantity independent of the coordinates. In general, for rotation about a fixed point, it is not possible to have three angles whose time derivatives give the angular velocities around the axes about which these angles are measured. These difficulties would appear to make the solution of problems pertaining to rotation about a fixed point almost impossible. However, there are simplifications which considerably overcome these difficulties.

* A. H. Shapiro, *Nature*, December 15, 1962.
† "Mémoire sur les équations du mouvement relatif des systèmes de corps," G. G. de Coriolis, *Journal de l'Ecole Polytechnique*, **15**, 142–154, 1835.

7.6   Moment of Inertia Tensor

Since the torque τ exerted on a body rotating about a fixed point is equal to the time rate of change of angular momentum $\mathbf{J}$, we have from Eq. 7.4

$$\tau = \dot{\mathbf{J}}$$

where from Eq. 7.3 for the ith particle in the body, Fig. 7.1,

$$\mathbf{J} = \sum_{1}^{n} m_i(\mathbf{r}_i \times \dot{\mathbf{r}}_i)$$

$$= \sum_{1}^{n} m_i \mathbf{r}_i \times (\boldsymbol{\omega} \times \mathbf{r}_i) \qquad (7.27)$$

In this double-vector product it is necessary first to evaluate the vector product $\boldsymbol{\omega} \times \mathbf{r}_i$. The vector product of $\mathbf{r}_i$ and the vector $\boldsymbol{\omega} \times \mathbf{r}_i$ is then found. As an exercise in algebraic manipulation, expand Eq. 7.27, using the determinant form for the vector product as in Eq. 7.2 and also the following equations.

$$\mathbf{r}_i = \mathbf{i}x_i + \mathbf{j}y_i + \mathbf{k}z_i$$

$$\boldsymbol{\omega} = \mathbf{i}\omega_x + \mathbf{j}\omega_y + \mathbf{k}\omega_z$$

For the component of $\mathbf{r}_i \times (\boldsymbol{\omega} \times \mathbf{r}_i)$ along the X axis, find that

$$\mathbf{i}J_x = \mathbf{i}\sum_{1}^{n} [\omega_x(y_i^2 + z_i^2) - \omega_y x_i y_i - \omega_z x_i z_i]m_i$$

The components along the Y and Z axes may be written from the above by cyclically changing the coordinates. Thus the angular momentum about a fixed point is

$$\mathbf{J} = \mathbf{i}\left[\omega_x \sum_{1}^{n} m_i(y_i^2 + z_i^2) - \omega_y \sum_{1}^{n} m_i x_i y_i - \omega_z \sum_{1}^{n} m_i x_i z_i\right]$$

$$+ \mathbf{j}\left[\omega_y \sum_{1}^{n} m_i(z_i^2 + x_i^2) - \omega_z \sum_{1}^{n} m_i y_i z_i - \omega_x \sum_{1}^{n} m_i y_i x_i\right]$$

$$+ \mathbf{k}\left[\omega_z \sum_{1}^{n} m_i(x_i^2 + y_i^2) - \omega_x \sum_{1}^{n} m_i z_i x_i - \omega_y \sum_{1}^{n} m_i z_i y_i\right]$$

In this expansion it is usual to abbreviate the terms in the above equation in the following manner:

$$I_x = \sum_1^n m_i(y_i^2 + z_i^2) \qquad I_{xy} = -\sum_1^n m_i x_i y_i \qquad I_{xz} = -\sum_1^n m_i x_i z_i$$

$$I_y = \sum_1^n m_i(z_i^2 + x_i^2) \qquad I_{yz} = -\sum_1^n m_i y_i z_i \qquad I_{yx} = -\sum_1^n m_i y_i x_i$$

$$I_z = \sum_1^n m_i(x_i^2 + y_i^2) \qquad I_{zx} = -\sum_1^n m_i z_i x_i \qquad I_{zy} = -\sum_1^n m_i z_i y_i$$

These nine quantities I_x, I_{xy}, ... , I_z are the components of the moment of inertia of the body about the fixed X, Y, Z axes. The components I_x, I_y, I_z are usually called the moments of inertia, and the components I_{xy}, I_{yz}, ... , I_{zx} are called the products of inertia. A quantity such as the moment of inertia is called a tensor of second rank. A vector can be considered as a tensor of first rank and a scalar as a tensor of zero rank.

Now it is possible to prove, although we shall not do so here, that any rigid body has three mutually perpendicular principal axes of inertia of such a nature that the products of inertia are zero when the principal axes are used as coordinate axes. This result is of considerable importance. In the following discussion we shall use the principal axes as the coordinate axes. For such axes the expression for the angular momentum becomes

$$\mathbf{J} = \mathbf{i}I_x\omega_x + \mathbf{j}I_y\omega_y + \mathbf{k}I_z\omega_z \qquad (7.28)$$

Notice that when the principal axes are used as coordinate axes the products of inertia vanish, and the tensor has only three terms, corresponding to the mathematical operation of diagonalizing the tensor. If we denote the moment of inertia tensor by ι then we can consider ι as an operator which, when it operates on the angular velocity vector $\boldsymbol{\omega}$, produces the angular momentum vector $\mathbf{J}$, that is,

$$\mathbf{J} = \iota\boldsymbol{\omega}$$

For axes in any general direction the tensor ι has nine elements (actually six different ones since $I_{xy} = I_{yx}$, etc.) whereas when the coordinate axes are along the principal axes the tensor has three non-zero elements as given in Eq. 7.28, and $J_x = I_x\omega_x$, etc. That $\mathbf{J}$ and $\boldsymbol{\omega}$ do not necessarily have the same magnitude and direction may be seen from the following example.

Suppose a thin uniform rod AB of mass M and length L is made to rotate with angular velocity $\boldsymbol{\omega}$, such that the rod makes an angle θ with the direction of $\boldsymbol{\omega}$ (Fig. 7.8). Choosing axes, fixed to the rod, as shown in the figure with the x axis along the rod and the z axis perpendicularly out of the page, it is seen that $I_x = 0$, $I_y = I_z = ML^2/12$. Since the direction of $\boldsymbol{\omega}$ is in the xy plane, it follows that

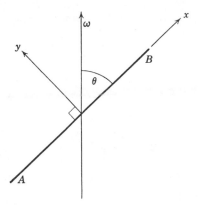

Fig. 7.8 Rod AB rotating about vertical axis with angular velocity ω.

$\omega_z = 0$, $\omega_x = \omega \cos \theta$ and $\omega_y = \omega \sin \theta$. Thus from Eq. 7.28

$$\mathbf{J} = \mathbf{j}J_y = \mathbf{j}I_y\omega_y = \frac{\mathbf{j}ML^2\omega \sin \theta}{12}$$

while

$$\boldsymbol{\omega} = \mathbf{i}\omega \cos \theta + \mathbf{j}\omega \sin \theta$$

Thus the angular momentum $\mathbf{J}$ and the angular velocity $\boldsymbol{\omega}$ are in different directions. Since in this case $\mathbf{J}$ is fixed relative to the rod, its direction is changing with time so there must be a torque acting, given by $\boldsymbol{\tau} = \dot{\mathbf{J}} = \boldsymbol{\omega} \times \mathbf{J}$. This torque is exerted by the bearings in which the rod is held, and such a rotating rod is not dynamically balanced.

The kinetic energy of the rotating body may be written

$$T = \tfrac{1}{2} \sum_1^n m_i \dot{r}_i^{\,2} = \tfrac{1}{2} \sum_1^n m_i (\boldsymbol{\omega} \times \mathbf{r}_i) \cdot (\boldsymbol{\omega} \times \mathbf{r}_i)$$

Using the principal axes and expanding, it follows that the kinetic energy is

$$T = \tfrac{1}{2}(I_x\omega_x^{\,2} + I_y\omega_y^{\,2} + I_z\omega_z^{\,2}) \qquad (7.29)$$

7.7 Equations of Motion of a Rotating Rigid Body Referred to a Fixed Point

We shall now find the expressions for the components of the torque acting on a rigid body rotating about a fixed point in terms of the corresponding components of angular momentum. To do this, let us suppose that the co-ordinate axes are the principal axes of the body, and that these coordinate axes are rigidly attached to the body. The coordinate axes then rotate with the body. Thus the moments of inertia about each of the axes are constants. From Eq. 7.4, $\tau = \dot{J}$, and, using Eq. 7.28 for J, it follows that

$$\tau = iI_x\dot{\omega}_x + I_x\omega_x\frac{di}{dt} + jI_y\dot{\omega}_y + I_y\omega_y\frac{dj}{dt} + kI_z\dot{\omega}_z + I_z\omega_z\frac{dk}{dt}$$

where $\dot{\omega}_x$ is the time rate of change of the angular velocity of the X axis and di/dt is the time rate of change of the unit vector i along the rotating X axis. Substituting for di/dt, dj/dt, dk/dt the values given in Eq. 7.13, it follows that

$$\tau_x = J_x = (I_x\dot{\omega}_x - I_y\omega_y\omega_z + I_z\omega_y\omega_z)$$
$$= [I_x\dot{\omega}_x + \omega_y\omega_z(I_z - I_y)]$$

and

$$\tau_y = J_y = [I_y\dot{\omega}_y + \omega_z\omega_x(I_x - I_z)]$$
$$\tau_z = J_z = [I_z\dot{\omega}_z + \omega_x\omega_y(I_y - I_x)]$$

(7.30)

The three equations given above for τ_x, τ_y, τ_z are known as Euler's equations. They relate the torque about each one of the principal axes to the angular velocity of the principal axes. Since the principal axes are fixed in the rotating body, it follows that the moments of inertia I_x, I_y, I_z are constants independent of time. We shall now apply this theory to the gyroscope.

7.8   The Motion of a Symmetrical Rigid Body. The Gyroscope

In this section we shall analyze the motion of a gyroscope a little more rigorously than was done in section 6.20. This theory applies to the motion of any rotating symmetrical body. By a symmetrical body is meant one that has two of its principal moments of inertia equal to each other. These are generally perpendicular to a third axis of rotation of the body. The gyroscope, the earth, and certain types of atomic and molecular structure are examples of such symmetrically rotating bodies.

Let us consider the gyroscopic motion of a symmetrical spinning top. To describe this motion we shall choose the principal axes of the top as the rotating coordinate axes, X, Y, Z. These rotating axes have a common origin O with a set of axes X_0, Y_0, Z_0 which are fixed in an inertial system. The common origin of the axes is taken as the fixed point of the top. In Fig. 7.9 the Y axis is the spin axis of the top about the origin O of the coordinate systems. Any motion of the top, other than its spin, is measured in terms of the angles θ and ϕ. These angles are called the Eulerian angles and serve as independent variables in the equations of motion.

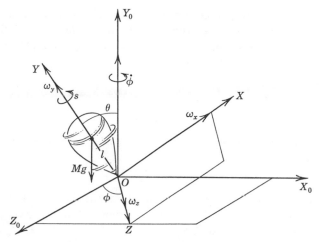

Fig. 7.9   The motions of a spinning top.

Let us first describe the motions of a spinning top. The top is spinning about the Y axis, which is rotating about the fixed Y_0 axis. In this motion the Y axis of the top describes a cone about the vertical Y_0 axis. This motion is known as the *precessional motion* and is measured by the time rate of change of the angle ϕ. In general the axis of the top changes its cone angle θ as the precession takes place. The angular velocity of θ is called *nutation*. Thus in general the top may be said to have three motions, spin, precession, and nutation.

The Eulerian angle ϕ is constructed in the following manner. The rotating OZ axis is perpendicular to the spin axis OY and also lies in the fixed $X_0 Z_0$ plane. Thus the precessional angle ϕ is the angle between the fixed OZ_0 axis and the rotating OZ axis. To complete the system of rotating axes, the OX axis is chosen perpendicular to the OY and OZ axes. From the nature of the construction it follows that OZ is perpendicular at every instant of time to the axes OY, OY_0, and OX. Thus the three axes OY, OY_0, and OX lie in the same plane.

We must try now to relate the Eulerian angles θ and ϕ to the components of the angular velocity of the rotating axes X, Y, Z. This we must do if we are to obtain the differential equations of motion of the top from Eq. 7.30. Let $\boldsymbol{\omega}$ be the vector angular velocity of rotation of the axes X, Y, Z with respect to the fixed axes X_0, Y_0, Z_0. The components of $\boldsymbol{\omega}$ along the axes X, Y, Z are ω_x, ω_y, and ω_z respectively.

Since OZ is perpendicular to OY and OY_0, it follows that the component ω_z of the angular velocity about the OZ axis is measured by the time rate of change of the angle θ between the OY and OY_0 axis, or

$$\omega_z = \dot{\theta} \tag{7.31}$$

Again, since OY_0 is perpendicular to OZ_0 and OZ, it can be seen from Fig. 7.9 that the rate of change of ϕ, the angle between the OZ_0 and OZ axes, is along the fixed OY_0 axis. This angular velocity $\dot{\phi}$ about the OY_0 axis can be resolved into the components ω_x and ω_y along the rotating OX and OY axes. From Fig. 7.9 it may be seen that

$$\omega_y = \dot{\phi} \cos \theta$$
$$\omega_x = \dot{\phi} \sin \theta$$

(7.31)

From the symmetry of the top the moments of inertia about the X and Z axes are equal, i.e.,

$$I_x = I_z$$

Also the total angular velocity of the top about the Y axis is the sum of the spin angular velocity s and the angular velocity of the Y axis. This total angular velocity will be called S.

$$\mathbf{S} = \boldsymbol{\omega}_y + \mathbf{s}$$

Now the only torque acting on the top is that due to its weight, and this acts about the Z axis.

$$\tau_z = Mgl \sin \theta$$

where l is the distance of the center of mass of the top from the point O. Also

$$\tau_x = \tau_y = 0$$

The equations of motion of the top are obtained from Eq. 7.30, but now the component of angular momentum along the Y axis is $I_y(\omega_y + s)$ rather than $I_y\omega_y$ used in Eq. 7.30. Hence, replacing I_z by I_x we have

$$I_x\dot{\omega}_x + \omega_y\omega_z(I_x - I_y) - I_ys\omega_z = 0$$
$$I_y(\dot{\omega}_y + \dot{s}) + \omega_z\omega_x(I_x - I_x) = 0$$
$$I_x\dot{\omega}_z + \omega_x\omega_y(I_y - I_x) + I_ys\omega_x = Mgl \sin \theta$$

(7.32)

From the middle equation above we have

$$I_y(\dot{\omega}_y + \dot{s}) = 0 \qquad I_y\dot{S} = 0 \qquad \text{or} \qquad I_yS = \text{Constant}$$

Since I_y is a constant, it follows that S, the total angular velocity about the Y axis, is a constant.

Next we turn to the total energy U of the top. This is the sum of its kinetic and potential energies. Using Eq. 7.29 and noting that the angular velocity about the Y axis is $\omega_y + s$ rather than ω_y, the total energy is

$$U = \tfrac{1}{2}I_x\omega_x^2 + \tfrac{1}{2}I_y(\omega_y + s)^2 + \tfrac{1}{2}I_z\omega_z^2 + Mgl \cos \theta$$

Substituting for ω_x, ω_y, ω_z from Eq. 7.31, we have

$$U = \tfrac{1}{2}I_x\dot{\phi}^2 \sin^2 \theta + \tfrac{1}{2}I_yS^2 + \tfrac{1}{2}I_x\dot{\theta}^2 + Mgl \cos \theta$$

We now require a further relationship between the angles θ and ϕ. This may be obtained by recognizing that there are no torques about the OY_0 axis and, hence, the component of angular momentum about this axis is a constant C.

$$\tau_{Y_0} = I_y S \cos \theta + I_x \omega_x \sin \theta = C$$

or

$$C = I_y S \cos \theta + I_x \dot{\phi} \sin^2 \theta \qquad (7.33)$$

Thus

$$\dot{\phi} = \frac{C - I_y S \cos \theta}{I_x \sin^2 \theta}$$

Substituting this value of $\dot{\phi}$ in the equation for the energy U given above, we obtain a relationship involving only θ and its derivatives. Thus

$$\dot{\theta}^2 = \frac{2}{I_x} \left[U - \frac{(C - I_y S \cos \theta)^2}{2 I_x \sin^2 \theta} - \frac{I_y S^2}{2} - Mgl \cos \theta \right] \qquad (7.34)$$

Theoretically at least we should be able to determine the angular velocities of precession and nutation from Eqs. 7.34 and 7.33 respectively. However, Eq. 7.34 is a difficult one to integrate and would take us much deeper into the analysis of gyroscopes than we wish to go. To avoid this, we shall introduce a new variable p where

$$p = \cos \theta$$

and

$$\dot{p} = -\sin \theta \dot{\theta}$$

It follows from the definition of θ that p must lie between -1 and $+1$. Substituting for θ and $\dot{\theta}$ in Eq. 7.34 gives

$$\dot{p}^2 = \frac{2U}{I_x}(1 - p^2) - \frac{(C - I_y Sp)^2}{I_x^2} - \frac{I_y S^2(1 - p^2)}{I_x} - \frac{2Mglp(1 - p^2)}{I_x}$$

$$= (1 - p^2)\left(\frac{2U - I_y S^2}{I_x} - \frac{2Mglp}{I_x}\right) - \frac{(C - I_y Sp)^2}{I_x^2}$$

$$= (1 - p^2)(A - Bp) - (D - Ep)^2 \qquad (7.35)$$

where A, B, D, E are the constants from the equation above. Equation 7.35 is a cubic and can vanish for no more than three values of p. The limits of the variable p are $+1$ and -1, and near these limits $\dot{p}^2$ is negative since

$$\dot{p}^2_{(p=1)} = -(D - E)^2 \qquad \text{and} \qquad \dot{p}^2_{(p=-1)} = -(D + E)^2$$

Thus the graph of $\dot{p}^2$ plotted against p must cross the p axis between -1 and $+1$, as shown in Fig. 7.10. In the physically possible motion the cone angle θ

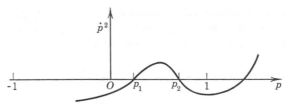

Fig. 7.10   Curve showing the variation of $\dot{p}^2$ with p, where $p = \cos \theta$.

must lie between 0 and $\pi/2$ so that p must lie between 1 and 0. That is, the curve in Fig. 7.10 must cross the p axis between 1 and 0 rather than between -1 and $+1$. These values at which $\dot{p}^2$ is zero are called p_1, p_2. They correspond to two values θ_1 and θ_2 of the cone angle and are called the *libration* limits. Notice that, when Bp is very large compared to any of the constants, then Eq. 7.35 becomes $\dot{p}^2 = Bp^3$, so that when Bp has a very large negative value $\dot{p}^2$ is negative, and when Bp has a very large positive value $\dot{p}^2$ is positive. It is from these considerations that Fig. 7.10 was qualitatively drawn. Since $p = \cos \theta$, it follows that the real or physically possible values of p must lie between $+1$ and -1. The large positive or negative values of p do not correspond to real angles.

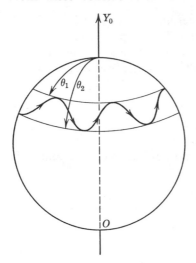

This analysis shows that the cone angle θ varies between the libration limits θ_1 and θ_2 as the precessional motion takes place. In this motion both the angular velocities $\dot{\theta}$ and $\dot{\phi}$ vary with time. The expression for these variations is given in treatises exclusively devoted to this subject.

A possible type of motion for the axis of the top is shown in Fig. 7.11. Here the axis of the top moves between the libration limits θ_1 and θ_2 at the same time that the precessional motion about the fixed vertical axis OY_0 is taking place.

Fig. 7.11   Nutation of a top between the libration limits θ_1 and θ_2.

7.9   The Condition for Precession without Nutation

In Section 6.20 we gave an approximate theory of the gyroscope. This applies to the special case where the spin angular momentum is very large compared with the precessional angular momentum and there is no nutational motion present. The top then precesses with its axis at a constant angle α with the fixed axis OY_0. In mathematical terms the conditions for this special case are

$$\theta = \alpha \qquad \dot{\theta} = 0$$

and from Eq. 7.31

$$\omega_z = \dot{\theta} = 0 \qquad \omega_y = \dot{\phi} \cos \alpha \qquad \omega_x = \dot{\phi} \sin \alpha$$

Applying these conditions to the equation of motion, Eq. 7.32, and setting $I_x = I_z$, we have for a top with a spin velocity s

$$I_x \dot{\omega}_x = 0$$

$$I_y(\dot{\omega}_y + \dot{s}) = 0$$

$$I_y(\omega_x \omega_y + s\omega_x) - I_x \omega_x \omega_y = Mgl \sin \alpha$$

These equations give

$$\dot{\omega}_y + \dot{s} = 0 \qquad \omega_y + s = \text{constant} = S$$

and

$$I_y \omega_x S - I_x \omega_x \omega_y = Mgl \sin \alpha$$

or substituting for ω_x and ω_y and dividing by $\sin \alpha$ gives

$$I_y S \dot{\phi} - I_x \dot{\phi}^2 \cos \alpha = Mgl$$

Solving for $\dot{\phi}$ gives

$$\dot{\phi} = \frac{I_y S \pm \sqrt{I_y^2 S - 4I_x \cos \alpha \, Mgl}}{2I_x \cos \alpha}$$

If $\dot{\phi}$ is to have real roots, i.e., the top precesses at a constant angle α without nutation, then

$$I_y^2 S^2 \geq 4I_x Mgl \cos \alpha \qquad (7.36)$$

There are two possible rates of precession given by the positive and negative values of the square root. The larger of these two rates, obtained by taking the positive sign for the square root, is difficult to produce physically. Its precessional angular velocity is approximately

$$\dot{\phi}_1 = \frac{I_y}{I_x \cos \alpha} \qquad (7.37)$$

and is independent of the torque but dependent on the angle of inclination of the top.

The smaller angular precessional velocity $\dot{\phi}_2$ may be obtained by expanding the square root using the condition that $I_y^2 S^2$ is very large compared to $4I_x Mgl \cos \alpha$. This gives

$$\dot{\phi}_2 = \frac{Mgl}{I_y S} \qquad (7.38)$$

This was investigated in Section 6.20. The angular velocity $\dot{\phi}_2$ given in Eq. 7.38 shows that, the smaller the spin velocity, the larger the precessional velocity. However, this does not hold for small values of the spin velocity since the conditions of the problem require a large spin velocity according to Eq. 7.36.

7.10   The Motion of a Top Started with a Spin at a Given Angle

In this section we shall investigate the motion of a top started with a given spin velocity s and with its axis making an angle α with the vertical. The initial conditions at zero time are

$$\theta = \alpha \qquad \dot{\theta} = 0 \qquad \dot{\phi} = 0$$

Thus at zero time the angular velocities ω_x, ω_y, ω_z of the rotating axes are zero. From Eq. 7.33 the angular momentum about the vertical fixed axis OY_0 is a constant. Hence

$$I_y S \cos \alpha = I_y S \cos \theta + I_x \dot{\phi} \sin^2 \theta$$

or

$$\dot{\phi} = \frac{I_y S(\cos \alpha - \cos \theta)}{I_x \sin^2 \theta} \tag{7.39}$$

Assuming the total energy of the top to be a constant, we have from the expression for the energy U in Section 7.8:

$$\tfrac{1}{2}I_x \sin^2 \theta \dot{\phi}^2 + \tfrac{1}{2}I_y S^2 + \tfrac{1}{2}I_x \dot{\theta}^2 + Mgl \cos \theta = \tfrac{1}{2}I_y S^2 + Mgl \cos \alpha$$

or

$$I_x(\sin^2 \theta \dot{\phi}^2 + \dot{\theta}^2) = 2Mgl(\cos \alpha - \cos \theta) \tag{7.40}$$

Substituting for $\dot{\phi}$ from Eq. 7.39 gives

$$\dot{\theta}^2 \sin^2 \theta = \frac{2Mgl \sin^2 \theta}{I_x}(\cos \alpha - \cos \theta) - \frac{I_y^2 S^2}{I_x^2}(\cos \alpha - \cos \theta)^2$$

For convenience we shall introduce a new constant γ, where

$$\gamma = \frac{I_y^2 S^2}{4I_x Mgl}$$

Then

$$\dot{\theta}^2 \sin^2 \theta = \frac{2Mgl \sin^2 \theta}{I_x}(\cos \alpha - \cos \theta) - \frac{4Mgl\gamma}{I_x}(\cos \alpha - \cos \theta)^2$$

From this equation, $\dot{\theta} = 0$ when

$$\theta = \alpha$$

and also when

$$\sin^2 \theta - 2\gamma(\cos \alpha - \cos \theta) = 0$$

or when

$$\cos^2 \theta - 2\gamma \cos \theta + 2\gamma \cos \alpha - 1 = 0$$

Thus $\dot\theta$ is zero when

$$\cos \theta = \gamma \pm \sqrt{\gamma^2 + 1 - 2\gamma \cos \alpha}$$

Both roots of this equation are real, but the larger of the two roots is greater than unity and corresponds to a physically impossible situation. The smaller root, involving the negative value of the square root, must be used. Hence the values of θ at which the nutational velocity is zero are

$$\theta_1 = \alpha \qquad \text{and} \qquad \cos \theta_2 = \gamma - \sqrt{\gamma^2 + 1 - 2\gamma \cos \alpha}$$

When the spinning top is initially set free at an angle α with the vertical, the top begins to descend and then the precessional and nutational motions set in. The axis of the top varies between the two values $\theta_1 = \alpha$ and the value of θ_2 given above. At the angle α, the values of the precessional and nutational motions are zero. Hence the axis of the top makes successive cusps with the horizontal circle around $\theta = \alpha$, as shown in Fig. 7.12. At the other libration limit where $\theta = \theta_2$, the nutational angular velocity $\dot\theta$ is zero and the precessional angular velocity may be obtained by dividing the energy equation, Eq. 7.40, by the angular momentum equation, Eq. 7.39, and setting $\dot\theta = 0$. This gives

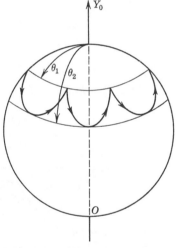

Fig. 7.12 The motion of a top started at a given angle.

$$\frac{I_x\dot\phi^2 \sin^2 \theta}{I_x\dot\phi \sin^2 \theta} = \frac{2Mgl(\cos \alpha - \cos \theta)}{I_yS(\cos \alpha - \cos \theta)}$$

or at $\theta = \theta_2$

$$\dot\phi = \frac{2Mgl}{I_yS}$$

Thus the upper end of the top touches the horizontal circle of angle θ_2, as shown in Fig. 7.12.

7.11   The Gyroscopic Compass

Possibly the most widespread application of the gyroscopic principle is in the gyroscopic compass. This is used for navigation on many ships and

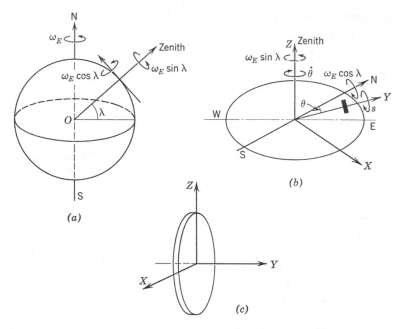

Fig. 7.13 The gyroscopic compass. (*a*) The motions on the earth. (*b*) The horizon circle. (*c*) The axes of the spinning wheel.

airplanes, and supplements the older magnetic compass. The theory of the gyroscopic compass was originally given by the French physicist Foucault. In its simplest form such a compass consists of a rapidly rotating circular disk whose axis is free to turn in a horizontal plane. We shall prove that owing to the earth's motion the system is in equilibrium when the axis of the rotating disk lies along a geographical north-south line or along a meridian.

Suppose that the earth is rotating on its axis with a constant angular velocity ω_E. At a place on the earth at north latitude λ, this angular velocity may be resolved into components $\omega_E \sin \lambda$ along the vertical and $\omega_E \cos \lambda$ about the horizontal, in a northerly direction, Fig. 7.13a. Let the angular velocity of spin of the disk be **s**. Take a system of axes X, Y, Z attached to the disk with the origin at the center of the disk, and let the Y axis be the axis of spin as in the case of the top. The X, Y, Z axes are as shown in Fig. 7.13b and c. Suppose that originally the Y axis of the disk makes an angle θ in the horizon circle with the meridian pointing north and that this angle is positive when measured east of the north line as shown in Fig. 7.13b. Thus the rotational velocity of the axis of the disk is measured by $\dot{\theta}$ along the negative Z

direction or by $-\dot{\theta}$ along the positive Z axis. Owing to the rotation of the earth there are components of angular velocity ω_x, ω_y, ω_z along the X, Y, Z axes respectively. From Fig. 7.13 these are

$$\omega_x = -\omega_E \cos \lambda \sin \theta$$

$$\omega_y = \omega_E \cos \lambda \cos \theta$$

$$\omega_z = \omega_E \sin \lambda - \dot{\theta}$$

If the axis of the disk is to place itself along a meridian, this implies motion about the vertical Z axis. We shall not need to have the X and Z axes rotate with the disk in order for their moments of inertia to be constant since the disk is symmetrical. Hence we can consider the Z axis to remain vertical and the components of angular velocity ω_x, ω_y, ω_z to be caused by the rotation of the earth. The torque about the Z axis is given in Eq. 7.30 with the modification that here the angular momentum about the Y axis is $I_y(\omega_y + s)$ rather than $I_y \omega_y$ as in Eq. 7.30. Since there is no torque about the Z axis and $I_z = I_x$, we have

$$I_z \dot{\omega}_z + \omega_x \omega_y (I_y - I_x) + s \omega_x I_y = 0$$

or by substitution

$$-I_x \ddot{\theta} - \omega_E^2 \cos^2 \lambda \sin \theta \cos \theta (I_y - I_x) - I_y s \omega_E \cos \lambda \sin \theta = 0$$

Since the angular velocity of the earth ω_E is approximately 7.3×10^{-5} radian/sec, it is negligibly small compared to the angular velocity of spin of the disk. Thus, without making any serious error, the second term containing ω_E^2 may be neglected. It then follows that

$$\ddot{\theta} + \frac{(I_y s \omega_E \cos \lambda)}{I_x} \sin \theta = 0$$

For small values of the angle θ we may replace $\sin \theta$ by θ, and for this the motion of the axis of the disk is simple harmonic in character about the geographic meridian. The period of the motion is

$$P = 2\pi \sqrt{\frac{I_x}{I_y s \omega_E \cos \lambda}}$$

The torque on the axis causing it to swing into the meridian is a maximum at the equator where λ is zero and zero at the poles. In practice the motion of the gyroscopic compass is damped so that the oscillations about the meridian do not continue indefinitely. A gyroscopic compass has to be constructed exceedingly carefully if it is to operate successfully. The next time you are on a ship, see this gyroscope if you can, for then you will appreciate how complicated such a compass can be.

7.12 The Lagrange Equations in Generalized and Polar Coordinates

In analyzing the motion of the planets about the sun, we found it convenient to express the equations of motion in polar coordinates rather than in Cartesian ones. Other problems may require other types of coordinates for their analysis. It can become very burdensome to derive the equations for the velocity and acceleration in terms of these new coordinates from the Cartesian expressions as was done for the polar coordinates. In 1760 Joseph Louis Lagrange (1736–1813), a French-Italian mathematician, gave a new formulation of Newton's Laws of motion and this we shall briefly present.

Suppose a mechanical system consists of N particles each with its own x, y, z coordinates. For such a system there would be three times as many coordinates as there are particles. In order to simplify the notation it is usual to replace x, y, z by x_1, x_2, x_3 for the first particle and so on for the other particles up to a total of, say, N coordinates corresponding to $N/3$ particles. What Lagrange did in part was to express the equations of motion not in terms of x_1, $x_2 \cdots x_N$ but in terms of n independent generalized coordinates $q_1, q_2 \cdots q_n$ which are functions of the x's, where n is less or equal to N.

In many problems the particles and their coordinates are connected by some form of constraint, for example, a rigid object in which the distance between the particles remains fixed, six coordinates are required to specify the position and orientation of the object. A bead constrained to move along a straight wire would require only one coordinate to specify its position. A constraint is said to be holonomic if the equations expressing the constraint can be used to reduce the number of independent generalized coordinates. With non-holonomic constraints it is not possible to directly reduce the number of independent coordinates, so we shall limit the discussion to holonomic constraints such as those involved in a rigid body. The generalized coordinates are suitably chosen for each problem and may be a distance, an angle, etc. In order to make the formal derivation of the Lagrange equations more meaningful, the same derivations are given alongside for the central force problem, with corresponding equations having the same numbers.

Generalized Coordinates

Consider a set of generalized coordinates $q_1, q_2 \cdots$ so that

$$x_i = x_i(q_1, q_2 \quad q_n) \qquad (7.41)$$

By differentiation

$$\dot{x}_i = \sum_j \frac{\partial x_i}{\partial q_j} \dot{q}_j \qquad (7.42)$$

where j is a dummy index and could be k, etc., but not i.
From Eq. 7.42 it follows that

$$\frac{\partial \dot{x}_i}{\partial \dot{q}_j} = \frac{\partial x_i}{\partial q_j} \qquad (7.43)$$

The kinetic energy of the system $T(q_i, \dot{q}_i)$ is given by

$$T = \sum_i \tfrac{1}{2} m_i \dot{x}_i{}^2 \qquad (7.44)$$

Hence

$$\frac{\partial T}{\partial \dot{q}_k} = \sum_i \frac{\partial}{\partial \dot{q}_k} (\tfrac{1}{2} m_i \dot{x}_i{}^2)$$

$$= \sum_i m_i \dot{x}_i \frac{\partial \dot{x}_i}{\partial \dot{q}_k}$$

$$= \sum_i m_i \dot{x}_i \frac{\partial x_i}{\partial q_k} \qquad (7.45)$$

Polar Coordinates: Central Force

For a single particle with coordinates r, θ; $n = 2$ and $q_1 = r$; $q_2 = \theta$.

$$x_i = x_i(r, \theta) \qquad (7.41)$$
$$x_1 = x = r \cos \theta$$
$$x_2 = y = r \sin \theta$$

$$\dot{x} = \frac{\partial}{\partial r}(r \cos \theta)\dot{r} + \frac{\partial}{\partial \theta}(r \cos \theta)\dot{\theta}$$

$$= \dot{r} \cos \theta - r\dot{\theta} \sin \theta \qquad \text{and}$$

$$\dot{y} = \dot{r} \sin \theta + r\dot{\theta} \cos \theta \qquad (7.42)$$

By differentiation of Eqs. 7.42

$$\frac{\partial \dot{x}}{\partial \dot{r}} = \cos \theta = \frac{\partial x}{\partial r} \quad \frac{\partial \dot{x}}{\partial \dot{\theta}} = -r \sin \theta = \frac{\partial x}{\partial \theta}$$

$$(7.43)$$

$$\frac{\partial \dot{y}}{\partial \dot{r}} = \sin \theta = \frac{\partial y}{\partial r} \quad \frac{\partial \dot{y}}{\partial \dot{\theta}} = r \cos \theta = \frac{\partial y}{\partial \theta}$$

The kinetic energy of the single particle

$$T = \tfrac{1}{2} m \dot{x}_1{}^2 + \tfrac{1}{2} m \dot{x}_2{}^2 = \tfrac{1}{2} m \dot{x}^2 + \tfrac{1}{2} m \dot{y}^2$$
$$(7.44)$$

As an exercise show by substitution in Eq. 7.42

$$T = \tfrac{1}{2} m(\dot{r}^2 + r^2 \dot{\theta}^2) \qquad (7.44a)$$

$$\frac{\partial T}{\partial \dot{r}} = \frac{\partial}{\partial \dot{r}}[\tfrac{1}{2} m(\dot{x}_2 + \dot{y}^2)]$$

$$= m\left(\dot{x} \frac{\partial \dot{x}}{\partial \dot{r}} + \dot{y} \frac{\partial \dot{y}}{\partial \dot{r}}\right) \qquad (q_k \equiv r)$$

$$= m\left(\dot{x} \frac{\partial x}{\partial r} + \dot{y} \frac{\partial y}{\partial r}\right)$$

$$\frac{\partial T}{\partial \dot{\theta}} = m\left(\dot{x} \frac{\partial x}{\partial \theta} + \dot{y} \frac{\partial y}{\partial \theta}\right) \qquad (q_k \equiv \theta)$$

$$(7.45)$$

Generalized Coordinates

using Eq. 7.43. Differentiating Eq. 7.45 with respect to time gives

$$\frac{d}{dt}\left(\frac{\partial T}{\partial \dot{q}_k}\right) = \sum_i \left[m_i \ddot{x}_i \frac{\partial x_i}{\partial q_k} + m_i \dot{x}_i \right.$$
$$\left. \times \frac{d}{dt}\left(\frac{\partial x_i}{\partial q_k}\right) \right] \quad (7.46)$$

The first term on the righthand side of Eq. 7.46 is defined as the generalized force Q_k, or

$$Q_k = \sum_i m_i \ddot{x}_i \frac{\partial x_i}{\partial q_k} \quad (7.47)$$

The second term of Eq. 7.46 containing $\partial x_i / \partial q_k$ is a function of time, so that

$$\sum_i m_i \dot{x}_i \frac{d}{dt}\left(\frac{\partial x_i}{\partial q_k}\right)$$
$$= \sum_{i,j} m_i \dot{x}_i \frac{\partial^2 x_i}{\partial q_k \partial q_j} \dot{q}_j \quad (7.48)$$

From Eq. 7.44 by differentiation

$$\frac{\partial T}{\partial q_k} = \sum_i m_i \dot{x}_i \frac{\partial \dot{x}_i}{\partial q_k}$$
$$= \sum_i m_i \dot{x}_i \frac{\partial}{\partial q_k}\left[\sum_j \frac{\partial x_i}{\partial q_j} \dot{q}_j\right]$$
$$= \sum_{i,j} \left[m_i \dot{x}_i \frac{\partial^2 x_i}{\partial q_k \partial q_j} \dot{q}_j \right] \quad (7.49)$$

Polar Coordinates: Central Force

Differentiating Eqs. 7.45;

$$\frac{d}{dt}\left(\frac{\partial T}{\partial \dot{r}}\right) = m\left[\ddot{x}\frac{\partial x}{\partial r} + \ddot{y}\frac{\partial y}{\partial r} \right]$$
$$\quad (7.46)$$
$$+ m\left[\dot{x}\frac{d}{dt}\left(\frac{\partial x}{\partial r}\right) + \dot{y}\frac{d}{dt}\left(\frac{\partial y}{\partial r}\right) \right]$$

As an exercise find the expression for the θ variable. The generalized force Q_r is given by

$$Q_r = m\left[\ddot{x}\frac{\partial x}{\partial r} + \ddot{y}\frac{\partial y}{\partial r} \right]$$

As exercises show (a) that

$$Q_r = m(\ddot{r} - r\dot{\theta}^2) \quad (7.47)$$

and (b) that

$$Q_\theta = m\left(\ddot{x}\frac{1}{r}\frac{\partial x}{\partial \theta} + \ddot{y}\frac{1}{r}\frac{\partial y}{\partial \theta} \right)$$
$$= m(r\ddot{\theta} + 2\dot{r}\dot{\theta}) \quad (7.47)$$

The second term of Eq. 7.46 is

$$m\dot{x}\frac{d}{dt}\left(\frac{\partial x}{\partial r}\right) + m\dot{y}\frac{d}{dt}\left(\frac{\partial y}{\partial r}\right)$$
$$= m\dot{x}\frac{d}{dt}(\cos\theta) + m\dot{y}\frac{d}{dt}(\sin\theta)$$
$$= -m\dot{x}\dot{\theta}\sin\theta + m\dot{y}\dot{\theta}\cos\theta \quad (7.48)$$

Differentiating Eq. 7.44 with respect to r

$$\frac{\partial T}{\partial r} = m\dot{x}\frac{\partial \dot{x}}{\partial r} + m\dot{y}\frac{\partial \dot{y}}{\partial r}$$
$$= m\dot{x}(-\dot{\theta}\sin\theta) + m\dot{y}(\dot{\theta}\cos\theta)$$
$$\quad (7.49)$$

Thus the second term of Eq. 7.46 is equal to $\partial T / \partial r$.

$$\left[\text{Note } \frac{d}{dt}\left(\frac{\partial x}{\partial r}\right) = \frac{\partial}{\partial \theta}\left(\frac{\partial x}{\partial r}\right)\dot{\theta} \right.$$
$$\left. = \frac{\partial^2 x}{\partial r\, \partial \theta}\dot{\theta} = -\sin\theta\,\dot{\theta} \right]$$

Generalized Coordinates

This is also Eq. 7.48, which is the second term of Eq. 7.46. Hence

$$\frac{d}{dt}\left(\frac{\partial T}{\partial \dot{q}_k}\right) = Q_k + \frac{\partial T}{\partial q_k} \quad (7.50)$$

If the system is conservative such that $Q_k = -\partial V/\partial q_k$ and $V = V(q)$ then we may introduce the Lagrangian function L where

$$L(q, \dot{q}) = T(q, \dot{q}) - V(q)$$

Thus

$$\frac{\partial T}{\partial \dot{q}_k} = \frac{\partial L}{\partial \dot{q}_k}$$

Since $Q_k = -\partial V/\partial q_k$ it follows that

$$Q_k + \frac{\partial T}{\partial q_k} = -\frac{\partial}{\partial q_k}(V - T) = \frac{\partial L}{\partial q_k}$$

Hence Eq. 7.50 may be written as

$$\frac{d}{dt}\left(\frac{\partial L}{\partial \dot{q}_k}\right) - \frac{\partial L}{\partial q_k} = 0 \quad (7.51)$$

This is known as the Lagrange equation. No physical principle other than Newton's second law has been used. However, it is the energy, which is a scalar, that is used rather than vector forces, as used in Newton's laws.

Polar Coordinates: Central Force

Thus from Eqs. 7.46 and 7.48 together with the expression for the generalized force Q_r it follows that

$$\frac{d}{dt}\left(\frac{\partial T}{\partial \dot{r}}\right) = Q_r + \frac{\partial T}{\partial r} \quad (7.50)$$

In a central force problem in which the potential energy V depends only on r, then $V = V(r)$ and $Q_r = -\partial V/\partial r$ and $Q_\theta = 0$. Thus we may introduce the Lagrangian function L given by

$$L = T - V = \tfrac{1}{2}m(\dot{r}^2 + r^2\dot{\theta}^2) - V(r)$$

Thus

$$\partial L/\partial \dot{r} = \partial T/\partial \dot{r} \quad \text{and} \quad \partial L/\partial \dot{\theta} = \partial T/\partial \dot{\theta}$$

Hence Eq. 7.50 can be expressed in terms of L in place of T, giving

$$\frac{d}{dt}\left(\frac{\partial L}{\partial \dot{r}}\right) - \frac{\partial L}{\partial r} = 0 \quad \text{and}$$

$$\frac{d}{dt}\left(\frac{\partial L}{\partial \dot{\theta}}\right) - \frac{\partial L}{\partial \theta} = 0$$

(7.51)

These are the Lagrange equations applied to the central force problem. We see that they give the equations of motion, namely $m\ddot{r} - mr\dot{\theta}^2 = F_r$ and since $Q_\theta = 0$ for central forces

$$\frac{d}{dt}(mr^2\dot{\theta}) = 0 \quad \text{or} \quad mr^2\dot{\theta} = \text{constant.}$$

That is, the angular momentum $mr^2\dot{\theta}$ is constant in a central force problem.

7.13   Cyclic or Ignorable Coordinates

The kinetic energy T of a system of particles is given as

$$T = \sum_k \tfrac{1}{2}m_k \dot{x}_k^2$$

For a single particle of mass m the kinetic energy T in polar coordinates is

$$T = \tfrac{1}{2}m(\dot{r}^2 + r^2\dot{\theta}^2)$$

Hence; $\partial T/\partial \dot{x}_k = m_k \dot{x}_k = p_k$ the component of momentum of the kth particle. The generalized momentum p_k is defined as

$$p_k = \frac{\partial L}{\partial \dot{q}_k} = \frac{\partial T}{\partial \dot{q}_k}$$

if the potential V does not contain $\dot{q}_k$. Suppose the Lagrangian L has one of the generalized coordinates q_s missing so that $\partial L/\partial q_s = 0$. Then from Eq. 7.51 it follows that

$$\frac{d}{dt}\left(\frac{\partial L}{\partial \dot{q}_s}\right) = 0 \quad \text{or} \quad \frac{d}{dt}(p_s) = 0$$

Thus the conjugate momentum p_s, corresponding to the missing coordinate q_s, is a constant. The definition of the generalized momentum is important and should be remembered.

$$p_k = \frac{\partial L}{\partial \dot{q}_k} \qquad (7.52)$$

Hence, $\partial T/\partial \dot{r} = m\dot{r} = p_r$, the linear momentum along r, and

$$\partial T/\partial \dot{\theta} = mr^2\dot{\theta} = P_\theta,$$

the angular momentum in the direction θ. Now in the expression for the kinetic energy the coordinate θ is missing, and the potential energy does not contain θ. Hence, from Eq. 7.51 it follows for the θ coordinate that

$$\frac{d}{dt}\left(\frac{\partial L}{\partial \dot{\theta}}\right) = 0 \quad \text{or} \quad \frac{d}{dt}(p_\theta) = 0$$

Thus it follows that p_θ, the angular momentum in the direction of the unit vector θ_1, is a constant. This result has been shown earlier for the central force problem. The two components of momenta are

$$p_r = m\dot{r} \quad \text{and} \quad p_\theta = mr^2\dot{\theta} \quad (7.52)$$

7.14 Hamilton's Equations in Generalized and Polar Coordinates

Sir William Rowan Hamilton (1805–1865) was an Irish mathematician and physicist who became a professor at Trinity College, Dublin, while still an undergraduate. In 1826 he introduced a "characteristic function" which was first applied to a system of light rays, and later in 1835 he applied this to dynamical systems.

The Hamiltonian function H is defined as

$$H = \sum_i p_i \dot{q}_i - L \qquad (7.53)$$

where $L = (T - V)$ and is the Lagrange function. We shall show that for conservative systems $H = (T + V)$ and when H is expressed in terms of the generalized momenta p and coordinates q, i.e., $H = H(p, q)$, then $\partial H/\partial q_i = -\dot{p}_i$ and $\partial H/\partial p_i = \dot{q}_i$. These latter two equations are known as Hamilton's **canonical equations**. It is important that the Hamiltonian function H be expressed in terms of the generalized *momenta* and *coordinates*.

Generalized Coordinates

$$H(p, q) = \sum_j p_j \dot{q}_j(p_k, q_k)$$
$$- L[\dot{q}_j(p_k, q_k)q_j] \quad (7.54)$$

From Eq. 7.44 the kinetic energy T is

$$T = \sum_i \tfrac{1}{2} m_i \dot{x}_i^2$$

Now

$$p_j = \frac{\partial T}{\partial \dot{q}_j} = \sum_i m_i \dot{x}_i \frac{\partial \dot{x}_i}{\partial \dot{q}_j}$$

Hence

$$\sum_j p_j \dot{q}_j = \sum_{i,j} m_i \dot{x}_i \frac{\partial \dot{x}_i}{\partial \dot{q}_j} \dot{q}_j \quad (7.55)$$

$$= \sum_i m_i \dot{x}_i \sum_j \frac{\partial x_i}{\partial q_j} \dot{q}_j$$

$$= \sum_i m_i \dot{x}_i^2,$$

using Eq. 7.42

$$= 2T \quad (7.56)$$

Thus

$$H = \sum_j p_j \dot{q}_j - L = (2T - T + V)$$
$$= (T + V) \quad (7.57)$$
$$= \text{Total Energy}$$

Differentiating Eq. 7.54 with respect to the q_i coordinate, it follows that

$$\frac{\partial H}{\partial q_i} = \sum_s p_s \frac{\partial \dot{q}_s}{\partial q_i} - \sum_s \frac{\partial L}{\partial \dot{q}_s} \frac{\partial \dot{q}_s}{\partial q_i} - \frac{\partial L}{\partial q_i} \quad (7.58)$$

Since $p_s = \partial L / \partial \dot{q}_s$ the first two terms on the right-hand side of Eq. 7.58 cancel. Also from the Lagrange equation we have

$$\frac{\partial L}{\partial q_i} = \frac{d}{dt}\left(\frac{\partial L}{\partial \dot{q}_i}\right) = \dot{p}_i$$

Hence

$$\frac{\partial H}{\partial q_i} = -\dot{p}_i \quad (7.59)$$

Polar Coordinates; Central Force

The coordinates q are r and θ, hence

$$H = \sum_j p_j \dot{q}_j - L = p_r \dot{q}_r + p_\theta \dot{q}_\theta$$
$$- T(r, \dot{r}, \theta, \dot{\theta}) + V(r) \quad (7.54)$$

The kinetic energy T is given as

$$T = \tfrac{1}{2} m(\dot{r}^2 + r^2 \dot{\theta}^2)$$

The momenta p_r and p_θ are given as

$$\frac{\partial L}{\partial \dot{q}_r} = \frac{\partial L}{\partial \dot{r}} = p_r = m\dot{r} \quad (7.55)$$

$$\frac{\partial L}{\partial \dot{q}_\theta} = \frac{\partial L}{\partial \dot{\theta}} = p_\theta = mr^2\dot{\theta}$$

$$\sum_j p_j \dot{q}_j = p_r \dot{r} + p_\theta \dot{\theta}$$
$$= m\dot{r}^2 + mr^2\dot{\theta}^2$$
$$= 2T \quad (7.56)$$

Thus

$$H = (2T - T + V)$$
$$= (T + V) = \text{Total Energy} \quad (7.57)$$

$$H(p, q) = \frac{p_r^2}{2m} + \frac{p_\theta^2}{2mr^2} + V(r)$$

Since the momenta and coordinates are considered as independent coordinates it follows that

$$\frac{\partial H}{\partial r} = -\frac{p_\theta^2}{mr^3} + \frac{\partial V}{\partial r} \quad (7.58)$$

$$= -\frac{m^2 r^4 \dot{\theta}^2}{mr^3} - F_r$$

$$= -m\ddot{r}; \quad \text{since } F_r = m(\ddot{r} - r\dot{\theta}^2)$$

Hence

$$\frac{\partial H}{\partial r} = -\frac{\partial(m\dot{r})}{\partial t} = -\frac{\partial p_r}{\partial t} = -\dot{p}_r \quad (7.59)$$

Generalized Coordinates

This is one of Hamilton's canonical equations, the second is obtained by finding $\partial H/\partial p_i$. From Eq. 7.57

$$\frac{\partial H}{\partial p_i} = \dot{q}_i + \sum_s p_s \frac{\partial \dot{q}_s}{\partial p_i} - \sum_s \frac{\partial L}{\partial \dot{q}_s}\frac{\partial \dot{q}_s}{\partial p_i}$$

Since $\partial L/\partial \dot{q}_s = p_s$, the last two terms on the right-hand side of the above equation cancel, giving

$$\frac{\partial H}{\partial p_i} = \dot{q}_i \qquad (7.60)$$

Polar Coordinates: Central Force

and

$$\frac{\partial H}{\partial \theta} = 0.$$

From the canonical equation $\partial H/\partial q_i = -\dot{p}_i$ it follows that p_θ is constant, i.e., the angular momentum in a central force system is constant. Now we shall calculate $\partial H/\partial p_r$.

$$\frac{\partial H}{\partial p_r} = \frac{p_r}{m} = \dot{r}$$

$$\frac{\partial H}{\partial p_\theta} = \frac{p_\theta}{mr^2} = \dot{\theta} \qquad (7.60)$$

These two equations agree with the one derived by generalized coordinates, namely, $\partial H/\partial p_i = \dot{q}_i$.

It should be noted that no physical principle was introduced in the derivations of the Lagrange and Hamilton equations other than what is contained in Newton's laws of motion. Newton's second law involves second order differentials between vector quantities whereas the Lagrange and Hamilton equations use energy, a scalar quantity. The Lagrange equations consist of n second order differential equations while those of Hamilton consist of $2n$ first order differential equations. In general, Hamilton's equations are not more convenient to use than the Lagrange equations for the solution of mechanical problems. However, the Hamilton equations have provided considerable insight in the most advanced branches of mechanics. Both statistical mechanics and quantum mechanics are essentially founded on Hamilton's equations. Hamilton's equations have been illustrated for a conservative system for which $H = T + V$. The canonical equations of Hamilton also apply to nonconservative systems and also for those in which the potential energy V depends on both q and $\dot{q}$, but in these cases H is not, in general, equal to the total energy $T + V$.

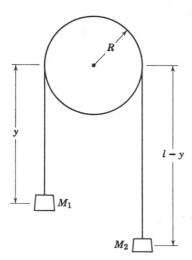

Fig. 7.14 Atwood machine.

Examples in the application of the Lagrange and Hamilton equations. *a. Atwood's machine.*
This problem is solved by Newton's laws in Section 2.9. We shall assume the moment of inertia of the pulley is I and its radius is R (Fig. 7.14). The kinetic energy T of the system is $T = \frac{1}{2}(M_1 + M_2 + I/R^2)\dot{y}^2$ since the kinetic energy of the

pulley is $\frac{1}{2}I\dot{\theta}^2$, and $\dot{\theta} = \dot{y}/R$. If the total length of cord hanging freely has a length l then the potential energy V of the masses relative to the center of the pulley is

$$V = -M_1 gy - M_2 g(l - y)$$

and

$$L = T - V = \tfrac{1}{2}(M_1 + M_2 + I/R^2)\dot{y}^2 + M_1 gy + M_2 g(l - y)$$

There is one variable y, so that the Lagrange equation is

$$\frac{d}{dt}\left(\frac{\partial L}{\partial \dot{y}}\right) - \frac{\partial L}{\partial y} = 0$$

or

$$\left(M_1 + M_2 + \frac{I}{R^2}\right)\ddot{y} = (M_1 + M_2)g$$

Thus

$$\ddot{y} = \frac{(M_1 - M_2)g}{M_1 + M_2 + I/R^2}$$

Now let us solve the problem of the Atwood's machine by using Hamiltonian's canonical equations. To do this we must express the energy in terms of the momentum and position coordinates.

For convenience let $k = (M_1 + M_2 + I/R^2)$. The total energy of the system is

$$T + V = \frac{k\dot{y}^2}{2} - M_1 gy - M_2 g(l - y)$$

From Eq. 7.52 the generalized momentum $p_k = \partial L/\partial \dot{q}_k$, hence $p = k\dot{y}$, and the Hamiltonian function $H(p, q)$ for the Atwood machine is

$$H(p, q) = \frac{p^2}{2k} - M_1 gy - M_2 g(l - y)$$

The canonical equations are $\partial H/\partial q_k = -\dot{p}_k$ and $\partial H/\partial p_k = \dot{q}_k$ so that

$$\frac{\partial H}{\partial y} = -M_1 g + M_2 g = -\dot{p} \qquad \text{and} \qquad \frac{\partial H}{\partial p} = \dot{y} = \frac{p}{k}$$

Hence

$$\dot{p} = (M_1 - M_2)g = k\ddot{y} = \left(M_1 + M_2 + \frac{I}{R^2}\right)\ddot{y}$$

b. *Cylinder rolling down inclined plane.* Assuming that there is no slipping, $r\, d\theta = ds$, Fig. 7.15. The kinetic energy can be given as the kinetic energy of the center of mass plus the kinetic energy about the center of mass:

$$T = \frac{1}{2} M\dot{s}^2 + \frac{1}{2}\frac{Mr^2}{2}\dot{\theta}^2 = \frac{M\dot{s}^2}{2} + \frac{M\dot{s}^2}{4}$$

The potential energy is $V = Mg(l - s)\sin\alpha$, so that

$$L = T - V = \frac{3M\dot{s}^2}{4} - Mg(l - s)\sin\alpha$$

The Lagrange equation is

$$\frac{d}{dt}\left(\frac{\partial L}{\partial \dot{s}}\right) - \frac{\partial L}{\partial s} = 0 \qquad \text{or} \qquad \frac{d}{dt}\left(\frac{3M\dot{s}}{2}\right) - Mg\sin\alpha = 0$$

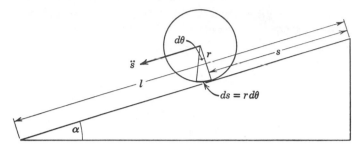

Fig. 7.15 Cylinder rolling down incline.

or the acceleration $a = \ddot{s} = (2/3)g \sin \alpha$, as given in section 6.12. As an exercise, find the acceleration $a = \ddot{s}$ for the rolling cylinder by the canonical equations of Hamilton and the definition of the generalized momentum.

7.15 Hamilton's Variational Principle

About 1834 Hamilton introduced a principle, which may be stated in his words as, "If a system of bodies is at P_1 at the time t_1 and at P_2 at the time t_2, it will pass from P_1 to P_2 by such a path that the mean value of the difference between the kinetic and potential energy of the system in the interval $t_2 - t_1$ is a minimum." The difference between the kinetic energy T and the potential energy V is known as the Lagrangian L. A dynamically possible motion is, according to Hamilton, one for which the line integral $\int_{t_1}^{t_2} L\, dt$, taken from some initial time t_1 to a final time t_2, has a minimum value. Now the line integral S, which has the dimensions of energy $\times$ time, is defined as

$$S = \int_{t_1}^{t_z} L\, dt$$

If δ represents a small variation from the dynamical possible path then Hamilton's variational principle may be stated as

$$\delta \int_{t_1}^{t_2} L\, dt = 0 \qquad \text{or} \qquad \delta S = 0$$

This is a problem in the calculus of variations, in which one is attempting to find the condition for an integral to have a minimum value, and can be compared to the relatively simple problem of finding the minimum of a known function $y = f(x)$ in differential calculus. In the latter, the condition for an extremum, maximum or minimum, is $df(x)/dx = dy/dx = 0$. What is wanted is to put the problem in the calculus of variations in such a form that a similar operation can be used to find the extremum of the integral. To do this one first represents the possible dynamical path as y and compares this path with nearby paths represented by $\tilde{y}$ (y tilde). All the paths must take place in the

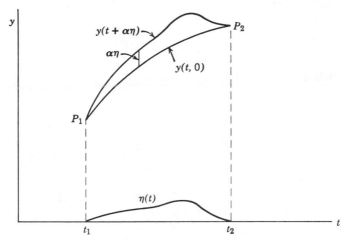

Fig. 7.16 Illustrating the dynamical path $y(t, 0)$ and another one $y(t + \alpha \eta)$ and also the arbitrary function $\eta(t)$ where $\eta(t_1) = \eta(t_2) = 0$.

same time interval $t_2 - t_1$. To put the problem in the calculus of variations in a form corresponding to that of finding the extremum of a function, we first consider some arbitrary function $\eta(t)$ which is zero at t_1 and at t_2. This arbitrary function is then multiplied by a parameter α such that $\alpha \eta(t)$ represents the difference between the paths y and $\tilde{y}$, as shown in Fig. 7.16. Thus:

$$\tilde{y}(t, \alpha) = y(t, 0) + \alpha \eta(t) \qquad (7.61)$$

where the arbitrary function η and its first two derivatives are continuous in the range of integration $t_1 - t_2$, and $\eta(t_1) = \eta(t_2) = 0$. Notice that the parameter α is a constant for any particular path but has a different value for a different path, that is, different paths correspond to different values of α. Thus the line integral S can now be expressed as $S(\alpha)$ in terms of y and $\alpha \eta$ for the different paths. Thus

$$S(\alpha) = \int_{t_1}^{t_2} L(t, y + \alpha \eta, \dot{y} + \alpha \dot{\eta}) \, dt \qquad (7.62)$$

As given above, Hamilton's principle states that the line integral $\int L \, dt = S$ has a minimum value over a possible dynamical path for which α must be equal to zero. The integral in α provides a way of finding its extremum value because what has to be done is to find $dS/d\alpha$, evaluated at $\alpha = 0$, and set this differential equal to zero. Since variations from the dynamically possible path are represented by δ, Hamilton's principle can be given as

$$\delta S = 0$$

We shall now illustrate the validity of this principle by showing that Lagrange's equations can be derived from it and also by using a simple problem that gives a result verifiable by Newton's second law.

Lagrange's Equations from Hamiltonian's Principle Variational

Expanding Eq. 7.62 for $S(\alpha)$ by Taylor's series keeping only the first order terms it follows that

$$S(\alpha) = \int_{t_2}^{t_1} \left[L(t, y, \dot{y}) + \alpha\eta \frac{\partial L}{\partial y} \right. $$
$$\left. + \alpha\dot{\eta} \frac{\partial L}{\partial \dot{y}} \right] dt$$

Only the first order terms are necessary since later the differential $dS/d\alpha$ is found at $\alpha = 0$.

The differential with respect to α is

$$\frac{\partial S(\alpha)}{\partial \alpha} = \int_{t_1}^{t_2} \left[\eta \frac{\partial L}{\partial y} + \dot{\eta} \frac{\partial L}{\partial \dot{y}} \right] dt \quad (7.63)$$

The second term on the right hand side of this equation can be integrated by parts as follows:

$$\int_{t_1}^{t_2} \dot{\eta} \frac{\partial L}{\partial \dot{y}} dt = \eta(t) \frac{\partial L}{\partial \dot{y}} \Big|_{t_1}^{t_2} - \int_{t_1}^{t_2} \eta \frac{d}{dt}\left(\frac{\partial L}{\partial \dot{y}}\right) dt$$

The first term on the right-hand side is zero, since $\eta(t_1) = \eta(t_2) = 0$.

Hamilton's principle states that the integral $\int L\, dt$ is an extremum for the dynamically possible path, that is,

$$\left(\frac{\partial S}{\partial \alpha}\right)_{\alpha=0} = 0$$

From Eq. 7.63 it follows that

$$\left(\frac{dS}{d\alpha}\right)_{\alpha=0} = 0$$
$$= \int_{t_1}^{t_2} \left[\frac{\partial L}{\partial y} - \frac{d}{dt}\left(\frac{\partial L}{\partial \dot{y}}\right) \right] \eta(t)\, dt$$

Since $\eta(t)$ is an arbitrary function of t, the only manner is which the integral can be zero is if

$$\frac{\partial L}{\partial y} - \frac{d}{dt}\left(\frac{\partial L}{\partial \dot{y}}\right) = 0$$

and this is the Lagrange equation.

Hamilton's Principle Applied to Vertical Motion in Earth's Gravitational Field

For an object of mass m shot vertically upwards the Lagrangian L is

$$L = \tfrac{1}{2}m\dot{y}^2 - mgy$$

and

$$S = \int_{t_1}^{t_2} (\tfrac{1}{2}m\dot{\tilde{y}}^2 - mg\tilde{y})\, dt$$

From the Taylor expansion for $S(\alpha)$, keeping only the first order terms, there results:

$$S(\alpha) = \int_{t_1}^{t_2} [\tfrac{1}{2}m\dot{y}^2 - mgy $$
$$- \alpha\eta mg + \alpha\dot{\eta}m\dot{y}]\, dt$$

The differential with respect to α is

$$\frac{\partial S(\alpha)}{\partial \alpha} = \int_{t_1}^{t_2} [-\eta mg + \dot{\eta}m\dot{y}]\, dt \quad (7.63)$$

The second term on the right-hand side of this equation can be integrated by parts as follows:

$$\int_{t_1}^{t_2} \dot{\eta}m\dot{y}\, dt = \eta(t)m\dot{y} \Big|_{t_1}^{t_2} - \int_{t_1}^{t_2} \eta \frac{d}{dt}(m\dot{y})\, dt$$

The first term on the right-hand side is zero, since $\eta(t_1) = \eta(t_2) = 0$, and the second term is $-\int_{t_1}^{t_2} \eta\, m\ddot{y}\, dt$.

Hamilton's principle states that the integral $\int L\, dt$ is an extremum for the dynamically possible path, that is,

$$\left(\frac{\partial S}{\partial \alpha}\right)_{\alpha=0} = 0$$

From Eq. 7.63 it follows that

$$\left(\frac{\partial S}{\partial \alpha}\right)_{\alpha=0} = 0 = \int_{t_1}^{t_2} [-mg - m\ddot{y}]\eta(t)\, dt$$

Since $\eta(t)$ is an arbitrary function of t, the only manner in which the integral can be zero is if

$$-mg - m\ddot{y} = 0 \quad \text{or} \quad m\ddot{y} = -mg$$

and this is Newton's second law for vertical motion in a uniform gravitational field.

As an example of the use of the calculus of variations let us find the shortest distance between two points on a plane. This is obviously a straight line but is to be proved by the calculus of variations.

An element of distance along the arc of a curve is given by

$$ds = \sqrt{dx^2 + dy^2} = dx\sqrt{1 + y'^2} \qquad \text{where} \qquad y' = dy/dx$$

The line integral which has to be minimized in this problem, is

$$I = \int_{x_1}^{x_2} \sqrt{1 + y'^2}\, dx$$

The condition for the minimum is given in this problem by the Euler-Lagrange equation, which is analogous to the Hamiltonian equation (for which the Lagrange equation was found). For this case let $F(x, y, y') = \sqrt{1 + y'^2}$; the condition for a minimum is

$$\frac{d}{dx}\left(\frac{\partial F}{\partial y'}\right) - \frac{\partial F}{\partial y} = 0$$

From the equation for F it follows that

$$\frac{\partial F}{\partial y'} = \frac{y'}{\sqrt{1 + y'^2}} \qquad \text{and} \qquad \frac{\partial F}{\partial y} = 0$$

Thus the condition for a minimum is

$$\frac{d}{dx}\left(\frac{y'}{\sqrt{1 + y'^2}}\right) = 0 \qquad \text{or} \qquad \frac{y'}{\sqrt{1 + y'^2}} = \text{constant } c$$

This solution is only valid if y' is a constant, hence

$$y' = \frac{dy}{dx} = \text{constant } m$$

Integrating this equation gives

$$y = mx + b$$

which is the equation of a straight line with a y intercept of b.

The Brachistochrone. This problem is to find the equation of the frictionless path along which an object moving from rest under gravity must travel in order to go from a higher point x_0, y_0 to a lower point x, y, in the least time. This is one of the early problems in the calculus of variations, worked on by the brothers James (1654–1748) and John (1667–1748) Bernouilli, Swiss mathematicians. It was solved by John Bernouilli about 1697 and this led to further work by Euler, at the time a student of the Bernouillis.

The time dt for an object to travel a distance ds with speed v is $dt = ds/v$, where $ds = \sqrt{dx^2 + dy^2} = \sqrt{1 + (dy/dx)^2}\, dx$ and therefore $dt = \sqrt{1 + (dy/dx)^2}\, dx/v$. Assuming there is conservation of energy along the path, the speed v after descending a vertical distance y is given by

$$\tfrac{1}{2}mv^2 = mgy \qquad \text{or} \qquad v = \sqrt{2gy}$$

and

$$dt = \sqrt{\frac{1 + (dy/dx)^2}{2gy}}\, dx = \sqrt{\frac{1 + y'^2}{2gy}}\, dx$$

where $y' = dy/dx$ and the time t for the object to go from $(0, 0)$ to (x_1, y_1) is

$$t = \int_0^{x_1} \sqrt{\frac{1 + y'^2}{2gy}}\, dx$$

The problem is now in the form required to apply the Euler-Lagrange equation; since we wish to find the equation of the path $y(x)$ for the minimum time, we must solve the equation

$$\frac{d}{dx}\left(\frac{\partial F}{\partial y'}\right) - \frac{\partial F}{\partial y} = 0$$

where now

$$F = \sqrt{\frac{1 + y'^2}{2gy}} \tag{7.64}$$

In this problem the function $F = F(y, y')$ and is explicitly independent of x so that the identity given below is valid, as shown by differentiating the left-hand side:

$$\frac{d}{dx}\left(F - y'\frac{\partial F}{\partial y'}\right) = y'\left(\frac{\partial F}{\partial y} - \frac{d}{dx}\frac{\partial F}{\partial y'}\right)$$

From the Euler-Lagrange equation the right-hand side of the above equation is zero, and hence for F independent of x it follows that

$$F - y'\frac{\partial F}{\partial y'} = \text{constant} = a$$

The above equation can be rewritten in terms of Eq. 7.64 for F as

$$a = \frac{1}{\sqrt{2gy(1 + y'^2)}}$$

which gives

$$y'^2 = \frac{1 - 2gya^2}{2gya^2} = \frac{b - y}{y}$$

where

$$b = \frac{1}{2ga^2}$$

Hence

$$dx = \sqrt{\frac{y}{b - y}}\, dy = \frac{y\, dy}{\sqrt{by - y^2}}$$

The integration of this equation can be found in mathematical tables or by breaking the integral into two parts, namely

$$x = \frac{b}{2} \int_0^y \frac{dy}{\sqrt{by - y^2}} - \frac{1}{2} \int_0^y \frac{(b - 2y)}{\sqrt{by - y^2}} \, dy$$

The integral of the second term on the right hand side is $-\sqrt{by - y^2}$, while the first term can be integrated by making the substitution

$$(b - 2y) = b \cos \theta \tag{7.65}$$

Thus

$$dy = \tfrac{1}{2}b \sin \theta \, d\theta \quad \text{and} \quad by - y^2 = \frac{b^2}{4} \sin^2 \theta$$

Hence the integral of the first term is

$$\frac{b}{2} \int_0^\theta d\theta = \frac{b}{2} \theta = \frac{b}{2} \cos^{-1} \left(\frac{b - 2y}{b} \right)$$

The complete equation of the path of quickest descent is

$$x = -\sqrt{by - y^2} + \frac{b}{2} \cos^{-1} \left(\frac{b - 2y}{b} \right) \tag{7.66}$$

From Eq. 7.65 it follows that

$$y = \frac{b}{2} (1 - \cos \theta)$$

and from Eq. 7.66 with the appropriate substitutions, the value of x is

$$x = \frac{b}{2} (\theta - \sin \theta)$$

These are the parametric equations of the brachistochrone and are the equations for a cycloid as given in Fig. 7.17.

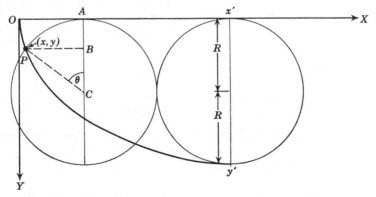

Fig. 7.17 The cycloid. $x = OA - PB = R\theta - R \sin \theta$; $y = AC - BC = R - R \cos \theta$. At the lowest point $\theta = \pi$ and $x' = R\pi$ and $y' = 2R$.

PROBLEMS

1. If **A**, **B**, **C** are three vectors having a common origin, show that $\mathbf{A} \times \mathbf{B} \cdot \mathbf{C}$ represents the volume of the parallelopiped formed by these three vectors. Show that the volume of the parallelopiped is also given by $\mathbf{A} \cdot \mathbf{B} \times \mathbf{C}$. Show analytically, by expressing the vectors in terms of their components along the X, Y, Z axes, that

$$\mathbf{A} \times \mathbf{B} \cdot \mathbf{C} = \mathbf{A} \cdot \mathbf{B} \times \mathbf{C}$$

2. A uniform horizontal rod 1 m long and of mass 1 kg is pivoted so as to turn about a vertical axis at 0.40 m from one end. A weight of 2 kg is placed on the end 0.40 m from the pivot, and a disk of radius 0.2 m and mass 4 kg is placed on the other end. The disk rotates freely about the horizontal axis of the rod at a rate of 100 rps. Find the rate of precession in revolutions per second, assuming the rod remains horizontal, and find the angular momentum of precession and that of the spin.

3. A top of the type shown in Fig. 7.9 has moments of inertia I_x equal to 6×10^{-4} kg m² and I_y equal to 2×10^{-4} kg m². The top is spinning so that the total angular velocity about the Y or spin axis is 37.95 radians/sec, the total energy U is equal to 0.1458 kg m²/sec², the total angular momentum C about the OY_0 axis is 3.06×10^{-3} kg m²/sec, and the maximum torque Mgl is 2.4×10^{-3} nt m. Show that the constants in Eq. 7.35, namely A, B, D, and E are 6/sec², 8/sec², 5.1/sec and 12.65/sec respectively, and also show that the limiting cone angles, Fig. 7.11, are 60° and 70° 30', and using Eq. 7.33, that the limiting rates of precession are 2.07 radians/sec and −1.63 radians/sec.

4. A top such as that shown in Fig. 7.9 has the moments of inertia I_x equal to 0.5 lb ft² and I_y equal to 0.1 lb ft². The weight of the top is 2 lbw, and the distance l of the center of mass from the point of rotation is 9 in. Show that steady motion is possible with the axis OY inclined at 60° to the vertical if the spin is greater than about 66 radians/sec.

5. A top is made by forcing a light pin through the center of a uniform circular disk whose radius is 1.5 in. The pin projects 3 in. below the disk. If the top is set into steady motion so that the rim of the disk just fails to touch the ground, show the minimum number of revolutions per second of the disk is 39.7.

6. An axle is placed through the center of a uniform circular disk so that the axle is inclined to the plane of the disk at an angle α. The system rotates with an angular velocity ω. Show that there is a torque on the bearings holding the axle which has a magnitude of $(I_y - I_x)\omega^2 \sin \alpha \cos \alpha$. This torque lies in a plane containing the axle and the perpendicular to the disk. The principal moments of inertia I_y, I_x are discussed in Section 7.6.

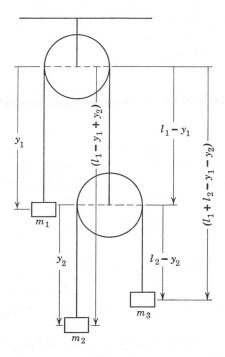

Fig. P-9   (Note velocity of m_1 is $\dot{y}_1$, of m_2 is $\dot{y}_2 - \dot{y}_1$, of m_3 is $-\dot{y}_1 - \dot{y}_2$.)

7. The axis of a gyroscope is mounted so that it is always horizontal but the frames containing the gyroscope can turn about a vertical axis. This system forms a simple gyroscopic compass. If a torque τ is applied to the frame about a vertical axis, show that the axis of the gyroscope comes to rest at an angle θ with the meridian where

$$\sin \theta = \frac{\tau}{I_y \omega_E s \cos \lambda}$$

The notation is that used in Section 7.11.

8. Show the period of the gyroscopic compass at latitude 40°N, and angular velocity of spin s of 1000 rad/sec, if the gyroscope is constructed such that $I_y = 2I_x$, is 18.8 sec.

9. Solve the following problems *a–i* by the Lagrange and Hamilton equations, that is, find the acceleration in terms of the displacement. (Calculate the kinetic and potential energies of the system in terms of suitable coordinates, q and $\dot{q}$ for L; and q and p for H.)

  a. An object of mass m is shot vertically upwards in a uniform gravitational field.

b. A mass m is held on the end of a vertical spring having negligible mass and a force constant k and undergoing simple harmonic motion.

c. A small mass m is held on the end of a light cord and set into oscillation as a simple pendulum.

d. A meter stick of mass m is suspended by a horizontal frictionless peg at a distance h from the center of mass. The stick is set into oscillation in a vertical plane. Show the equation of motion is $I\ddot{\theta} = -Mgh \sin \theta$.

e. A solid uniform sphere of mass m, radius R, rolls without slipping down an incline plane of angle of inclination α with the horizontal.

f. A bullet is shot with an initial speed v_0 at an angle ϕ with the horizontal. If the coordinate axes have their origin at the point of firing, then show by the Lagrange method that the equation of the trajectory is $y = x \tan \phi - (gx^2)/(2v_0^2) \cos^2 \phi$

g. Using the Lagrange method derive Eqs. 5.67 and 5.68 for the coupled pendulums.

h. Problems 11 and 13, Chapter 5.

i. The system shown in the figure consists of two light pulleys with masses connected by light cords of lengths l_1 and l_2. Show by the Lagrange method that the acceleration of m_3 is $g[4m_2m_3 + m_1(m_3 - 3m_2)]/[4m_2m_3 + m_1(m_2 + m_3)]$. (See Fig. P-9.)

10. Show that the time to go to the lowest point $(b\pi/2, b)$ or $\theta = \pi$, for an object starting from rest and moving along the path traced out by a cycloid is $\pi\sqrt{b/2g}$ assuming it starts from the origin, $(0, 0)$. Suppose the object starts from rest at some point (x_0, y_0) on the brachistochrone show the time to descend to the lowest point is again $\pi\sqrt{b/2g}$, where $v = \sqrt{2g(y - y_0)}$ and the limits of the integral are θ_0 and π. (To carry out the integral let $u = \cos(\theta/2)$ and the integral $du/\sqrt{a^2 - u^2} = \sin^{-1}(u/a) + C$.)

11. A body is dropped from the top of a tower 200 ft high at a latitude of $40°$ N. Show the easterly displacement of the point directly below the point of dropping is 0.316 in. Neglect air resistance but take into account the rotation of the earth.

12. A body is projected vertically upward with an initial velocity v at a latitude λ. Show that the distance between the point of projection and the point of return to the ground is $1.33\omega v^3 \cos \lambda/g^2$. Is the point of return east or west of the point of projection? In the above, ω is the angular velocity of the earth and air resistance is neglected.

PROPERTIES OF SOLIDS
AND LIQUIDS

8.1 Elasticity and Hooke's Law

In the previous chapters we have been concerned with the action of forces on rigid bodies. These bodies have been translated or rotated as a whole, but with the exception of the spring they have not been considered as changing in size or shape. Any changes in size or shape were of secondary importance, and their effects were assumed to be zero. We shall now drop this assumption.

Any change from the normal configuration of a body in either size or shape is a deformation, and the corresponding fractional change is a *strain*. In general such a change is the result of a force that acts either on the surface or on the body as a whole. If a force ΔF acts on area ΔA at some point of a body, then the *stress* at the point within this area is

$$\text{Stress} = \lim_{\Delta A \to 0} \frac{\Delta F}{\Delta A} = \frac{dF}{dA}$$

Where the force F is uniformly distributed over an area A, then the stress over the area A is F/A.

It is necessary in discussing stress-strain relationships in different materials to divide these into two main classes called *isotropic* and *anisotropic*. In isotropic materials the strain at any point is independent of direction, or the elastic properties of the material are the same in every direction. For anisotropic materials the elastic properties at any point are different in different directions about the point. Quartz and many other crystalline substances are anisotropic. They also show different optical properties in different directions.

Crystalline substances are built up from a characteristic unit cell with atoms of the substance so placed as to outline the shape of the cell. Large single crystals having a great number of unit cells in orderly arrangement can be made. It has been found that the elastic, thermal, and magnetic properties of these single crystals depend on the direction relative to the crystal axes. Most solids are polycrystalline, i.e., they consist of microscopic crystals, called grains, which are joined together at their boundaries. If the grains are oriented at random in the solid, then the variation in properties with crystal direction averages out and the polycrystalline solid behaves as an isotropic substance. Here we shall limit the discussion to isotropic materials.

We shall also assume that the materials are homogeneous, i.e., they have the same elastic properties at all points in the material. If this were not assumed, the analysis would be impossible. Thus the analyses given in this chapter apply only to *homogeneous isotropic materials*. Furthermore, unless the stress is relatively small, the strain is not proportional to it. Within the limits in which this proportionality holds, the modulus of elasticity is defined as the ratio of the stress to the corresponding strain. This region of proportionality is usually designated as the region in which Hooke's law is valid. Robert Hooke (1635–1703) showed experimentally for coiled and flat springs and the bending of wooden rods that, if "one power stretch or bend it one space, two will bend it two and three will bend it three, and so forward."

In the following discussion we shall assume that Hooke's law holds and that when the load is removed the body returns to its original shape or there is no permanent set. The modulus of elasticity of a substance is defined as the ratio of the stress to the strain produced by the stress. Corresponding to different kinds of strains, there are three simple moduli of elasticity.

8.2   Young's Modulus E and Poisson's Ratio σ

This modulus of elasticity E is concerned with the stretch produced by a simple tension. A convenient method of measuring Young's modulus for a substance is to prepare a long wire of uniform cross section from some of the substance. This wire is rigidly held at its upper end, and its elongation is measured for various loads placed on the lower end. The strain is measured by the increase in length divided by the original length, and the stress by the stretching force divided by the area of cross section of the wire. If a wire of length l and cross section A is subjected to a force F that produces an elongation Δl, then Young's modulus E for the material of the wire is

$$E = \frac{\text{Stress}}{\text{Strain}} = \frac{F}{A}\frac{l}{\Delta l}$$

When a wire is extended there is in general an accompanying decrease in the area of cross section. Suppose a wire of original length l_0 and diameter d_0 is subjected to a tension F, changing the length to l and the diameter to d. The longitudinal strain is

$$\frac{l - l_0}{l_0} = \frac{\Delta l}{l_0}$$

and the transverse strain is

$$\frac{d - d_0}{d_0} = -\frac{\Delta d}{d_0}$$

The ratio of these two strains is called Poisson's ratio σ, or

$$\sigma = \frac{\text{Transverse strain}}{\text{Longitudinal strain}} = -\frac{\Delta d}{\Delta l}\frac{l_0}{d_0} \tag{8.1}$$

For most substances the value of σ lies between 0.25 and 0.5. As may be appreciated from a numerical example, problem 1 at end of chapter, the change in cross-sectional area of a wire for any load within the elastic limit is negligible. Thus, in the expression for Young's modulus, the area A can be taken as a constant.

8.3 Bulk Modulus k

A second type of simple strain is the change in volume per unit volume when a substance is subjected to normal stresses over its whole surface. These normal stresses are most easily produced by immersing the substance in a fluid subjected to a pressure p. If under these conditions a normal inward force F acts on each element of area A of a substance of volume V to produce a decrease in volume of $-\Delta V$, then the bulk modulus k for this substance is

$$k = \frac{\text{Stress}}{\text{Strain}} = \frac{F}{A}\frac{V}{-\Delta V} = -\frac{pV}{\Delta V} \tag{8.2}$$

where p is the increase in pressure producing the decrease in volume $-\Delta V$. The value of the bulk modulus for most common metals is of the order of 10^{12} dynes/cm² $= 10^{11}$ N/m^2 (Newtons per square meter).

8.4 Shear Modulus or Modulus of Rigidity n

This third and final type of simple strain is that in which successive layers of the material are moved or sheared by tangential surface forces. This is

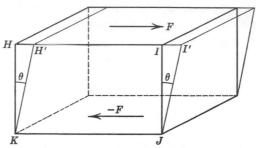

Fig. 8.1 Shearing of a rectangular block by a tangential force F.

represented in an exaggerated manner in Fig. 8.1. To the upper surface of the rectangular block a tangential force is applied, and an equal and opposite force $-F$ is applied to the lower surface. As a result, the upper surface is sheared relative to the lower. If the lower surface is considered fixed, then the upper edge is moved from the position HI to $H'I'$. Each layer is moved or sheared by an amount proportional to its distance from the lower fixed surface. The strain is measured by the ratio of the horizontal displacement of any layer to its distance from the fixed surface:

$$\text{Shearing strain} = \frac{HH'}{HK} = \tan\theta$$

where θ is called the angle of shear and is always very small if the elastic limit is not to be exceeded.

The shearing stress is the tangential force F divided by the area A over which it acts. Thus the modulus of rigidity n is given for Fig. 8.1 as

$$n = \frac{\text{Stress}}{\text{Strain}} = \frac{F}{A\tan\theta}$$

Table 8.1 gives the values of E, N, and k in dynes per square centimeter and the corresponding values of σ.

Theoretically the three moduli of elasticity for a homogeneous isotropic body are related by the expressions

$$E = 3k(1 - 2\sigma) = 2n(1 + \sigma) = \frac{9kn}{3k + n} \tag{8.3}$$

As may be checked from the values in the table, these relationships are only approximately correct. We shall not give the proofs since they add little to the understanding of elasticity.

For anisotropic substances no such simple relationships exist. It is shown in more advanced treatises that for anisotropic substances there are 21 elastic constants. The general expressions for stresses and strains are called dyadics

TABLE 8.1   Approximate Values of Elastic Moduli

Substance	Young's Modulus dynes/cm²	Shear Modulus, dynes/cm²	Bulk Modulus, dynes/cm²	Poisson's Ratio
Aluminum	7.0×10^{11}	2.4×10^{11}	7.5×10^{11}	0.34
Cast bronze	8.1×10^{11}	3.4×10^{11}	9.6×10^{11}	0.18
Copper	12.3×10^{11}	4.5×10^{11}	13.1×10^{11}	0.34
Gold	8.0×10^{11}	2.8×10^{11}	16.6×10^{11}	0.42
Lead	1.6×10^{11}	0.54×10^{11}	5.0×10^{11}	0.45
Silver	7.8×10^{11}	2.8×10^{11}	10.9×10^{11}	0.37
Steel	20.6×10^{11}	8.9×10^{11}	18.1×10^{11}	0.33
Tin	4.5×10^{11}	1.67×10^{11}	5.1×10^{11}	0.31
Quartz fiber	5.2×10^{11}	3.0×10^{11}	1.4×10^{11}	0.37
Glass, crown	7.0×10^{11}	3.0×10^{11}	5.0×10^{11}	0.24
Phosphor bronze	12.0×10^{11}	4.3×10^{11}	—	0.36

or tensors, and each contains nine terms. These, of course, reduce to the simple expressions we have obtained for isotropic substances.

8.5   Young's Modulus as Obtained from the Bending of Beams

We have already shown how Young's modulus for a wire can be obtained by measuring the extension produced by a known load. A less obvious method for obtaining Young's modulus utilizes the deflection of a beam under a given load.

Suppose that a rod such as a meter stick is slightly bent into a circular arc of radius R. There is a surface passing through CC' in Fig. 8.2 which does not undergo any change in length with the bending, whereas surfaces above and below CC' respectively undergo extension and contraction. The surface through CC' that remains the same length before and after bending is called the *neutral surface* or neutral plane. It will now be shown that the neutral surface passes through the center of mass of the rod.

Consider a surface DD' at a distance y above the neutral surface CC'. From Fig. 8.2 it follows that

$$CC' = R\phi \quad \text{and} \quad DD' = (R + y)\phi$$

so that the elongation of DD' is $y\phi$. Thus the linear strain of the surface DD', or its elongation divided by its original length, is $y\phi/R\phi$ or y/R. At a distance y below the neutral surface, there is similarly a linear strain of contraction of $-y/R$. These strains increase from zero at the neutral surface to a maximum at the surface of the rod. Corresponding to each of these

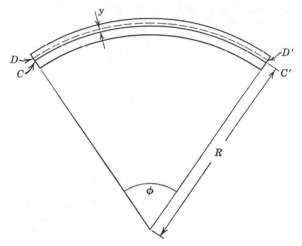

Fig. 8.2   The bending of a rod.

strains are stresses on the different layers. In order to evaluate these stresses
consider a thin layer of area dA in a cross section of the rod at a distance y
above the neutral surface as shown in Fig. 8.3. If F is the force acting perpendicular to the area dA, then by definition of Young's modulus

$$E = \frac{\text{Stress}}{\text{Strain}} = \frac{F}{dA}\frac{R}{y}$$

or

$$F = \frac{Ey\,dA}{R} \tag{8.4}$$

In Fig. 8.3b these forces are shown in a cross section of the rod. They act in
opposite directions above and below the neutral surface and produce elonga-
tions and contractions respectively. These forces are in translational equilibrium, so that their vector sum is zero or

$$\sum F = 0$$

and from Eq. 8.4 it follows that

$$\frac{E}{R}\sum y\,dA = 0$$

Since neither E nor R is zero, then $\sum y\,dA$ must be equal to zero. If the rod is
of uniform density, the masses dm are proportional to the areas dA so that
$\sum y\,dm = 0$. It thus follows from the definition of center of mass that the
neutral surface from which y is measured must pass through the center of
mass of the beam.

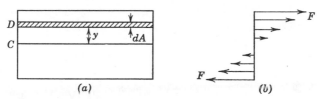

Fig. 8.3 Forces acting on a cross section of a bent rod.

Now in order for the beam to be bent, there must be external torques applied to the beam. These must be balanced by or in equilibrium with the moments of the internal stress forces. The moment of the stress force acting on dA at a distance y from the neutral surface is

$$Fy = \frac{E}{R} y^2 \, dA$$

and the total moment of all the forces acting over the whole cross section of the rod is

$$\sum \frac{E}{R} y^2 \, dA = \frac{E}{R} \sum y^2 \, dA = \frac{E}{R} I_A$$

where

$$I_A = \sum y^2 \, dA$$

The quantity I_A is very similar to the corresponding moment of inertia $\sum y^2 \, dm$. For this reason I_A is often called the *second moment of area*. Like the moment of inertia, its value depends on the axis about which it is taken.

Let us consider a rod whose second moment of area is I_A and whose value for Young's modulus is E. If an external torque τ acts on this rod so as to bend it in an arc of a circle whose radius is R, then

$$\tau = \frac{E}{R} I_A \tag{8.5}$$

This is the fundamental equation for the bending of beams.

It is often useful to express the radius of curvature R in terms of the derivatives in an XY coordinate system. Thus, as is proved in most elementary calculus books,

$$R = \frac{[1 + (dy/dx)^2]^{3/2}}{d^2y/dx^2} \tag{8.6}$$

In all the cases that we shall consider the deflections are within the elastic limit, and the slope dy/dx at any point in the beam is so small that it will be sufficiently accurate to take R as

$$R = \frac{1}{d^2y/dx^2}$$

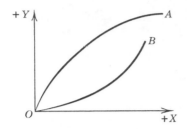

Fig. 8.4 Curves having positive and negative curvatures.

Thus from Eq. 8.5 the external torque τ is given by

$$\tau = EI_A \frac{d^2y}{dx^2} \qquad (8.7)$$

In applying this equation, one has to be careful of the signs of the terms, and for this purpose the curves in Fig. 8.4 are drawn. Along OA in Fig. 8.4 the slope dy/dx of the curve is continuously decreasing so that along this curve d^2y/dx^2 is negative. Similarly, along OB the slope dy/dx is continuously increasing so that along OB d^2y/dx^2 is positive. The torque τ about any point is positive whenever it has a direction such as to tend to cause the rod to have a positive curvature, i.e., to make d^2y/dx^2 positive. Conversely, τ is negative if it tends to produce a negative curvature.

8.6 The Shape of a Heavy Beam Clamped at One End and Loaded at the Other End

Let us determine the shape and depression of a uniform heavy beam clamped at one end and loaded at the free end. Suppose that the beam has a length of l and a weight of W'. A load W is placed on the free end, as shown in Fig. 8.5. Consider any point x, y on the beam with the X, Y axes drawn as shown in the figure. The torque exerted by the forces to the right of the point x, y is balanced by the internal torque in the beam exerted at the point x, y

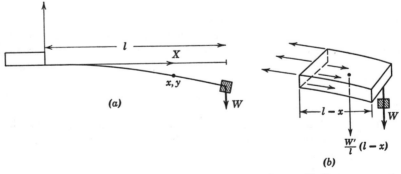

Fig. 8.5 (a) Deflection of a horizontally clamped heavy beam. (b) Forces on section of length $l - x$.

as represented in Fig. 8.5b. From Eq. 8.7 for the moments about the point x, y

$$EI_A \frac{d^2y}{dx^2} = -W(l - x) - \frac{W'}{2l}(l - x)^2$$

In this case d^2y/dx^2 is negative along the bent rod, and the moments that tend to bend the rod in the shape shown must be negative. The second term on the right-hand side of the equation is that due to the weight $[W'(l - x)]/l$ of the portion of the rod of length $l - x$ acting at its center of gravity a distance of $(l - x)/2$ from the point x, y. Integration gives

$$EI_A \frac{dy}{dx} = \frac{W}{2}(l - x)^2 + \frac{W'}{6l}(l - x)^3 + C_1$$

where C_1 is the constant of integration and is evaluated from the condition $dy/dx = 0$ at $x = 0$. Hence

$$C_1 = -\frac{Wl^2}{2} - \frac{W'l^2}{6}$$

Integrating a second time gives

$$EI_A y = -\frac{W}{6}(l - x)^3 - \frac{W'(l - x)^4}{24l} - \frac{Wl^2x}{2} - \frac{W'l^2x}{6} + C_2$$

where C_2, the constant of integration, is evaluated from the condition $y = 0$ at $x = 0$. Hence

$$C_2 = \frac{Wl^3}{6} + \frac{W'l^3}{24}$$

Thus the deflection y of the beam at any point x is given by

$$EI_A y = -\frac{W}{6}(l - x)^3 - \frac{W'(l - x)^4}{24l} - \frac{Wl^2x}{2} - \frac{W'l^2x}{6} + \frac{Wl^3}{6} + \frac{W'l^3}{24} \quad (8.8)$$

and the deflection y at the free end where $x = l$ is

$$y = -\frac{l^3}{EI_A}\left(\frac{W}{3} + \frac{W'}{8}\right) \quad (8.9)$$

Thus the effect of the weight W' of the beam on the deflection at the end of the beam is the same as if the beam were weightless and a weight of $3W'/8$ were added to the end.

8.7 The Shape of a Heavy Beam Clamped at One End and Supported at the Other End

The problem is to determine the shape and the depression at the center of a uniform heavy beam clamped horizontally at one end and supported at

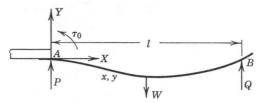

Fig. 8.6   Beam clamped at one end and supported at the other end.

the same level at the other end. Let us suppose that a beam having a length l and a weight W is supported as shown in Fig. 8.6. Let the clamped end A be the origin of the X, Y coordinates. At A there is an upward supporting force P and a torque τ_0 due to the clamp; at end B there is only an upward supporting force Q. Owing to the nature of the supports the beam is not symmetrical about the mid-point. For translational equilibrium

$$P + Q - W = 0$$

The torque about A is

$$Ql + \tau_0 - \frac{Wl}{2} = 0$$

From Eq. 8.7 for a point x, y we have

$$EI_A \frac{d^2y}{dx^2} = Px - \tau_0 - \frac{Wx^2}{2l}$$

where E is Young's modulus for the beam and I_A is the second moment of area of a cross section. The term $Wx^2/2l$ is the torque due to the weight Wx/l of the portion of the beam between A and x, whose center of gravity is at $x/2$.

By substitution for P and τ_0, we have

$$EI_A \frac{d^2y}{dx^2} = Wx - Qx - \frac{Wl}{2} + Ql - \frac{Wx^2}{2l}$$

Integrating this equation, using the boundary conditions, $x = 0$, $y = 0$, and $dy/dx = 0$, gives

$$EI_A \frac{dy}{dx} = \frac{Wx^2}{2} - \frac{Qx^2}{2} - \frac{Wlx}{2} + Qlx - \frac{Wx^3}{6l}$$

and

$$EI_A y = \frac{Wx^3}{6} - \frac{Qx^3}{6} - \frac{Wlx^2}{4} + \frac{Qlx^2}{4} - \frac{Wx^4}{24l}$$

To determine Q in terms of W, we must use another boundary condition, namely, $y = 0$ at $x = l$. This gives

$$Q = \frac{3W}{8}$$

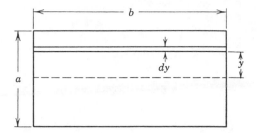

Fig. 8.7   Second moment of area of a rectangle.

Thus the deflection y at any point x is

$$y = \frac{Wx^2}{48EI_A l} (5xl - 3l^2 - 2x^2)$$

and the deflection at the mid-point $x = l/2$ is

$$y_{l/2} = -\frac{Wl^3}{192EI_A}$$

As an example, let us suppose that the beam is a meter stick having a length l of 100 cm, weight W of 150 gm, width of 2.5 cm, and thickness of 0.5 cm. Young's modulus E for the beam is 0.5×10^{11} dynes/cm². If the meter stick is placed with the 2.5-cm side horizontal, then from Fig. 8.7

$$I_A = \sum y^2 \, dA = \int_{-a/2}^{a/2} y^2 b \, dy$$

$$= \frac{ba^3}{12} = \frac{2.5 \times (0.5)^3}{12} = \frac{0.625}{24} \text{ cm}^4$$

Hence

$$y_{l/2} = -\frac{150 \times 980 \times 10^6 \times 24}{192 \times 5 \times 10^{10} \times 0.625} \frac{\text{dynes cm}^3}{\text{(dynes/cm}^2)\text{ cm}^4} = -0.588 \text{ cm}$$

or the center of the beam is depressed a distance of 0.588 cm below the horizontal owing to its own weight.

8.8   Period of Vibration of a Light Horizontal Beam Carrying a Load

The problem is to determine the period of oscillation of a light beam of length l clamped at one end and carrying a load W at the other, as shown in Fig. 8.8. If the beam is horizontal before the load W is added, then this load causes a depression AB, given by Eq. 8.9 as $Wl^3/3EI_A$, since the beam is

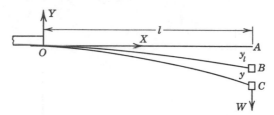

Fig. 8.8 Period of oscillation of horizontally clamped light beam carrying a load W.

assumed to be weightless or $W' = 0$. The equilibrium position of the beam with the load W on it is OB.

Suppose now that the end of the beam is pulled down from B to C, a distance y below the equilibrium position. If F is the applied force, then there is a total deflection of $y_l + y$ due to the total force $W + F$ where, by Eq. 8.9,

$$y_l + y = - \frac{l^3}{EI_A}\left(\frac{W + F}{3}\right)$$

By substitution of the value of y_l, it follows that

$$y = - \frac{l^3 F}{3EI_A}$$

The force F required for a displacement y below the equilibrium position is

$$F = - \frac{3EI_A y}{l^3}$$

If the beam is released when in the displaced position, it starts upward with an acceleration $\ddot{y}$ given by Newton's second law as

$$\ddot{y} = \frac{F}{M} = \frac{F}{W/g} = - \frac{3EI_A g y}{W l^3} \tag{8.10}$$

where the mass of the body moved is W/g. The negative sign appears because the acceleration is in the opposite direction to the displacement y. The acceleration $\ddot{y}$ in Eq. 8.10 is proportional to and in the opposite direction from the displacement y. This equation then represents simple harmonic motion, the period of which is

$$P = 2\pi \sqrt{\frac{W l^3}{3EI_A g}} = 2\pi \sqrt{\frac{M l^3}{3EI_A}} \tag{8.11}$$

In this analysis the mass of the beam has been neglected. If this is to be included, then one has to take into account the fact that different portions of the beam move with different velocities. The clamped end has zero velocity whereas the free end has maximum velocity. An analysis of this problem shows that the period of oscillation of a heavy beam of weight W' clamped at one end and carrying a load W at the other end is

$$P = 2\pi \sqrt{\frac{[(W/g) + (33W'/140g)]l^3}{3EI_A}}$$

8.9   Measurement of the Modulus of Rigidity n by a Static Method

One of the simplest means of determining the modulus of rigidity is to measure the angle of torsion in a cylindrical rod produced by a known torque. Suppose that a uniform cylindrical rod is hung vertically with its upper end rigidly clamped. The lower end is subjected to a known torque, and the resulting angle of twist is measured, Fig. 8.9a. Each cross section of the cylindrical rod is twisted through some angle proportional to the distance of the cross section from the upper fixed end.

Suppose that the cylindrical rod has a length l and a radius R, Fig. 8.9b, and that a straight line AB on the unstrained rod is twisted into position AC by the torque at the lower end. The angle BAC on the rod is ϕ, and the angle through which the point B is turned is θ. From the geometry of the figure

$$l\phi = R\theta = \text{arc } BC$$

Consider now a magnified cross section of the lower end of the rod, Fig. 8.9c. The shear on this is measured by the angle ϕ. If F is the shearing force on an annulus of radius r and thickness dr, then the coefficient of rigidity n is

$$n = \frac{F}{2\pi r \, dr} \frac{1}{\phi}$$

or

$$F = 2\pi n \phi r \, dr$$

The moment of F about the center O is

$$Fr = 2\pi n \phi r^2 \, dr$$

For this annulus

$$l\phi = r\theta$$

where θ is constant over the cross section. Hence

$$Fr = \frac{2\pi n \theta}{l} r^3 \, dr$$

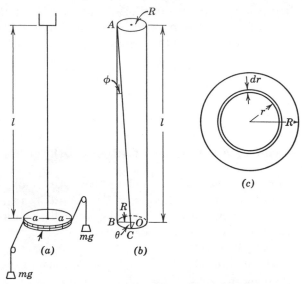

Fig. 8.9   Measurement of modulus of rigidity n.

These shearing moments act over the whole cross section, and their total sum is equal to the applied moment of force or torque τ. Thus

$$\sum_0^R Fr = \tau = \int_0^R \frac{2\pi n\theta}{l} r \; dr$$

or

$$\tau = \frac{\pi n\theta R^4}{2l} \tag{8.12}$$

The torque required to produce an angle of twist of 1 radian is often called τ_0, the torsion constant, and is equal to τ/θ, or

$$\tau_0 = \frac{\pi n R^4}{2l}$$

Thus, by measuring the quantities τ_0, R, and l, the modulus of rigidity n may be obtained. In Fig. 8.9a the torque τ is equal to $2mga$, and θ is measured on a graduated scale so that τ_0 is readily obtained.

8.10   Measurement of the Modulus of Rigidity n by a Dynamic Method

Suppose that the cylindrical rod in Fig. 8.9a is fixed at its upper end and has a concentric circular disk attached to its lower end. If the lower end and disk are twisted through some angle θ, then the system is subjected to a torque given by Eq. 8.12. This torque produces an angular acceleration $\ddot{\theta}$. If I is the moment of inertia of the system about an axis through the longitudinal axis of the rod, then the equation of motion of the system is

$$\frac{\pi n R^4 \theta}{2l} = -I\ddot{\theta} \tag{8.13}$$

The negative sign appears because the angular acceleration is toward the origin while θ increases away from the origin. Equation 8.13 represents angular harmonic motion whose period P is

$$P = 2\pi \sqrt{\frac{2lI}{\pi n R^4}} \tag{8.14}$$

The moment of inertia I is the sum of the moments of the inertia of the disk and of the cylindrical rod about the central axis. In general, the moment of inertia of the disk is so much larger than the moment of inertia of the rod that the latter may be neglected. From Eq. 8.14 it follows that the modulus of rigidity of the material used in the rod shown in Fig. 8.9 is given by

$$n = \frac{8\pi lI}{P^2 R^4}$$

By placing two similar bodies on the disk at equal distances from the axis and then determining the new period, it is readily seen that the moment of inertia of the bodies can be obtained. This is a fairly common method of obtaining the moment of inertia of unsymmetrical bodies since it is not easy to calculate them.

8.11   A Qualitative Account of some Properties of Solids

The advance in knowledge in the present century has nowhere been more evident than in the understanding of the properties of matter. Solid state physics is now a thriving part of physics and a basis of a large technology. We have only to consider that transistors, colored television with its solid state phosphors, and many other smaller but important advances have come into being since the end of World War II. These advances have come about through the understanding of the microscopic properties of the solid state. It is obvious that this book cannot do justice to the properties of matter as this term is now understood.

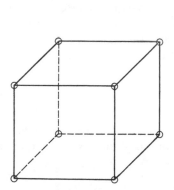

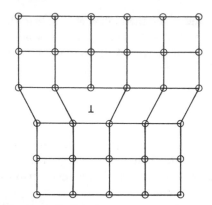

Fig. 8.10   Unit cell of a simple cubic crystal.

Fig. 8.11   A dislocation in a row of atoms.

During the nineteenth century, physicists regarded matter as essentially structure-less and measured elastic constants that could be put into increasingly complicated equations. The simple theories for isotropic substances used scalar quantities for the various moduli of elasticity: shear, Young's, and bulk moduli. For anistropic substances, in which the moduli change with direction, the scalar quantities were replaced by tensors resulting in much more complicated relationships. It was the x-ray diffraction experiments of Freidrich, Knipping, and von Laue in 1912 and a short time later by the Braggs, father and son, that allowed physicists to "look inside" crystals and measure the distances between the regularly spaced atomic planes. As time went on, other methods were used among which were the electron microscope, neutron diffraction, electrical and thermal conductivities of crystals, etc., so that the investigation of the solid state has become a separate science with its theoretical and experimental investigations.

The diffraction experiments showed that all metals and most solids are crystalline, that is, the atoms or ions are located in a regular geometrical pattern (represented in Fig. 8.10) for a conventional unit cell of a cubic crystal. These unit cells are re-peated so as to form a lattice, in which the unit cells are frequently represented as points. A single crystal is composed of a huge number of unit cells or lattice points, although it is almost impossible to produce a perfect crystal without any imperfec-tions.

That the atoms or ions in a crystal can move was shown by an English metal-lurgist, Roberts-Austin, in 1896. In order to show the diffusion of one metal into another he fused a thin disk of gold on the top of an inch-long cylinder of lead. This gold-lead cylinder was kept at about 200°C for ten days after which it was cut into thin sections. The amount of gold in each section was measured, and it was found that this progressively decreased from the end originally having the gold disk. There was even a measureable amount of gold at the end of the lead cylinder. It was also found that lead had diffused into the gold.

One of the problems that the early investigators encountered was that the strength of materials was much less than that theoretically expected from quantum mechanical considerations. For iron the measured values were of the order of 1500 psi (lb/in^2) whereas the theoretical values were about 1,600,000 psi. An attempt to resolve this discrepancy was made in 1934 independently by G. I. Taylor in England and E. Orowan in Germany, who suggested that the strength of crystalline materials depended not on their average properties but on imperfections within the crystal called *dislocations*. The two most important types of dislocations are the edge and screw dislocations, though in many cases these appear to merge into one another. Such dislocations have been observed on the surface of a crystal in the process of deposition of atoms in the crystal growth. A representation of a dislocation is shown in Fig. 8.11 in which there is a mismatched region or there is a plane of atoms missing from the crystal lattice. If shear forces are applied above and below the dislocation plane, marked by an inverted T, it is assumed that the dislocation would move or the material would undergo shear. Calculations on this model showed that the theoretical strength was less than the measured value, that is, the theory worked too well. With further development of theory it became apparent that this could be accounted for if it was assumed that dislocations, being regions of high potential energy, could immobilize one another in crystals. In general the strengthening of metals is interpreted as the immobilization of the dislocations. For instance, the empirical metallurgical processes have been explained in terms of the modified dislocation theory. The addition of carbon in iron from one to two parts of carbon in a million parts of iron increases the strength of the metal by about four times. Hammering, cold rolling, or bending metals can strengthen them. If a piece of ductile copper strip is bent around it is found to be considerably strengthened—the dislocations having largely immobilized one another. If such a piece of workhardened copper is heated it is found that at a few hundred degrees centigrade it gives out heat, and this exothermic energy is interpreted as the energy released by the dislocations. Copper exposed to a large dose of neutrons is found to be hardened. It is considered that neutrons introduce lattice defects which immobilize the dislocations so the copper becomes so hard that it can produce a sound similar to that of a tuning fork.

Most solids are made up of a large number of lattice points regularly arranged in units called grains with the grains having irregular boundaries. In some materials grains can be seen with the unaided eye. With the grains oriented at random the material behaves isotropically with no preferred directions. It is considered that dislocations cannot readily pass across grain boundaries, so that controlling the size of the grains is one of the processes used in strengthening steel, the smaller the grain size the larger the strength. By heating, rolling, and cooling, the grains in iron can be approximately aligned so that the iron has desirable magnetic properties. It is found that the degree to which iron can be magnetized depends on the direction within the unit cell of the iron, as shown in Fig. 8.12. These properties have been made use of in transformers and other electromagnetic equipment.

It is estimated that there may be several million dislocation lines in each square centimeter of a crystal. However, it is possible to have small, so-called "whiskers" grow on iron that are free of dislocations; and the strength of the whiskers approaches the theoretical value. Silicon crystals have been drawn from a melt in a way that the crystals have no dislocations. Nevertheless, it is found that dislocations do develop with time, and these are thought to come from minute imperfections on the surface which eventually moved into the whole crystal. Very small imperfections on the surface of metals, which are usually too small to be seen without considerable magnification, are considered to grow in the crystal lattice, somewhat as if a wedge were being driven in, and eventually to cause considerable weakness in the metal. This is one of the processes that accounts for fatigue and brittleness in metals.

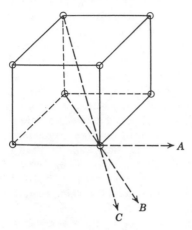

Fig. 8.12 Magnetization of iron crystal depends on direction. Along direction A the magnetization is the largest, along B it is intermediate, and along C it is least.

Investigations on the thermal and electrical conductivities of metals and other solids, especially at very low temperatures, have yielded much information on the structure of substances. It is a well-known experimental fact that metals are good conductors of electricity and heat. As one proceeds from the metals to the semiconductors to the insulators, one finds a striking change between the conductivity of electricity and the conductivity of heat in these substances as shown in Fig. 8.13.

In this figure the number of free electrons per unit volume in the substances is plotted against the electrical and thermal conductivities. An interpretation of these results has been given in terms of the quantum theory, and in this brief survey only a qualitative discussion can be given.

A metal has an abundance of free or conduction electrons that do not belong to

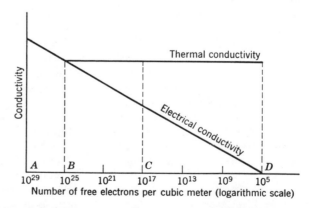

Fig. 8.13   Thermal and electrical conductivities plotted against the number of free electrons. Region A–B represents the metals; B–C the semiconductors; C–D the insulators.

individual atoms but to the solid as a whole. These free electrons account for the relatively large conductivities of heat and electricity. In moving through a metal the conduction electrons are diverted or scattered by the thermal vibrations of the atoms in the lattice and by any imperfections such as substitution or foreign atoms in the lattice. According to the theory, a perfect crystal, having no thermal motions, would produce no electron scattering and would have an infinite conductivity. In a real metal lattice the important factor is the distance the electron can travel before being scattered, that is, the mean free path for scattering. The greater this is, the greater the conductivity. Although it is the electric charge of the electron which is involved in electrical conductivity and the kinetic or thermal energy of the electron in thermal conductivity the scattering of electrons applies to both kinds of conductivities so that they should be proportional to each other. This proportionality of the thermal and electrical conductivities was given the name many years ago of the Wiedemann-Franz law and is shown in the upper left corner of Fig. 8.13.

The changes in the two conductivities with temperature are markedly dissimilar. The electrical conductivity of most metals varies inversely as the absolute temperature, whereas the thermal conductivity is approximately constant and independent of temperature. This may be accounted for theoretically by the increase in atomic vibrations and electron scattering with increase in temperature. Thus the electron mean free path for scattering is decreased with increase in temperature, which at first sight would imply that both conductivities would decrease with increase in temperature. This is the case for the electrical conductivity. For the thermal conductivity it is the kinetic energy of the electrons that is involved, and this increases with temperature and balances the decrease resulting from electron scattering so that the thermal conductivity is approximately constant.

Consider now the effect of the change in the free electron density as one proceeds from metals to semiconductors to insulators. As shown in Fig. 8.13 the electrical conductivity decreases approximately linearly with the decrease in the free electron density whereas the thermal conductivity, after the initial drop, remains approximately constant and independent of the free electron density. There are thus no effective solid heat insulators. Some other process must be introduced to account for the approximately constant thermal conductivity with the comparatively low density of the free electrons. This new process is concerned with the vibrations of the atoms in the lattice under the action of the interatomic forces. These vibrations have frequencies of the order of 10^{13} per sec. The waves associated with these vibrations travel with the speed of sound in the solid. Quantum theory assumes that these vibrations are quantized having energy of $h\nu$ where ν is the frequency of the vibrations and h is Planck's constant. This quantum unit of sound waves, having energy $h\nu$, is called a *phonon*, being a first cousin to the *photon* (a quantum unit of energy of electromagnetic radiation). It should be noted that Planck's constant is so small (6.62×10^{-34} joule-sec) that it is practically impossible to detect phonons in audible sounds in air. The number of phonons in a solid increases rapidly with the temperature of the solid since the kinetic energy of the vibrating atoms increases with temperature. However, it is the mean free path that a phonon travels before being scattered by imperfections that controls the thermal conductivity of the solid. There is also phonon-phonon scattering that becomes important at high temperatures. Heavy atomic imperfections also produce relatively large scattering, so that it has been found possible to detect and measure, at liquid helium temperatures, heavy foreign atoms in concentrations of about 1 part in 10^6 by the change in thermal conductivity produced by these foreign atoms.* As exercises, consider the following statements.

   a. On what principle or principles do the practical heat insulators operate? Consider a blanket, eiderdown, and vacuum flask.

   b. If the frequency of the sound waves in a solid is 10^{13} per sec and the speed of sound in this solid is 5000 m/sec, find the wavelength of the phonons and their energy.

8.12   The Elastic Properties of the Three States of Matter

   The three states of matter—solid, liquid, and gaseous—are characterized by very different physical properties which for our present purposes can be summarized in the following manner. Large forces are required to change the shape or volume of a solid. With liquids, although a large force is required to change their volume, practically no force is required to change their shape. In other words, liquids have a large bulk modulus but practically no rigidity or shear modulus. Almost no force is required to change the shape of a gas, and only a small force to change the volume. Thus gases have practically no rigidity or shear modulus and only a small bulk modulus.

   When we attempt to place every substance into one of these three classes, we experience some difficulty. For example, ice is considered to be a solid. However, this substance is known to flow very slowly downhill under the action of gravity. The flow of glaciers is well known to geologists. Since the flow of so-called solids is rarely met, we need not be concerned with such solids here.

   In many cases the properties of gases and liquids are similar and can be discussed under the single heading of fluids. For the present we shall deal with perfect fluids, which exhibit no forces opposing changes in shape, or have no shear modulus. In a later section this assumption will be modified and the results discussed under the subject of viscosity. It will be further assumed that solids, liquids, and gases are continuous throughout their volume. This, of course, contradicts the atomic or molecular theory of matter. However, it is easy to show that any very small element of volume contains a very large number of molecules; for example, 1 mm³ of air contains about 3×10^{16} molecules. Thus for our present discussion, we may consider matter

---

* *Sci. Am.* September 1967. This issue is completely devoted to articles on the properties of materials.

to be continuous, as we have already done in the discussion of elasticity of solids.

First we shall discuss the theory of fluids in equilibrium, i.e., hydrostatics, and later some of the phenomena exhibited by fluids in motion, i.e., hydrodynamics.

8.13   Pressure in Hydrostatics

Let us consider a small plane area ΔA inside a perfect fluid at rest. A force ΔF is exerted by the fluid on one side of this area, and an equal and opposite force is exerted by the fluid on the other side. The pressure p at this area is defined as the force per unit area of

$$p = \frac{\Delta F}{\Delta A}$$

At the point at which this area is situated the pressure is given by

$$p = \lim_{\Delta A \to 0} \frac{\Delta F}{\Delta A} = \frac{dF}{dA} \tag{8.15}$$

By writing Eq. 8.15 as $d\mathbf{F} = \mathbf{n}_1 p \, dA$, where $\mathbf{n}_1$ is a unit vector along the outward drawn normal, it appears that the pressure p is a scalar quantity. Actually, pressure is a tensor, though in the simple problems we can consider it to be a scalar just as we consider temperature to be.

The force exerted by a fluid at rest on any area must always be at right angles to the area. If the force is not at right angles, then it has a component along the surface, and the equal and opposite force exerted on the fluid would cause the fluid to move, which is contrary to the assumption of the fluid at rest.

To obtain the variation of pressure with depth in a fluid, consider an imaginary right circular cylinder of cross-sectional area A and height $y_2 - y_1 = \Delta y$ drawn in the fluid. The cylinder is in equilibrium under the vertical forces of the weight of the fluid in the cylinder and the forces on the bottom and top due to pressure, Fig. 8.14. The weight of fluid in the cylinder is $\rho A g \, \Delta y$, where ρ is the density of the fluid. The vertical forces on the faces of the cylinder are $(p_2 - p_1)A = \Delta p A$. Hence for equilibrium

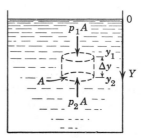

Fig. 8.14   Pressure increases with depth.

$$\Delta p = \rho g \, \Delta y$$

or in the limit

$$dp/dy = \rho g \tag{8.16}$$

If ρ is constant, then

$$p = \rho g y + \text{constant}$$

where y increases downward. If the boundary conditions are such that at $y = 0, p = p_0$, then the pressure at depth y is

$$p = p_0 + \rho g y \tag{8.17}$$

The pressure increases linearly with downward depth in a liquid of constant density.

The common units of pressure are dynes per square centimeter, pounds weight per square foot, and pounds weight per square inch. In meteorological work a unit of pressure called the *bar* is introduced. It is equivalent to a million dynes per square centimeter. Since normal atmospheric pressure is the pressure equal to that exerted by a column of mercury 76 cm high and 13.6 gm/cm³ in density, it is given by Eq. 8.17 as:

$$
\begin{aligned}
p_{\text{atm}} &= 13.6 \times 980 \times 76 \\
&= 1.013 \times 10^6 \text{ dynes/cm}^2 \\
&= 1.013 \text{ bars}
\end{aligned}
$$

For a large mass of gas such as exists in our atmosphere the density ρ is not a constant independent of the height above sea level and the pressure is not a simple linear function of the height.

To consider the forces exerted by a liquid let us think of a vessel filled with water, as shown in Fig. 8.15. The problem is to find the forces exerted by the water on the various portions of the vessel. The force on the base is

$$
\begin{aligned}
F &= pA = hg\rho A \\
&= 2 \times 62.4 \times 2 \times 1.5 = 374.4 \text{ lbw}
\end{aligned}
$$

We have taken ρg as equal to 62.4 lbw/ft³ for water.

To determine the force on any of the sides consider a horizontal strip along one of the sides at a depth y from the top and a width dy. The pressure at this strip produced by the water is $y\rho g$, and the force exerted on such a strip is

$$
\begin{aligned}
F &= 1.5 \int_0^2 y g \rho \, dy = \left[1.5 g \rho \frac{y^2}{2} \right]_0^2 \\
&= 1.5 \times 62.4 \times \frac{2^2}{2} = 187.2 \text{ lbw}
\end{aligned}
$$

Next let us examine the vertical height at which this force acts. Suppose that this is a distance y_0 from the top of the vessel. Then the moment of the resultant force, about the top edge of this wall as the axis, must be equal to the sum of the

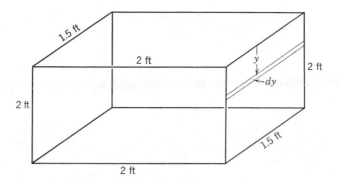

Fig. 8.15 Forces exerted by a liquid on the sides of a vessel.

moments of all the strips making up this wall about the same axis. Thus

$$y_0 = \frac{1.5\rho g \displaystyle\int_0^2 y^2 \, dy}{1.5\rho g \displaystyle\int_0^2 y \, dy} = \frac{8}{3} \times \frac{2}{4} = \frac{4}{3} \text{ ft}$$

Therefore the point at which this resultant force acts, sometimes called the center of pressure, is $\frac{4}{3}$ ft from the top and $\frac{2}{3}$ ft from the bottom of the liquid.

8.14 Archimedes' Principle and Stability of Floating Objects

An important principle in hydrostatics was first presented by Archimedes (287–212 B.C.). This principle states that when a body is immersed in a fluid there is a resultant upward vertical force exerted on the body by the fluid equal to the weight of the fluid displaced by the body. The bouyant force on a body of volume V immersed in a fluid of density ρ is then

$$F_y = \int_0^V g\rho \, dV = g\rho V \tag{8.18}$$

Now ρV is the product of the density of the fluid and the volume of the body and is, therefore, the mass of fluid displaced by the body. Thus Eq. 8.18 is a mathematical statement of *Archimedes' principle*.

When a body is floating in a liquid, the weight of the liquid displaced must equal the weight of the body. In other words, for a body to float in a liquid the average density of the body must be less than the density of the liquid. This, of course, must be the case for a ship whose hull is made of steel. Not only must the ship float but it must also be stable in the water.

If the ship is disturbed from its equilibrium position, there must be forces present which, when brought into play, will tend to return the ship to its equilibrium position.

The upward buoyant force on the ship acts through a point in the ship called the *center of buoyancy* which is at the center of gravity of the displaced liquid before its displacement. To show that this is true, let us consider what would happen if the ship were removed and the displaced water were back in its former position. The water surrounding the ship is in equilibrium with either the ship or with the water replacing the ship. In the latter case the force is exerted at the center of gravity of the displaced water and, consequently, the center of gravity of the displaced water must be the same as the point in the ship where the buoyant force acts.

Consider the ship shown in Fig. 8.16a, whose center of gravity is at G and whose center of buoyancy is at B. In Fig. 8.16b the ship is shown slightly displaced so that the center of buoyancy is now at B'. Let the vertical line through B' intersect the line of symmetry BG at M, called the *metacenter*. Figure 8.16b shows that, if M is above G, the equilibrium is stable; if M is below G, the equilibrium is unstable; and if M coincides with G, the equilibrium is neutral. The distance GM is called the *metacentric height h*. Thus the equilibrium is stable, unstable, or neutral, according to whether h is positive, negative, or zero.

In practice, the point M and the metacentric height h depend on the angle of displacement θ (see Fig. 8.16b), but for relatively small displacements the point M can be considered to be fixed. We shall now determine the period of roll of the ship in terms of the metacentric height h, assuming it to be positive. Consider the motion about the center of mass of the ship. The torque about G is

$$V\rho gh \sin \theta$$

since the buoyant force is $V\rho g$, where V is the volume of displaced water whose density is ρ. This torque produces an angular acceleration $\ddot{\theta}$ about the

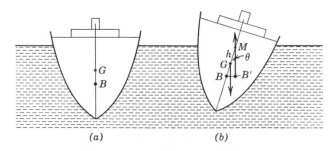

(a)   (b)

Fig. 8.16 Equilibrium of a floating body.

center of mass of the ship given by

$$I\ddot{\theta} = -V\rho gh \sin \theta$$

where I is the moment of inertia of the ship about a horizontal axis through its center of mass.

If θ is sufficiently small so that it may be substituted for $\sin \theta$, then

$$\ddot{\theta} = -\frac{V\rho gh\theta}{I}$$

This is the equation for simple harmonic motion whose period P is

$$P = 2\pi \sqrt{\frac{I}{V\rho gh}}$$

Thus the period is inversely proportional to the square root of the meta-centric heght h. A very stable ship has a large metacentric height and, consequently, a relatively small period. This may be very safe but gives somewhat uncomfortable riding.

The positions of G and M depend to a large extent on the load the ship is carrying. When a ship has no pay load, it usually carries ballast purely for reasons of safety. The angle through which a ship can roll safely depends, of course, on how the ship is built, but for most ships it is of the order of $40°$. Long before this angle of roll is reached, most passengers would be sure that the safety angle had been passed.

8.15 Hydrodynamics. The Equation of Continuity

The problems associated with fluids in motion can be very complex, so we shall analyze only some of the simplest of them. Let us consider first what is meant by the stationary or *steady flow* of a fluid. In this type of flow the velocity of the fluid at any one point is always the same. Of course the velocity will probably be different at different points along a line of flow. Any small element of volume of the fluid follows a particular path called a *stream line*. The tangent at any point in the stream line gives the direction of the velocity of the fluid at the point. If V is the velocity and ds is a distance measured along the steam line whose components along the X, Y, Z axes are dx, dy, dz respectively, then the equation of the stream line in differential form is

$$\frac{ds}{V} = \frac{dx}{u} = \frac{dy}{v} = \frac{dz}{w} \tag{8.19}$$

where u, v, w are the components of V along the X, Y, Z axes respectively.

The stream lines in a liquid can be made visible by inserting in the liquid a very small amount of colored liquid and following the lines taken by the colored liquid. It follows from the definition of stream lines that an element of fluid can never pass from one stream line to another. Suppose that the flow of the fluid is such that the velocities of the fluid elements over any cross section perpendicular to the flow are constant. The stream lines then form a tube of flow which may be considered a bundle of stream lines. Let Fig. 8.17 represent a tube of flow which might be an ideal liquid flowing in a tube with a constriction. Over the cross-sectional area A_1 the velocity is v_1 and, similarly, over A_2 it is v_2. If ρ_1 is the density of the fluid at A_1, then $A_1 v_1 \rho_1$ represents the mass of fluid crossing the area A_1 each second. Similarly, if ρ_2 is the density of the fluid at the cross section A_2, then $A_2 v_2 \rho_2$ is the mass of fluid crossing the area A_2 per second. If there are no sources or sinks within the volume between A_1 and A_2, then the mass of fluid flowing into the tube per second at A_1 is equal to the mass of fluid flowing out of the tube per second at A_2. This is a statement of the principle of conservation of matter. Thus we have

$$A_1 v_1 \rho_1 = A_2 v_2 \rho_2 \tag{8.20}$$

If the fluid is incompressible or if its density is constant, as is very approximately true for all liquids, then

$$A_1 v_1 = A_2 v_2 \tag{8.21}$$

Thus where the cross section is large, the velocity is small, or "still water runs deep."

The principle of continuity expressed in Eqs. 8.20 and 8.21 can be given in a differential form. Let $\mathbf{V}$ be the velocity at some point x, y, z with components u, v, w parallel to the X, Y, Z axes respectively. Around the point x, y, z construct a parallelopiped whose sides have the lengths dx, dy, dz respectively, Fig. 8.18. Consider the fluid flowing through this element of volume. Into the left-hand face whose area is $dy\,dz$, perpendicular to the X axis, there is a mass of fluid flowing per second equal to $\rho u\,dy\,dz$. Through the right-hand face there, to a first approximation, is a mass of fluid flowing out per second equal to

$$\left[\rho u + \frac{\partial(\rho u)}{\partial x}\,dx \right] dy\,dz$$

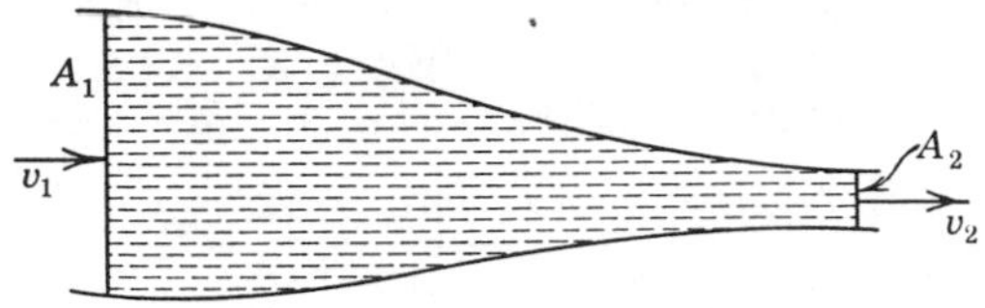

Fig. 8.17 The principle of continuity in a liquid.

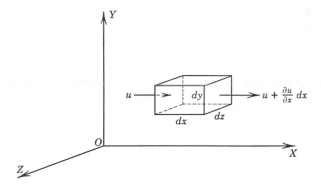

Fig. 8.18   The principle of continuity for a small parallelopiped of liquid.

The net outward flow per second through the parallelopiped in the X direction is

$$\frac{\partial(\rho u)}{\partial x} \, dx \, dy \, dz$$

Similar expressions hold for the net outward flow in the Y and Z directions. Hence the total excess mass of fluid flowing per second out of the parallelopiped over that entering each second is

$$\left[\frac{\partial(\rho u)}{\partial x} + \frac{\partial(\rho v)}{\partial y} + \frac{\partial(\rho w)}{\partial z}\right] dx \, dy \, dz$$

This expression gives the rate of loss of mass of fluid within the volume $dx \, dy \, dz$. Hence, if there are no sources or sinks within this volume, the equation of continuity can be written in the following manner:

$$\left[\frac{\partial(\rho u)}{\partial x} + \frac{\partial(\rho v)}{\partial y} + \frac{\partial(\rho w)}{\partial z}\right] dx \, dy \, dz = -\frac{\partial \rho}{\partial t} \, dx \, dy \, dz$$

or dividing by $dx \, dy \, dz$

$$\frac{\partial(\rho u)}{\partial x} + \frac{\partial(\rho v)}{\partial y} + \frac{\partial(\rho w)}{\partial z} + \frac{\partial \rho}{\partial t} = 0 \qquad (8.22)$$

If the density of the fluid is a constant, then the *equation of continuity* becomes

$$\frac{\partial u}{\partial x} + \frac{\partial v}{\partial y} + \frac{\partial w}{\partial z} = 0 \qquad (8.23)$$

Eqs. 8.22 and 8.23 can be written in an abbreviated form by means of the vector differential operator del written as

$$\nabla = \mathbf{i}\frac{\partial}{\partial x} + \mathbf{j}\frac{\partial}{\partial y} + \mathbf{k}\frac{\partial}{\partial z}$$

where $\mathbf{i}$, $\mathbf{j}$, $\mathbf{k}$ are the unit vectors along the X, Y, Z axes. The velocity $\mathbf{V}$ of the fluid has the components u, v, w so that

$$\mathbf{V} = \mathbf{i}u + \mathbf{j}v + \mathbf{k}w \qquad (8.24)$$

Writing the dot product of $\mathbf{V}$ and $\rho\mathbf{V}$, we may express the equation of continuity, Eq. 8.22, as follows:

$$\nabla \cdot (\rho\mathbf{V}) + \frac{\partial \rho}{\partial t} = 0$$

If the density ρ is constant as in Eq. 8.23. then this may be written

$$\nabla \cdot \mathbf{V} = 0$$

The quantity $\nabla \cdot \mathbf{V}$ (del dot $\mathbf{V}$) is called the divergence of the vector $\mathbf{V}$. It is a measure of how much more of a vector quantity leaves a unit volume than enters the unit volume per second. The equation of continuity for electric charges is similar to that given above. If j is the electric current density or the amount of electric charge crossing unit area per second and ρ is the volume density of charge at the point, then the law of conservation of electric charge requires that

$$\nabla \cdot j + \frac{\partial \rho}{dt} = 0 \qquad (8.25)$$

8.16   The Eulerian Equations of Motion of a Liquid

In this mode of analysis, first given by Euler about 1755, we choose a particular point in space and try to determine the velocity, density, and pressure of the liquid at that point for all instants of time. First we find the expression for the acceleration at a point in the liquid.

Suppose that a fluid element is at P, the point x, y, z, at time t, and in an interval of time dt moves to Q, the point whose coordinates are $x + dx$, $y + dy$, $z + dz$. Let $\mathbf{V}$ be the velocity of the fluid element at P which has components parallel to the coordinate axes of u, v, w respectively. Then each of the components u, v, w are functions of the independent variables x, y, z, and t. The increase in the X component of the velocity of the fluid element in moving from P to Q is

$$du = \frac{\partial u}{\partial t} dt + \frac{\partial u}{\partial x} dx + \frac{\partial u}{\partial y} dy + \frac{\partial u}{\partial z} dz \qquad (8.26)$$

The X component of the velocity u is equal to dx/dt. Similarly

$$v = \frac{dy}{dt} \qquad w = \frac{dz}{dt}$$

Thus from Eq. 8.26 the acceleration of the fluid element at P in the X direction is

$$\frac{du}{dt} = \frac{\partial u}{\partial t} + u\frac{\partial u}{\partial x} + v\frac{\partial u}{\partial y} + w\frac{\partial u}{\partial z} \qquad (8.27)$$

Similar expressions hold for the components of acceleration along the Y and Z axes. In moving from P to Q there is a step forward in time and a step forward in distance. The expression for the acceleration, Eq. 8.27, contains these two factors: the first term on the right is caused by the time factor; the remaining three terms are caused by the space factor. Notice that the partial derivative of the component velocity u with respect to time is the rate of change of the velocity u with respect to time but not with respect to distance.

To relate the acceleration to the forces, we shall choose a small rectangular parallelopiped drawn about the point P in the liquid, shown in Fig. 8.19. Let p be the pressure, ρ the density, and $\mathbf{F}$ the extraneous *force per unit mass* at the point P. Suppose that the force $\mathbf{F}$ has components F_x, F_y, F_z along the X, Y, Z axes respectively.

The rate of change of momentum of the fluid element in the X direction is

$$\rho\, dx\, dy\, dz\, \frac{du}{dt}$$

The body or extraneous force in the X direction on this fluid element is

$$\rho\, dx\, dy\, dz F_x$$

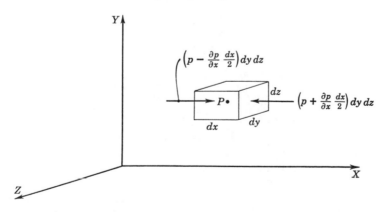

Fig. 8.19   Horizontal forces on a small parallelopiped of liquid.

Finally there is a force due to the pressure. On the left-hand face of the fluid element, which has an area $dy\,dz$, there is a force of

$$\left(p - \frac{\partial p}{\partial x}\frac{dx}{2}\right)dy\,dz$$

Similarly, on the right-hand face there is a force of

$$\left(p + \frac{\partial p}{\partial x}\frac{dx}{2}\right)dy\,dz$$

The resultant force in the X direction due to the pressure change is

$$-\frac{\partial p}{\partial x}\,dx\,dy\,dz$$

Thus from Newton's second law the equation of motion of the point P in the fluid in the X direction is

$$\rho\,dx\,dy\,dz\frac{du}{dt} = \rho\,dx\,dy\,dzF_x - dx\,dy\,dz\frac{\partial p}{\partial x}$$

and, dividing by $\rho\,dx\,dy\,dz$, it follows that

$$\frac{du}{dt} = F_x - \frac{1}{\rho}\frac{\partial p}{\partial x} = \frac{\partial u}{\partial t} + u\frac{\partial u}{\partial x} + v\frac{\partial u}{\partial y} + w\frac{\partial u}{\partial z} \qquad (8.28)$$

using Eq. 8.27. Similar expressions can be immediately written for the Y and Z components. If the force $\mathbf{F}$ per unit mass can be derived from a potential Ω which is independent of time, then

$$F_x = -\frac{\partial \Omega}{\partial x} \qquad F_y = -\frac{\partial \Omega}{\partial y} \qquad F_z = -\frac{\partial \Omega}{\partial z}$$

and the equation of motion, Eq. 8.28, may be written

$$\frac{du}{dt} = \frac{\partial u}{\partial t} + u\frac{\partial u}{\partial x} + v\frac{\partial u}{\partial y} + w\frac{\partial u}{\partial z} = -\frac{\partial \Omega}{\partial x} - \frac{1}{\rho}\frac{\partial p}{\partial x} \qquad (8.29)$$

Similar equations can be written for dv/dt and dw/dt.

Now these equations are a starting point for many relatively difficult derivations which we shall have to omit here. Instead we shall apply the equations of motion to the derivation of Bernoulli's theorem.

Let us transform Eq. 8.29 to an expression for steady motion along a stream line in a direction s. At the point P the velocity V of the fluid element is tangential to the stream line and V is a function of s and t. The acceleration along the stream line may be obtained from Eq. 8.27, and is given as

$$\frac{dV}{dt} = \frac{\partial V}{\partial t} + V\frac{\partial V}{\partial s}$$

Now in the Bernoulli theorem the motion is considered to be steady, or the velocity at any point in space is constant. Hence for steady flow

$$\frac{\partial V}{\partial t} = 0$$

and the acceleration along the stream line is

$$\frac{dV}{dt} = V\frac{\partial V}{\partial s}$$

Similarly the resultant force per unit mass at the point P along the stream line can be obtained from Eq. 8.29 and is

$$-\frac{\partial \Omega}{\partial s} - \frac{1}{\rho}\frac{\partial p}{\partial s}$$

Thus the equation of motion for steady flow along a stream line is

$$V\frac{\partial V}{\partial s} = -\frac{\partial \Omega}{\partial s} - \frac{1}{\rho}\frac{\partial p}{\partial s} \tag{8.30}$$

By integration of Eq. 8.30, we have

$$\int \frac{dp}{\rho} + \Omega + \frac{V^2}{2} = \text{constant}$$

If the fluid is incompressible so that ρ is a constant, then for steady flow along a stream line we have Bernoulli's theorem:

$$\frac{p}{\rho} + \Omega + \frac{V^2}{2} = \text{constant} \tag{8.31}$$

If we assume the extraneous force per unit mass to be that due to gravity, the potential energy per unit mass Ω is given by

$$\Omega = gh$$

where h is measured vertically in the Y direction. Thus Eq. 8.31 becomes

$$\frac{p}{\rho} + hg + \frac{V^2}{2} = \text{constant}$$

which is known as Bernoulli's equation. This equation represents conservation of energy per unit mass along a stream line. The terms in this equation can be physically interpreted in the following manner: p/ρ is the pressure head or the potential energy per unit mass of the liquid due to the pressure; hg is the elevation head or the potential energy per unit mass of the liquid due to gravity; and $v^2/2$ is the kinetic energy per unit mass of the liquid.

8.17   Applications of Bernoulli's Theorem

The Venturi meter.   This application of Bernoulli's theorem is extensively used for measuring the velocity of liquids and gases. The Venturi meter usually consists of a horizontal pipe with a constriction in it. The pressure in the wide and the constricted portions can be measured by means of liquid manometers, as shown in Fig. 8.20, if gas is flowing in the tube. If a liquid is flowing in the tube, the manometers will consist of three vertical tubes attached to the top of the main tube. If there is no friction between the tube and the moving fluid and the cross-sectional areas A_1 and A_3 are equal, then the pressure p_1 is equal to p_3 and the velocity V_1 is equal to V_3. It thus follows that the liquid in the manometer tubes at A_1 and A_3 stands at the same level. From the equation of continuity, Eq. 8.21, the velocities and cross sections at the wide and constricted portions are given by

$$A_1 V_1 = A_2 V_2 \tag{8.21}$$

Bernoulli's theorem, Eq. 8.31, assuming stream-line flow, applies to the Venturi tube. From this and the relation $h_1 = h_2$, we have

$$\frac{1}{2} V_1{}^2 + \frac{p_1}{\rho} = \frac{1}{2} V_2{}^2 + \frac{p_2}{\rho}$$

where ρ is the density of the fluid and is assumed to be constant. Substituting for V_2 from Eq. 8.21 in the above equation, it follows that the velocity V_1 is given by

$$V_1 = A_2 \sqrt{\frac{2(p_2 - p_1)}{\rho(A_2{}^2 - A_1{}^2)}}$$

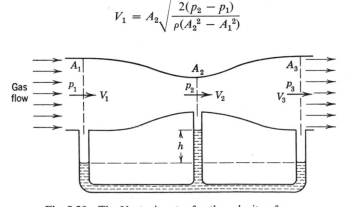

Fig. 8.20   The Venturi meter for the velocity of a gas.

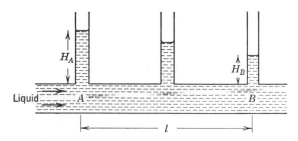

Fig. 8.21 Energy loss in a pipe carrying a liquid.

If the liquid in the monometer tubes in Fig. 8.20 has a density ρ' and the difference in levels of the liquid in the two tubes is h, then

$$p_2 - p_1 = h\rho'g$$

In using the Venturi meter to measure the speed of a gas, there is some error introduced due to the difference in density of the gas in the wide and constricted portions of the tube. This is less than 1 % if the velocity of the gas does not exceed about 50 m/sec.

In the above example we have assumed that there is no friction between the moving liquid and the stationary tube. This, of course, is not possible in practice. Its effect can be shown by allowing a liquid to flow through a horizontal tube of constant diameter, as shown in Fig. 8.21. The pressure in the liquid decreases in the direction of the motion, as indicated by the height of the liquid in the manometer tubes. From Eq. 8.21 the energy per unit mass of the liquid at A is

$$\frac{p_A}{\rho} + gh_A + \frac{v_A^2}{2}$$

Similarly for the point B the energy per unit mass is

$$\frac{p_B}{\rho} + gh_B + \frac{v_B^2}{2}$$

If there were no energy loss along the tube, these two expressions would be equal according to Bernoulli's theorem. However, there is an energy loss. Let E be the energy loss per unit mass per unit length along the tube. The energy loss per unit mass of fluid between A and B, Fig. 8.21, is El and is given by

$$\frac{p_A}{\rho} + gh_A + \frac{v_A^2}{2} - \frac{p_B}{\rho} - gh_B - \frac{v_B^2}{2} = El$$

From the conditions of the problem it follows that

$$h_A = h_B \quad \text{and} \quad v_A = v_B$$

Hence

$$\frac{p_A - p_B}{\rho} = El$$

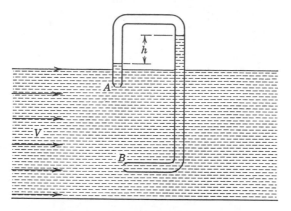

Fig. 8.22 The Pitot tube for a liquid.

If H_A is the height of the liquid in the manometer tube A and H_B the height in manometer tube B, then

$$p_A = H_A g \rho \qquad p_B = H_B g \rho$$

The energy loss per unit mass per unit distance along the tube is then

$$E = (H_A - H_B) \frac{g}{1}$$

The Pitot tube. Another device for measuring the speed of fluids is the Pitot tube. One type for measuring the speed of liquids is shown in Fig. 8.22. In this the opening A is perpendicular to the flow and the opening B faces the flow. As a result the levels of the liquid in the two arms are separated by a height h. Along the stream line, which is in line with the tube B, the velocity of the liquid falls rapidly to zero, and the manometer reading gives the value of $p + \rho V^2/2$. The tube A, past which the stream lines flow, measures the static pressure p in the liquid. Since the two manometer tubes are connected together and have a difference in levels of h, then it follows that

$$g \rho h = \frac{\rho V^2}{2} \qquad \text{or} \qquad V = \sqrt{2gh}$$

A Pitot tube is frequently attached to an airplane for measuring the velocity of the air relative to the plane, or the so-called air speed of the plane. It is represented diagrammatically in Fig. 8.23. It is placed on the leading edge of a wing or in the nose. The air rushes through tube B, giving the value of $p + \rho V^2/2$; a second tube A has openings on its sides giving the value of the pressure p. The two tubes are usually connected to a differential aneroid barometer which measures the difference in these pressures or $\rho V^2/2$. By proper calibration the pointer on the aneroid gives the velocity V of the plane, though, of course, this is only possible for one particular value of the density ρ of the air. Since the density varies with altitude, this is hardly an accurate instrument for measuring air speed.

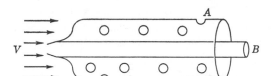

Fig. 8.23   The Pitot tube for a gas.

Torricelli's theorem.   This theorem, given by Torricelli in 1643, concerns the velocity of flow of a liquid from an orifice in a vessel. Consider a stream line starting from A and moving to B, as shown in Fig. 8.24. At A the velocity of the stream line is zero if the height of the liquid is kept constant, or approximately zero if the vessel has a large diameter. Also, the pressure p at both A and B is regarded as atmospheric. From Eq. 8.31 Bernoulli's theorem for a stream line moving from A to B gives

$$\frac{p}{\rho} + gy_A = \frac{p}{\rho} + gy_B + \frac{V^2}{2}$$

where gy_A and gy_B are respectively the potential energy per unit mass at A and B, and V is the velocity of emergence of the liquid at B. Thus

$$V = \sqrt{2g(y_A - y_B)}$$
$$= \sqrt{2gh}$$

where h is the height of A above B. This result, known as Torricelli's theorem, was originally derived not by the above method but rather along the lines of reasoning given by Galileo for falling bodies.

Now the result given above is only approximately correct, for observation shows that the jet of liquid does not emerge from the orifice in a cylindrical form. The liquid coming out of the orifice is made up of a large number of stream lines converging towards the orifice. This convergence continues outside the orifice until there is a minimum cross section at some place C, Fig. 8.24, called the *vena contracta*.

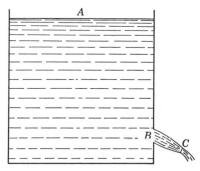

Fig. 8.24   Torricelli's theorem for a liquid emerging from a hole in a vessel.

The ratio of the area of cross section at C to the area of the orifice is called the coefficient of contraction. This coefficient depends on the shape of the orifice. For a plane circular orifice the coefficient of contraction is experimentally found to be about 0.62.

Furthermore, the convergence of the stream lines at the orifice implies that the pressure there is not atmospheric. At the vena contracta the stream lines are approximately parallel and the pressure there is equal to atmospheric. Hence the height h in Torricelli's theorem should be $y_A - y_C$ rather than $y_A - y_B$, as given above.

8.18 The Shape of the Surface of a Rotating Liquid

When a liquid is set in rotation, as for example by stirring a cup of coffee with a spoon, the free surface is no longer horizontal but is parabolic in shape.

Consider a liquid in a cylindrical vessel rotating with a constant angular velocity ω about the Y axis, as shown in Fig. 8.25. A fluid element of mass m at the point x, y on the surface of the liquid is in equilibrium under the action of its weight mg and the centrifugal reaction $mx\omega^2$ outward. If the tangent to the surface at x, y makes an angle θ with the horizontal, then from Fig. 8.25 we see that

$$\tan \theta = \frac{mx\omega^2}{mg} = \frac{x\omega^2}{g}$$

Since $\tan \theta = dy/dx$, we obtain the differential equation for the shape of the curve as

$$\frac{dy}{dx} = \frac{x\omega^2}{g}$$

By integration

$$y = \frac{x^2\omega^2}{2g} + C \tag{8.32}$$

If the X axis is drawn through the lowest point in the surface, then $x = 0$ at $y = 0$ and the constant of integration is zero. Thus the shape of the surface is parabolic.

Suppose now that the rotation of the cylindrical vessel is suddenly stopped with the liquid still rotating. Gradually the motion of the liquid stops, and, as it does so, we see that the fluid elements near the stationary wall of the vessel lag behind those near the axis. Thus there are tangential forces between contiguous fluid elements. These tangential forces do not exist when the liquid is rotating as a solid body but only when there is a velocity gradient

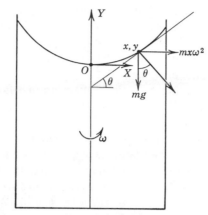

Fig. 8.25 The shape of the surface of a rotating liquid.

from the axis outward in the liquid. Up to the present we have assumed that we have been dealing with ideal fluids in which these tangential forces could not exits. We shall now investigate some of the properties of real liquids in connection with these tangential forces.

8.19   Viscosity

To simplify the understanding of these tangential forces, we shall consider two plain horizontal plates with liquid placed between them. The lower plate is kept at rest while the upper plate, which is resting on the upper surface of the liquid, moves with a small constant velocity v. This is represented diagrammatically in Fig. 8.26.

We shall assume that the liquid in contact with the upper plate moves with this plate and that the liquid in contact with the lower plate is at rest.

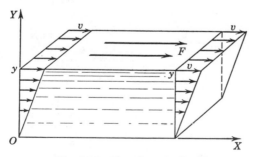

Fig. 8.26 Viscosity of a liquid.

In other words, there is no slippage between the plates and the liquid. If the velocity v of the upper plate is sufficiently small, then the liquid moves in horizontal stream lines or with laminar motion where each stream line moves relative to a contiguous one. The magnitude of the velocity v depends on the type of liquid, the distance between the plates, and the force F. If the velocity exceeds a critical amount, the regular laminar flow is broken up and turbulence occurs. We shall here assume that the velocity is below the critical velocity and that the liquid flows with this laminar motion.

In order for the motion to take place, a tangential force F must be applied to the plates. Suppose that the plates have an area A and are separated by a vertical distance y. It is found experimentally that the force F required to produce laminar flow, with the upper layer moving with the velocity v, is proportional to the velocity v and to the area A, and inversely proportional to the vertical distance y between the plates. Thus

$$F \propto \frac{vA}{y}$$

The constant of proportionality in this equation is called the *coefficient of viscosity* η of the liquid. (See Table 8.2.) Hence

$$F = \eta \frac{vA}{y}$$

or

$$\eta = \frac{Fy}{Av} \tag{8.33}$$

The quantity F/A is the shearing stress, and the quantity v/y is called the velocity gradient.

TABLE 8.2   Coefficients of Viscosity η

Substance	Temperature, °C	Coefficient of Viscosity, poises
Water	10	0.01308
Water	20	0.01005
Glycerin	0	121.1
Glycerin	20	14.9
Castor oil	10	24.2
Castor oil	20	9.86
Air	0	170.8×10^{-6}
Air	18	182.7×10^{-6}
Hydrogen	0	83.5×10^{-6}
Hydrogen	20	87.5×10^{-6}

The coefficient of viscosity of a liquid is the tangential force per unit area per unit velocity gradient. For two contiguous layers having a velocity gradient dv/dy the shearing stress F/A is given by

$$\frac{F}{A} = \eta \frac{dv}{dy} \tag{8.34}$$

From Eq. 8.33 it follows that the dimensions of the coefficients of viscosity are $[MT^{-1}L^{-1}]$. The cgs unit of coefficient of viscosity is the dyne second per square centimeter which is also called a *poise*. It follows that *a coefficient of viscosity of 1 poise means that 1 dyne of force is required to maintain a tangential velocity difference of 1 cm per sec between two surfaces each a square centimeter in area and 1 cm apart.*

It is found experimentally that η decreases with temperature increase for liquids and increases with temperature increase for gases. A few values of η for liquids and gases are given in Table 8.2.

8.20 Measurement of the Coefficient of Viscosity by the Capillary-Tube Method

About 1840 the French physicist Poiseuille carried out experiments on the flow of water through tubes of very small diameter. He found that the volume of water flowing per unit time through a horizontal capillary tube varied as the fourth power of the diameter of the tube, as the difference in pressure at the two ends, and inversly as the length of the tube. We shall now derive these experimental results from theoretical considerations.

Consider a horizontal capillary tube of length l and radius R, Fig. 8.27. Suppose that a liquid having a coefficient of viscosity η is moving with laminar flow in the tube. The velocity of the liquid is a maximum along the axis and is zero at the walls of the tube. Let the X axis lie along the axis of the tube, and let the pressure at $x = 0$ be p_0 and at $x = l$ be p_l.

Let us now consider the forces acting on an imaginary cylinder of radius

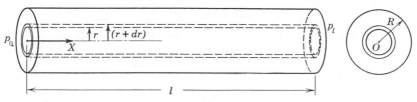

Fig. 8.27   The measurement of the viscosity of a liquid by a capillary tube.

r described about the axis of the tube. The radial velocity gradient at all points on the surface of this imaginary cylinder is dv/dr. Then from Eq. 8.34 the force F exerted on the liquid inside the cylinder by the liquid outside is

$$F = 2\pi r l \eta \frac{dv}{dr}$$

where $2\pi r l$ is the area of the surface of the imaginary cylinder. The liquid inside the imaginary cylinder is moving with a larger velocity than that outside and, therefore, the force F is to the left. This force is supplied by the pressure difference between the ends of the imaginary cylinder and is equal to

$$\pi r^2(p_0 - p_l)$$

Since the flow is steady, the forces acting on the liquid in the imaginary cylinder must be in equilibrium or

$$\pi r^2(p_0 - p_l) + 2\pi r l \eta \frac{dv}{dr} = 0$$

Thus

$$\frac{dv}{dr} = -\frac{r}{2\eta} \frac{(p_0 - p_l)}{l} \tag{8.35}$$

where $(p_0 - p_l)/l$ is the pressure gradient of the liquid in the tube. There is no radial pressure gradient and hence no radial flow since there is laminar flow of the liquid. The radial velocity gradient dv/dr is negative as shown by Eq. 8.35 and also by the fact that the velocity is a maximum along the axis and zero at the walls. By integration of Eq. 8.35, using the boundary conditions $v = 0$ at $r = R$, we have for the velocity v at any distance r from the axis

$$v = \frac{p_0 - p_l}{4\eta l}(R^2 - r^2) \tag{8.36}$$

To obtain the quantity of liquid flowing through the tube per unit of time, we shall first consider a thin cylindrical shell whose radius is r and whose thickness is dr. The area of cross section of this shell is $2\pi r\, dr$, and the velocity of flow at the distance r is given by Eq. 8.36. Thus the volume passing this cross section per second is

$$dQ = 2\pi v r\, dr$$

and the volume of liquid passing any cross section of the tube per second is

$$Q = \int_0^R 2\pi \frac{p_0 - p_l}{4\eta l}(R^2 r - r^3)\, dr$$

$$= \frac{\pi(p_0 - p_l)}{8\eta l} R^4 \tag{8.37}$$

This is known as Poiseuille's formula and agrees with his experimental results. From Eq. 8.37 the coefficient of viscosity of a liquid can be obtained experimentally.

8.21 Measurement of the Coefficient of Viscosity by the Revolving Cylinder Method

In this method the liquid whose coefficient of viscosity is to be determined is placed between two concentric cylinders, as shown in Fig. 8.28. The outer cylinder is rotated with a constant angular velocity ω_0 while the inner is at rest supported by a torsion fiber. Owing to the viscosity of the liquid, a torque is transmitted by the liquid to the inner cylinder which is balanced by the torque in the torsion fiber.

To calculate the torque on the inner cylinder, consider an imaginary concentric cylinder of radius r and length l drawn within the liquid, where l is the length of the inner cylinder immersed in the liquid. If ω is the angular velocity of the surface of this cylinder of radius r, then the radial velocity gradient at this radius in the liquid is $(r \, d\omega)/dr$. At the outer cylinder where $r = r_2$ the angular velocity of the liquid is ω_0, and at the inner cylinder where $r = r_1$ the angular velocity of the liquid is zero. The liquid on the outer surface of this imaginary cylinder exerts a tangential force and a torque on the liquid on the inner surface of the cylinder. Since the surface area of the

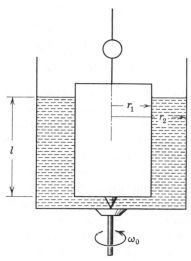

Fig. 8.28 The measurement of the viscosity of a liquid by a rotating cylinder.

imaginary cylinder is $2\pi rl$, the total tangential force F exerted on this surface is given by Eq. 8.34 as

$$F = 2\pi rl\eta r \frac{d\omega}{dr}$$

The torque τ exerted by this tangential force is

$$\tau = Fr = 2\pi l\eta r^3 \frac{d\omega}{dr}$$

Since the liquid between the imaginary cylinder of radius r and the stationary cylinder of radius r_1 is in steady motion or undergoing no acceleration, the torque τ is exerted on the stationary cylinder and is equal to that exerted by the imaginary cylinder.

By integrating the above equation, applying the boundary conditions of $\omega = 0$ at $r = r_1$ and $\omega = \omega_0$ at $r = r_2$, we have

$$\int_0^{\omega_0} 2\pi l\eta \, d\omega = \int_{r_1}^{r_2} \tau \frac{dr}{r^3}$$

or

$$2\pi l\eta\omega_0 = \frac{\tau}{2}\left(\frac{1}{r_1^2} - \frac{1}{r_2^2}\right)$$

Now the torque τ exerted on the inner cylinder is balanced by the torque in the supporting fiber. If the fiber has a torsion constant τ_0 and is twisted through an angle of θ radians

$$\tau = \tau_0\theta$$

and

$$\eta = \frac{\tau_0\theta}{4\pi l\omega_0}\left(\frac{1}{r_1^2} - \frac{1}{r_2^2}\right) \tag{8.38}$$

In the above derivation we have neglected the torque exerted by the liquid on the under surface of the stationary cylinder. If there were regular stream-line motion between the two horizontal surfaces, it would be possible to calculate an expression for it. However, at the edge of the inner cylinder the stream lines would not be concentric with those nearer the axis. We can overcome this difficulty by making a determination of the angle of torsion θ for two different heights of the liquid. We must, of course, use the same angular velocity of rotation of the outer cylinder in both experiments.

The viscosity of liquids increases very much as their temperature is decreased. We are all familiar with this phenomenon since we change the oil in our car from one having a high viscosity in summer to one having a relatively low viscosity in winter. The grade number of car oil is proportional to the number of seconds a given quantity of oil takes to go through a capillary tube. The smaller the number, the less viscous or the "thinner" the oil.

8.22   The Velocity of Fall of a Sphere through a Viscous Liquid

This problem was first solved by Sir George Stokes about the middle of the last century. Since the method he used is beyond the scope of this book, we shall give only a partial solution by dimensional analysis. As was mentioned earlier, both sides of an equation between physical quantities must have the same dimensions. This statement is the foundation of dimensional analysis. In applying dimensional analysis to this problem, we must first attempt to ascertain the physical quantities that determine the force exerted on a sphere moving with constant velocity in a viscous liquid. This must be done by intelligent guesswork or intuition.

Suppose that a sphere moves through a viscous liquid with a constant small velocity so that there is stream-like motion. The applied force must be exactly equal and opposite to the friction force exerted by the liquid on the sphere, since we have assumed constant velocity. This frictional force F might be expected to depend on the size or radius r of the sphere, the viscosity η of the liquid, and the velocity v with which the sphere moves. If these are considered to be the only quantities on which the force depends, then we may say that the friction force is a function of the radius r, the viscosity η, and the velocity v. We can then write

$$F = Kr^a\eta^bv^c$$

where K is a dimensionless constant and a, b c are dimensionless numbers to which r, η, and v must be raised respectively for the result to have the same dimensions as that of a force.

Writing out the dimensions of these quantities in terms of mass $[M]$, length $[L]$, and time $[T]$, we have

$$[MLT^{-2}] = [L]^a[ML^{-1}T^{-1}]^b[LT^{-1}]^c$$

Equating the powers for $[M]$, $[L]$, and $[T]$ respectively, we have:

For $[M]$: $\qquad\qquad 1 = b$

For $[L]$: $\qquad\qquad 1 = a - b - c$

For $[T]$: $\qquad\qquad -2 = -b - c$

The solution of these three equations gives

$$a = b = c = 1$$

Thus

$$F = Kr\eta v$$

The dimensionless constant K could be obtained by experiment by actually measuring the force required to pull a sphere of known radius through a liquid of known viscosity at a known speed. If this were done accurately, then a value of 6π should be obtained. This is the value obtained theoretically by Stokes, and the expression

$$F = 6\pi r\eta v \qquad (8.39)$$

is known as Stokes' law.

If we now return to the problem of a small sphere falling at a constant velocity through a viscous liquid, the effective weight of the sphere in the liquid is balanced by the frictional force. This effective weight is given by Archimedes' principle as

$$\tfrac{4}{3}\pi r^3(\rho_s - \rho_l)g$$

where r is the radius of the sphere of density ρ_s, and ρ_l is the density of the liquid. Thus we have for the velocity of fall v, using Eq. 8.39,

$$F = 6\pi r\eta v = \tfrac{4}{3}\pi r^3(\rho_s - \rho_l)g$$

Hence

$$v = \frac{2g}{9\eta}\, r^2(\rho_s - \rho_l) \qquad (8.40)$$

From this equation one can experimentally determine the viscosity of a liquid since all the remaining quantities in Eq. 8.40 can be easily measured. However, one must be careful to see that the motion is stream line or laminar. This type of motion depends very much on the viscosity of the liquid and the shape of the body. A marble falling through glycerin does so with stream-line flow, and the motion is in a straight line with constant speed. If, however, the same marble is dropped through water, then the motion may be irregular because turbulence has been created in the liquid. Turbulent motion is very difficult to analyze, and we shall again have to revert to the method of dimensional analysis.

8.23 Turbulent Motion and Reynolds' Number

Toward the end of the last century Sir Osborne Reynolds investigated the change from laminar to turbulent flow in liquids in cylindrical tubes as the velocity of flow was increased. He found that for any liquid there was a critical velocity at which there was a sudden change from the laminar to the turbulent type of motion. His experiment consisted in inserting a narrow thread of colored liquid into the main body of liquid flowing through the tube. With laminar flow the narrow thread followed a straight line, but once the critical velocity was passed, the thread broke up and the colored liquid ultimately diffused into the main body of the liquid. To determine the point at which turbulence sets in, we must ask ourselves on what quantities we might expect this critical velocity v_c to depend. These would seem to be the viscosity η and the density ρ of the liquid and the diameter d of the cylindrical tube into which the liquid is flowing. If these are the correct and only factors on which the critical velocity depends, we may write for dimensional purposes

$$v_c = R\eta^a\rho^b d^c$$

where R is a dimensionless number called Reynolds' number. Writing the dimensions of these quantities we have

$$[LT^{-1}] = [ML^{-1}T^{-1}]^a[ML^{-3}]^b[L]^c$$

Then

For $[L]$: $1 = -a - 3b + c$

For $[M]$: $0 = a + b$

For $[T]$: $-1 = -a$

Hence from these equations we have

$$a = 1 \qquad b = -1 \qquad c = -1$$

Thus we have for the critical velocity v_c

$$v_c = R\,\frac{\eta}{\rho d} \tag{8.41}$$

Reynolds' number R is then given by

$$R = \frac{\rho\, d\, v_c}{\eta}$$

If the critical velocity is determined experimentally for a given liquid and diameter of pipe, then Reynolds' number can be found for a cylindrical pipe. Of course this analysis is only approximate for, as we have seen, the velocity of a liquid in a cylindrical tube varies from a maximum along the axis to zero at the edges of the tube when there is laminar flow. The critical velocity is the average velocity over the cross section of the tube.

Now when an object is pulled through a liquid so that there is laminar motion, then, as we know, the force varies with the velocity, but as soon as the motion becomes turbulent, the force varies with the square of the velocity. Eddies are set up behind the moving body so that the pressure behind the

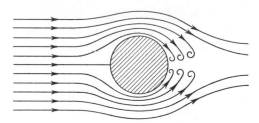

Fig. 8.29 Turbulent motion at an obstacle.

moving body is less than that in front, Fig. 8.29. Evidence of these eddies is seen when a car moves at a high speed over light objects that are whirled around in the air behind the car.

With laminar or stream-line motion no eddies are set up and there is no pressure difference between the front and back of the body moving through the fluid. There the resistance force is entirely caused by viscosity, whereas with turbulence the resistance force is largely caused by the pressure difference and only slightly by the viscosity.

The force due to pressure difference in the turbulent motion depends on the area of cross section A of the body perpendicular to the direction of the motion, on the velocity v of the body relative to that of the fluid, and on the density ρ of the fluid, which is either a gas or a liquid. Thus we may write the pressure-resistance force F for dimensional purposes as

$$F = KA^a v^b \rho^c$$

where K is a dimensionless quantity. Writing out the dimensions of these quantities we have

$$[MLT^{-2}] = [L^2]^a [LT^{-1}]^b [ML^{-3}]^c$$

Equating the powers of $[M]$, $[L]$, and $[T]$ respectively, it follows that

For $[M]$: $1 = c$

For $[L]$: $1 = 2a + b - 3c$

For $[T]$: $-2 = -b$

Solving these equations, we have

$$a = 1 \qquad b = 2 \qquad c = 1$$

Hence

$$F = KA\rho v^2 \tag{8.42}$$

The value of the dimensionless constant K in Eq. 8.42 depends on the shape of the body, which is moving relative to the fluid. For any particular body it can be measured experimentally. It is about 0.7 for a flat disk and about 0.03 for a pear-shaped body.

PROBLEMS

1. A wire 2 m in length and 2×10^{-4} m in radius is clamped at its upper end and has a load of 4 kg on its lower end. If Young's modulus is 9.8×10^{10} N/m² and Poisson's ratio is 0.3, find the extension of the wire and the decrease in the radius and the cross-sectional area due to the lateral strain.

2. Find the work done by the stretching force, and the potential energy of the stretched wire in problem 1.

3. Show that for a homogeneous isotropic substance a volume strain is approximately three times the corresponding linear strain. As a method of proof, consider a cube whose length of side is l, subjected to a uniform pressure p producing a decrease in length of the sides of Δl.

4. Show that, when a shearing stress T shears a body through an angle θ, the work done per unit volume is $\frac{1}{2}T\theta$, and that, when a uniform stress p produces a volume strain v, the work done per unit volume is $\frac{1}{2}pv$.

5. A cubical block of material is subjected to a uniform tension perpendicular to one pair of faces. (a) Show that the change in volume per unit volume is approximately $\alpha(1 - 2\sigma)$, where α is the longitudinal strain in the cube and σ is Poisson's ratio for the material. (b) Show that the fractional decrease in area of a cross section perpendicular to the tension is approximately $2\alpha\sigma$.

6. A cylindrical bar of copper 10^{-2} m² in cross section and 50 cm long at 0°C is rigidly held between two end blocks. If the coefficient of linear expansion of copper is 1.7×10^{-5}/°C and the copper is heated through 50°C, find the force in newtons on the end blocks.

7. Find the increase in density of water at 99 ft below the surface of a lake if the pressure increases by 1 atm for each 33 ft of depth. The bulk modulus of water is about 2×10^4 atm, and the normal density of water is about 62.5 lb/ft³. (Using 1 lb of water, find its decrease in volume, its new volume and density, and then the increase in density.)

8. Show that the bulk modulus k may be expressed $k = (\rho\, dp)/d\rho$, where ρ is the density of the substance. If the rate of change of pressure p with depth y below the surface of a lake is given by $dp/dy = \rho g$, find the expression for the density at any depth y below the surface, assuming ρ_0 is the density at the surface where $y = 0$.

9. If the pressure p of a gas varies with volume v so that $pv = $ constant when the temperature is constant, show that the bulk modulus of a gas under these conditions is equal to its pressure.

10. When a gas is thermally insulated, its pressure and volume are related by $pv^\gamma = $ constant, where γ is the ratio of the specific heat at constant pressure to that at constant volume for the gas. Show that for a thermally insulated gas or for adiabatic changes in the gas the bulk modulus is $k = \gamma p$.

11. A light rod of length l is supported at its two ends in a horizontal position, and a load W is placed on the center of the rod. Show that the deflection of the center of the rod is $Wl^3/48EI_A$, where E is Young's modulus and I_A is the second moment of area of the beam.

12. A light rod of length l is clamped at its two ends in a horizontal position, and a load W is placed on the center of the rod. If the coordinate axes have their origin at the end of the rod, show that the equation for the shape of the rod is

$$y = \frac{W}{EI_A}\left(\frac{x^3}{12} - \frac{lx^2}{16}\right)$$

13. A uniform beam of length l and weight W has a weight W' placed at its center and is supported at each end at the same horizontal level. Show that the deflection of the center of the beam is $[-l^3(5W + 8W')]/384EI_A$.

14. A uniform beam of length l and weight W has a weight W' placed at its center, and the two ends are clamped at the same horizontal level. Find the equation of the shape of the beam in the region of positive x for axes whose origin is at the left-hand end of the beam.

15. A light beam of length l is supported at each end at the same horizontal level. A load W is placed on the beam at a distance of a third of the length of the rod from the left-hand support. Show that the equation for the shape of the beam between O and $l/3$ is

$$y_1 = \frac{Wx}{81EI_A}(9x^2 - 5l^2)$$

and between $l/3$ and l it is

$$y_2 = \frac{W}{162EI_A}(27x^2l - 9x^3 - 19l^2x + l^3)$$

16. A light cylindrical wooden rod 0.5 in. in diameter and 24 in. long is horizontally supported at its two ends so as to be used as a clothes hanger. When a suit weighing 5 lb is suspended at the mid-point, it is found that this point is depressed a distance of 0.25 in. Find Young's modulus for the rod in pounds weight per square foot.

17. A torque of 200 lbw in. applied to one end of a cylindrical rod 0.6 in. in diameter and 4 ft in length produces a twist of 2° when the other end is fixed. Find the modulus of rigidity of the rod in pounds per square inch and the work done by the torque in foot pounds weight.

18. The top of a vertical gate is 50 ft below the level of the water at a dam. If the gate is 10 ft wide and 4 ft high and is hinged along its upper horizontal edge, find the torque exerted by the water about the line of hinges as an axis.

19. A horizontal cylindrical pipe 9 ft in radius has its ends closed with vertical slabs of glass. Find the resultant force on the glass ends and the distance of this force from the center of the pipe when the pipe is filled to a height of 6 ft with water.

20. The cross section of a wooden trough is an equilateral triangle, 3 ft on a side, with the upper side horizontal. Assuming that such a trough will hold water, find the force on one of the ends when the trough is filled with water.

21. A U-tube has a uniform diameter of 3 cm and contains 200 cm³ of oil of density 0.9 gm/cm³. The oil originally stands at the same horizontal level in the two vertical arms. By blowing into one of the arms, the column of oil is set into oscillation with an amplitude of 1 cm. Assuming no damping, find the period of the simple harmonic vibrations of the oil and the force originally exerted by blowing on the oil.

22. Show that for any point in a liquid at rest having a density ρ and a pressure p, the force per unit mass in the X direction is

$$ F_x = -\frac{1}{\rho}\frac{\partial p}{\partial x} $$

23. If p_0 is the pressure at sea level and p is the pressure at the top of a column of air h meters in height and having a uniform temperature of T degrees absolute, show that

$$ \log_{10} p_0 - \log_{10} p = 0.0149\,\frac{h}{T} $$

Take the density of air at 0°C as 0.001293 gm/cm³ and the pressure p_0 of air at sea level as 1.01×10^6 dynes/cm².

24. Water flows out of a circular hole in the side of a large cylindrical tank 1 m in diameter. Initially the water level is 5 m above the vena contracta which has a circular cross section of 2-cm radius. Find the volume of water escaping from the tank per second when the water level is 4 m above the vena contracta, taking into account the velocity of fall of the top of the liquid in the tank. Find how long it takes to empty the vessel from the height of 5 m.

25. At a fixed point the velocity of the water in a streamline is 50 cm/sec and the rate of change of the velocity with distance along the streamline is 8 cm/sec/cm. Find the acceleration of the liquid at the fixed point.

26. Find the volume of water flowing per minute through a horizontal pipe 6 in. in diameter if there is a difference in height of 1 ft in the levels of the liquid in the two manometers, one of which is placed at the wide portion and the other at the constricted portion. The constricted portion has a diameter of 2 in.

27. A horizontal tube 0.8 mm in diameter and 16 cm long is connected at one end to a constant-level water tank whose height is 200 cm. If 600 cm^3 of water flows through the tube in 8 min, find the coefficient of viscosity of water.

28. A horizontal capillary tube with radius R and length l is connected to an airtight vessel of volume V. Enough air escapes from the vessel through the capillary tube during a time of t sec to reduce the pressure of the air in the vessel from P_1 to P_2 without any change in temperature. If atmospheric pressure if P_0, prove that the coefficient of viscosity η of air is given by

$$\eta = \frac{\pi R^4 P_0 t}{8lV \log_e \dfrac{(P_2 + P_0)(P_1 - P_0)}{(P_2 - P_0)(P_1 + P_0)}}$$

(In order to prove this relationship consider a small section of the capillary tube of length dx where the pressure is p and then integrate over the length l of the tube. If P is the pressure at the entrance to the capillary and P_0 is atmospheric pressure at the end of the tube, then show that the time rate dV/dt at which air enters the capillary is

$$P \frac{dV}{dt} = \frac{\pi R^4}{16\eta l} (P^2 - P_0{}^2)$$

From Boyle's law

$$P \frac{dV}{dt} = -V \frac{dP}{dt}$$

where V is the constant volume of the vessel and dP/dt is the rate of change of pressure in the vessel.)

29. A capillary tube 0.04 cm in radius and 100 cm in length is connected to a vessel having a volume of 4000 cm^3. The original pressure of the air in the vessel is 80.5 cm of mercury, and after 20 sec the pressure has fallen to 79.4 cm of mercury. If the temperature of the air is constant throughout the experiment and the atmospheric pressure is 76 cm of mercury, find the coefficient of viscosity of air.

30. A liquid having a coefficient of viscosity η is flowing through a vertical capillary tube whose length is l and whose radius is R. If the exit end of the capillary is a vertical height H below a constant level tank, then show that the volume of liquid Q emerging from the capillary per second may be given as

$$Q = \frac{\pi R^4 g \rho}{8\eta l} \left(H - \frac{2Q^2}{g \pi^2 R^4} \right)$$

31. Assuming that Reynolds' number is 1000 for cylindrical pipes, calculate the critical velocity of water and of glycerine at 20°C in tubes of 2-cm diameter. The density of glycerin is 1.36 gm/cm^3.

32. In the determination of the coefficient of viscosity by the rotating cylinder method as described in Section 8.21, two heights of liquid l_1 and l_2 are used and the corresponding angles of torsion θ_1 and θ_2 are observed. If the radii of the inner and outer cylinders are r_1 and r_2 respectively, and Ω is the angular velocity of rotation of the outer cylinder in both cases, and τ_0 is the torsion constant of the suspension, show that according to Eq. 8.38 the coefficient of viscosity η of the liquid is

$$\eta = \frac{\tau_0(\theta_1 - \theta_2)}{4\pi\Omega(l_1 - l_2)}\left(\frac{1}{r_1^{\,2}} - \frac{1}{r_2^{\,2}}\right)$$

33. The coefficient of viscosity η of a gas or a liquid can be found by measuring the torsion $\tau_0\theta$ exerted on a horizontal circular plate placed a small distance d above a concentric and parallel plate which is rotating with an angular velocity Ω. If R is the radius of the circular plates, show that $\eta = 2d\tau_0\theta/\pi R^4\Omega$.

WAVE MOTION

9.1 Wave and Particle Motion

Classical physics has two main branches: the one deals with the motion of particles or rigid bodies, and the other with wave motion. These two branches developed independently up to about the beginning of the present century. Until then electric charges had, along with many other concepts, been explained in terms of particles, whereas light and sound phenomena were explained in terms of wave theory. Very briefly we may say that the motion of particles can be interpreted by Newton's laws, together with the additional developments of the eighteenth and nineteenth centuries made by Laplace, Lagrange, Hamilton, and others. In the case of a particle, energy is carried by the moving particle, whereas the energy of a spherical wave spreads out in all directions. The course of physics in the present century has shown that natural phenomena cannot be explained quite so simply. Light or electromagnetic radiation has both wave and particle properties. The properties of electrons are also interpreted in terms of both particles and waves. These two distinct theories, each formerly used independently to explain certain natural phenomena. have, in a sense, been merged into the modern theory of wave mechanics.

In this chapter we shall be concerned with some of the fundamental characteristics of the propagation of waves in continuous media such as solids, liquids, and gases. When an oscillatory motion is impressed on some part of a body, this motion is propagated as a wave motion to all parts of the body. Up to now our only discussion of vibratory motion impressed on a body has been concerned with the oscillations of a spring. We may properly ask whether waves were not propagated in the spring? Actually, waves were set up in the spring, but they were so small that they could be neglected, since we were dealing with a spring whose mass was negligible compared to the mass of the body attached to its end. When the mass of a spring is assumed to be zero, no inertial force is required to set any portion of the spring in motion, so that any force applied at one end of the spring is passed on to the other end undiminished by each elementary portion of the spring. This force remains the same throughout the spring, and every elementary length of the spring is stretched an amount proportional to its length, so that the resultant extension of the whole spring is proportional to the applied force.

Let us drop the assumption that the spring has zero mass. Now a force is required to accelerate each elementary portion of the spring. The force at any instant is no longer the same throughout the spring, and waves are propagated in the spring. Wave motion occurs in any medium in which mass and restoring forces are associated with each elementary portion of the medium. In our previous discussion of springs, we considered the restoring force as evenly distributed throughout the spring, whereas the mass undergoing acceleration was considered to be concentrated at the lower end of the spring. A similar situation occurs in electric circuits. So long as we assume that inductance and capacitance are separate or lumped parameters, we deal with forced oscillations or alternating currents and not with true waves. Although in practice there is always some capacitance associated with an inductance and some inductance with a capacitance, these small effects can generally be neglected if the frequency of the alternations is small. When the frequency is large, these effects cannot be neglected and waves are propagated. Here, again, the parameters—inductance and capacitance—are distributed throughout the circuit.

In most practical cases the medium in which the wave is propagated is limited and not infinite. At the boundaries of the medium the waves are reflected so that the medium is acted on by both the incident and the reflected waves. With suitable geometrical configurations, standing waves can be set up in the medium. These may occur in a string fastened at both ends, in a membrane such as a drumhead clamped around its edges, and so forth. Before discussing the phenomenon of standing waves, let us find an expression for the velocity of propagation of transverse waves in an infinitely long cord.

9.2   Transverse Waves in a Cord

When a stretched cord is caused to vibrate either by giving sinusoidal vibrations to one end or by plucking it, transverse waves travel along the cord. In this transverse wave the motion of the elementary portions of the cord is at right angles or transverse to the direction of the traveling wave. The wave is a motion of a geometrical pattern transmitting energy and not a mass motion of the cord as a whole.

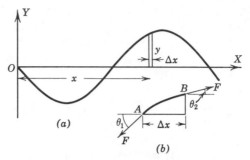

Fig. 9.1   Transverse wave in a cord.

To determine the velocity of a wave in a cord, let us consider a uniform cord whose mass per unit length is μ. The cord is subjected to a constant tension F. We shall assume that the deflections of the cord are so small that the tension remains constant throughout the deflections and that damping can be neglected. Suppose that at some instant of time the cord assumes the configuration shown in Fig. 9.1a. An element of the cord whose horizontal length is Δx at position x, y is shown enlarged in Fig. 9.1b. On the ends of this element the forces make angles θ_1, θ_2 respectively with the horizontal X axis. The problem is to find the motion of the element AB under the action of these forces. We shall assume that the displacements are so small that $\sin \theta \approx \tan \theta$, where the notation $\approx$ means approximately equal to. The mass of the element AB may be approximately taken as $\mu \, \Delta x$ since $\cos \theta \approx 1$. The resultant vertical force on the element is $F(\sin \theta_2 - \sin \theta_1)$, and this may be set equal to $F(\tan \theta_2 - \tan \theta_1)$. Now

$$F \tan \theta_1 = F \left(\frac{\partial y}{\partial x} \right)_A \tag{9.1}$$

where $(\partial y / \partial x)_A$ is the slope of the tangent at A at the instant of time under consideration. It is necessary to use partial derivatives since y is a function of both the position x and the time t. The resultant vertical force on the element of length Δx is then

$$F \left(\frac{\partial y}{\partial x} \right)_B - F \left(\frac{\partial y}{\partial x} \right)_A$$

By Taylor's theorem, using only the first-order terms,

$$\left(\frac{\partial y}{\partial x} \right)_B = \left(\frac{\partial y}{\partial x} \right)_A + \frac{\partial}{\partial x} \left(\frac{\partial y}{\partial x} \right) \Delta x$$

The resultant vertical force F_y on an element of length Δx is

$$F_y = F \frac{\partial^2 y}{\partial x^2} \Delta x \tag{9.2}$$

and by Newton's second law this is equal to the mass $\mu \Delta x$ of the element multiplied by the vertical acceleration. Hence

$$F \frac{\partial^2 y}{\partial x^2} \Delta x = \mu \, \Delta x \frac{\partial^2 y}{\partial t^2}$$

or

$$\frac{\partial^2 y}{\partial x^2} = \frac{\mu}{F} \frac{\partial^2 y}{\partial t^2} \tag{9.3}$$

This is the wave equation for waves propagated along the X axis. You should be able to show from the assumptions that the resultant horizontal force is zero. From Eq. 9.3 it follows that $\sqrt{F/\mu}$ has the dimensions of a velocity which we shall see is the velocity of the transverse waves in the cord. The wave equation above is a linear, homogeneous partial-differential equation of the second order. There are several methods of obtaining a solution of such an equation. Because of its importance, we shall illustrate two of these methods. It may be readily verified by substitution that

$$y = f(x - vt) \tag{9.4}$$

is a solution of the wave equation. In this equation f is any function—sine. cosine, square, cube, etc. By substituting Eq. 9.4 in Eq. in Eq. 9.3, it is easily shown that $v = \sqrt{F/\mu}$. Similarly, $y = f(x + vt)$ is also a solution. Let us ask what $y = f(x - vt)$ signifies physically. It represents a wave traveling in the positive X direction, and $y = f(x + vt)$ represents a wave traveling in the negative X direction. Suppose that a wave is traveling along the positive X direction (Fig. 9.2). If there is a displacement y at position x and time t. then, if the wave is traveling in the positive X direction, there is the same displacement y at some position $x + vP$ to the right of x at the time $t + P$ or a time P later. Mathematically this is given as

$$y = f(x - vt)$$
$$= f(x + vP - vt - vP)$$

In order to understand this, let $y = f(x)$ be the equation of a curve, shown

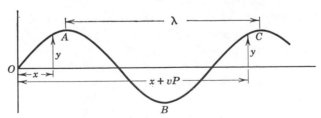

Fig. 9.2   Equal displacements on a transverse wave a wavelength $\lambda = vP$ apart.

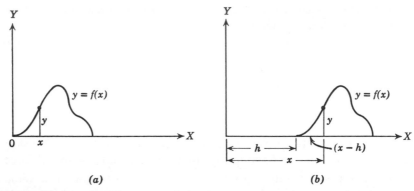

(a)   *(b)*

Fig. 9.3 Displacement of curve $y = f(x)$ a distance h along the positive X axis giving the curve whose equation is $y = f(x - h)$.

in Fig. 9.3a. Then $y = f(x - h)$ is the equation of the same curve translated along the positive x axis a distance h (Fig. 9.3b).

  If the function f is to represent a wave, then it must be periodic, and an important special case of the function is the sine or cosine function. Consider the solution

$$y = A \sin k(x - vt) \tag{9.5}$$

where A is the maximum displacement called the *amplitude*. It gives the value of the height of the crest at A or C or the depth of the trough at B. Fig. 9.2. The quantity $k(x - vt)$ is called the *phase* of the vibration, and the distance between two points in the same phase, such as AC, is called the *wavelength* λ. If P is the *period* of the wave, then, if the wave is traveling to the right, there is the same displacement at x and t as there is at $x + \lambda$ and $t + P$. Hence

$$A \sin k(x - vt) = A \sin k[x + \lambda - v(t + P)]$$

or, since the arguments of the sine function are equal,

$$\lambda = vP \qquad v = \frac{\lambda}{P} \tag{9.6}$$

The quantity v is the ratio of the distance λ traveled by the wave to the period P. It is called the *phase velocity* of the wave since it gives the velocity at which any point of any phase travels along. The *frequency f* of the waves is $1/P$ as indicated in the discussion of simple harmonic motion. This phase velocity may be understood as follows: If a given value is taken for y, then, for this to be constant, the quantity $x - vt$ must also be constant. The derivative of this latter quantity with respect to time gives $v = dx/dt$. This is the velocity of a point of constant phase and is therefore called the phase velocity of the wave.

It may be seen that every element of the cord executes simple harmonic motion with a period of P and a frequency of f. For any particular point where x is a constant

$$y = A \sin (k_1 t + k_2)$$

where k_1 and k_2 are constants depending on the period P and the position x.

The quantity k in Eq. 9.5 may be obtained in terms of the wavelength by noting that the distance λ corresponds to an increase in the argument of the sine function of 2π radians:

$$k(x - vt) + 2\pi = k(x + \lambda - vt)$$

or

$$k = \frac{2\pi}{\lambda} \tag{9.7}$$

Thus the displacement y at any point x and time t on the rope along which a simple harmonic wave is traveling in the positive X direction may be written as

$$y = A \sin k(x - vt) = A \sin \frac{2\pi}{\lambda} (x - vt) = A \sin 2\pi \left(\frac{x}{\lambda} - \frac{t}{P} \right)$$

$$= A \sin (kx - \omega t) = A \sin \omega \left(\frac{x}{v} - t \right) \tag{9.8}$$

where $\omega = 2\pi/P = 2\pi v/\lambda = kv$.

Since the wave equation is linear and homogeneous, it follows that

$$y = f_1(x - vt) + f_2(x + vt)$$

is also a solution, as may be readily verified by substitution. Now this solution deals with waves traveling in opposite directions along the string. If these waves are simple harmonic and have the same amplitude, they give rise to what are called standing waves. The equation of such standing waves is given by

$$y = A \sin (kx - \omega t) + A \sin (kx + \omega t)$$
$$= 2A \sin kx \cos \omega t$$

and this satisfies the wave equation, Eq. 9.3. These standing waves will be discussed in a later section.

9.3 Solution of the Wave Equation by Separation of the Variables

The one-dimensional form of the wave equation for a transverse wave in a cord of mass per unit length μ and tension F placed along the X axis is given by Eq. 9.3 as

$$\frac{\partial^2 y}{\partial x^2} = \frac{\mu}{F} \frac{\partial^2 y}{\partial t^2}$$

In this partial-differential equation, the dependent variable is the displacement y. It is a function of the two independent variables x, the position along the cord, and t, the time under consideration. A relatively simple method of solving such a partial-differential equation is to find a solution which is the product of functions of each of the independent variables. Let us assume that a product solution is possible of the type

$$y = X(x)T(t)$$

where $X(x)$ is a function of x alone and $T(t)$ is a function of t alone. From this equation it follows by differentiation that

$$\frac{\partial^2 y}{\partial x^2} = T \frac{d^2 X}{\partial x^2} \qquad \text{and} \qquad \frac{\partial^2 y}{\partial t^2} = X \frac{d^2 T}{dt^2}$$

The wave equation becomes

$$T \frac{d^2 X}{dx^2} = \frac{\mu X}{F} \frac{d^2 T}{dt^2}$$

Dividing by XT gives

$$\frac{1}{X} \frac{d^2 X}{dx^2} = \frac{\mu}{F} \frac{1}{T} \frac{d^2 T}{dt^2}$$

In this equation the left-hand side is a function of x alone and the right-hand side of t alone. These two sides cannot be equal unless each is equal to some constant k'. Thus we have

$$\frac{d^2 X}{dx^2} = k'X \qquad \text{and} \qquad \frac{d^2 T}{dt^2} = k'T \frac{F}{\mu}$$

These two equations are homogeneous differential equations of the second order. We have encountered such equations in the discussion of simple harmonic motion where k' was intrinsically negative. Since any point x on the cord is undergoing simple harmonic motion, we are led to believe that k' must be negative in this problem. To emphasize the negative character of k', we shall put $k' = -c^2$. The equations resulting from the separation of the variables then become

$$\frac{d^2 X}{dx^2} = -c^2 X \qquad \text{and} \qquad \frac{d^2 T}{dt^2} = -\frac{F}{\mu} c^2 T \qquad (9.9)$$

From the procedure outlined in Chapter 5 the solutions of these equations are

$$X = A \cos (cx + \alpha) \qquad \text{and} \qquad T = B \cos \left(\sqrt{\frac{F}{\mu}} \, ct + \beta \right)$$

where A, B, α and β are constants. It would have been equally valid to use the sine function in place of the cosine. The displacement y of any point x on the cord at a time t is given by

$$y = XT = AB \cos(cx + \alpha) \cos\left(\sqrt{\frac{F}{\mu}}\, ct + \beta\right)$$

By a trigonometric transformation this may be written

$$y = \frac{AB}{2} \cos\left(cx + \sqrt{\frac{F}{\mu}}\, ct + \alpha + \beta\right) + \frac{AB}{2} \cos\left(cx - \sqrt{\frac{F}{\mu}}\, ct + \alpha - \beta\right)$$

To interpret this equation, we shall set $\sqrt{F/\mu}$ equal to the phase velocity v of the wave and the quantity c equal to $2\pi/\lambda$. The latter follows since cx and vct must be angles, so that c must have the dimensions of an inverse length. This length is the wavelength λ of the wave in the cord. Thus

$$y = \frac{AB}{2} \cos\left(2\pi\frac{x}{\lambda} + \frac{2\pi vt}{\lambda} + \alpha + \beta\right) + \frac{AB}{2} \cos\left(\frac{2\pi x}{\lambda} - \frac{2\pi vt}{\lambda} + \alpha - \beta\right)$$

$$= \frac{AB}{2} \cos 2\pi\left(\frac{x}{\lambda} + \frac{t}{P} + \frac{\alpha + \beta}{2\pi}\right) + \frac{AB}{2} \cos 2\pi\left(\frac{x}{\lambda} - \frac{t}{P} + \frac{\alpha - \beta}{2\pi}\right)$$

where $P = \lambda/v$ is the period of the waves. The solution represents two progressive waves: the one traveling in the negative X direction and the other in the positive X direction. Their amplitudes are $AB/2$, and their phases at $x = 0$, $t = 0$ are respectively $(\alpha + \beta)/2\pi$ and $(\alpha - \beta)/2\pi$. These quantities are determined from the initial or boundary conditions of the particular problem.

From the above we see that the wave equation, Eq. 9.3, is satisfied by any of the above solutions representing a wave traveling along the positive X axis or similar expressions for a wave traveling along the negative X axis.

9.4   Energy of a Wave in a Vibrating String

The energy of a vibrating string is partly kinetic and partly potential energy. An element of length dx and mass $\mu\, dx$, having a transverse velocity $\dot{y}$ has kinetic energy $(\mu\dot{y}^2\, dx)/2$. The kinetic energy $W_{\text{K.E}}$ of a length L is

$$W_{\text{K.E}} = \frac{\mu}{2} \int_0^L \left(\frac{\partial y}{\partial t}\right)^2 dx$$

The potential energy of the string is the work done in stretching it. Consider the element AB of the string, Fig. 9.1, of original length dx which is stretched to length $AB = ds$. If dy is the vertical height of the triangle in the figure, then

$$(ds)^2 = (dx)^2 + (dy)^2$$

and

$$ds = dx\left[1 + \left(\frac{dy}{dx}\right)^2\right]^{\frac{1}{2}}$$

If dy/dx is very small compared to unity, then it follows from the binomial theorem that approximately

$$ds = dx\left[1 + \frac{1}{2}\left(\frac{dy}{dx}\right)^2\right]$$

The work done in stretching this element of length is $F(ds - dx)$, which from the above equation is equal to $[F(\partial y/\partial x)^2\, dx]/2$. Thus the potential energy of a length L of the string is

$$W_{\text{P.E}} = \frac{F}{2}\int_0^L \left(\frac{\partial y}{\partial x}\right)^2 dx$$

The velocity v of a progressive wave in the string is given by $v = \sqrt{F/\mu}$ or $v^2\mu = F$, and the potential energy $W_{\text{P.E}}$ becomes

$$W_{\text{P.E}} = \frac{\mu v^2}{2}\int_0^L \left(\frac{\partial y}{\partial x}\right)^2 dx$$

The total wave energy of the vibrating string of length L is

$$W = \frac{\mu}{2}\int_0^L \left[\left(\frac{\partial y}{\partial t}\right)^2 + v^2\left(\frac{\partial y}{\partial x}\right)^2\right] dx \tag{9.10}$$

For a progressive wave in a cord, this energy is traveling with the speed v of the wave

9.5   Longitudinal Waves in a Solid Rod

A *longitudinal* wave is one in which the displacement of any part of the medium is along the direction of propagation of the wave. Sound waves are longitudinal waves in air. A perfect fluid, i.e., a perfect gas or a perfect liquid, can transmit longitudinal waves but not transverse waves since a perfect fluid has no shear modulus.

As an introduction to longitudinal waves we shall first determine the velocity of propagation of a longitudinal wave in a uniform rectangular rod. Figure 9.4 shows such a rod of a uniform cross section A and a density ρ oriented along the X axis. This rod is subjected to a longitudinal force which

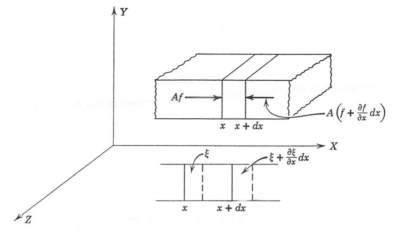

Fig. 9.4 Analysis of a longitudinal wave in a solid rod.

varies with position and time as would occur if a longitudinal wave were passing through the rod. This varying force causes any cross section of the rod to undergo small longitudinal displacements about a mean position. It is these varying displacements that constitute the wave motion. We shall assume that the displacements are within the elastic limit of the material and that any damping can be neglected.

Let us now consider a thin slab of this rod at position x and of a thickness dx. At some particular instant of time, let the force per unit area transmitted to the slab by the material of the rod at the left of the slab be f so that the total force is Af. The force exerted on the slab at $x + dx$ by the material to the right may be expressed to a sufficient degree of accuracy by the first term in Taylor's expansion as

$$-A\left(f + \frac{\partial f}{\partial x}\,dx\right)$$

The resultant force acting on the slab is

$$-A\frac{\partial f}{\partial x}\,dx$$

Suppose that ξ is the displacement from its equilibrium position of the cross section in the X direction. This is the displacement of the slab that has a mass $\rho A\,dx$. From Newton's second law the acceleration $\partial^2\xi/\partial t^2$ is given as

$$\rho A\,dx\,\frac{\partial^2\xi}{\partial t^2} = -A\frac{\partial f}{\partial x}\,dx$$

or

$$\rho \frac{\partial^2 \xi}{\partial t^2} = -\frac{\partial f}{\partial x} \qquad (9.11)$$

Since the stress on the rod is different at different points and times, it follows that the strain must likewise vary with position and time. At position x the displacement is ξ, and at $x + dx$ the displacement is $\xi + (\partial \xi / \partial x)\, dx$. The strain on the slab is the resultant displacement $(\partial \xi / \partial x)\, dx$ divided by the original length dx, or the strain is $\partial \xi / \partial x$. From the definition of Young's modulus E for a rod it can be seen that

$$E = \frac{\text{Stress}}{\text{Strain}} = \frac{-f}{\partial \xi / \partial x}$$

since, if $\partial \xi / \partial x$ is positive, then the forces must be those of extension and not of compression, as used in Fig. 9.3. Hence

$$-f = E \frac{\partial \xi}{\partial x}$$

and

$$-\frac{\partial f}{\partial x} = E \frac{\partial^2 \xi}{\partial x^2}$$

Thus from Eq. 9.11 the equation of motion of the slab becomes

$$\rho \frac{\partial^2 \xi}{\partial t^2} = E \frac{\partial^2 \xi}{\partial x^2} \qquad (9.12)$$

This is the wave equation for a longitudinal wave traveling in the rod with a phase velocity v of

$$v = \sqrt{\frac{E}{\rho}} \qquad (9.13)$$

This longitudinal wave is a plane wave inasmuch as the displacement ξ is the same at all points on any given cross section.

9.6   Longitudinal Waves in a Gas

The analysis of the motion of plane waves in a gas is almost the same as that for the rod discussed in the previous section. Suppose that the plane wave is traveling through the gas in the X direction. If the gas is considered to be composed of molecules, then, as the wave passes through the gas, the molecules vibrate about a mean position along the X axis. This motion is

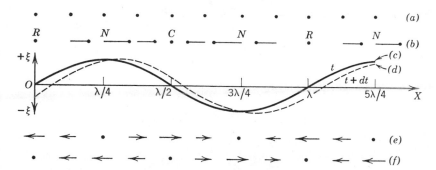

Fig. 9.5 Longitudinal wave in a gas. (*a*) Equilibrium positions; (*b*) displaced positions; (*c*) longitudinal displacements replaced by equivalent vertical displacements at time *t*, (*d*) at time $t + dt$; (*e*) velocity of particles at time *t*; (*f*) accelerations of particles at time *t*. Note that $\partial\xi/\partial x$ is positive at a rarefaction *R*, is negative at a compression *C*, and is zero at normal pressure *N*.

superimposed on the random motion of the molecules. All molecules in the *YZ* plane having a given value of *x* on the average have the same displacement at any instant, or the plane wave front of the wave is in the *YZ* plane. The displacement of the molecules about their mean positions produces rarefactions and compressions in the gas. These are shown in Fig. 9.5, where *R* indicates the position of a rarefaction, *N* the position of normal pressure, and *C* the position of a compression. In Fig. 9.5*a* are shown the normal positions of the molecules; (*b*) shows their positions when displaced any amount ξ at some instant of time. This displacement varies with *x*, the position of the molecule. Where the molecules crowd together there is a *compression*, and where they are far apart there is a *rarefaction*. Notice that at the center of a rarefaction or compression there is no displacement, whereas normal pressure occurs at maximum displacement. In Fig. 9.5*c* the longitudinal wave is represented as a transverse wave by rotating the displacements through 90°, thus making displacements to the right positive and vertically up, and displacements to the left negative and vertically down. According to this convention, $\partial\xi/\partial x$ has a *maximum negative* value at a *compression*, a *maximum positive* value at a *rarefaction*, and a value of *zero* at a position of *normal pressure*.

Suppose that the wave is traveling to the right. After a short time *dt*, the wave has the configuration shown by the dotted curve (*d*). From this the velocities of the molecules may be obtained. The displacement, velocity and acceleration of the molecules can be also obtained from the equation of the wave

$$\xi = A \sin \frac{2\pi}{\lambda}(x - vt)$$

The curve c is drawn for $t = 0$, so for this time

$$\xi = A \sin \frac{2\pi x}{\lambda}$$

The velocity of the particles at any time t and position x is

$$\frac{\partial \xi}{\partial t} = -Av \frac{2\pi}{\lambda} \cos \frac{2\pi}{\lambda}(x - vt)$$

which for $t = 0$ becomes

$$\frac{\partial \xi}{\partial t} = -Av \frac{2\pi}{\lambda} \cos \frac{2\pi x}{\lambda}$$

This is shown in curve (e). Similarly the accelerations are given by $\partial^2 \xi / \partial t^2$, which at time $t = 0$ is

$$\frac{\partial^2 \xi}{\partial t^2} = -\frac{4\pi^2}{\lambda^2} Av \sin \frac{2\pi x}{\lambda}$$

These are shown in (f).

Let us now return to the determination of the velocity of sound in a gas. Consider a thin section of the gas of thickness dx located at x, as shown in Fig. 9.4. At some instant of time the displacement of the particles at x is ξ and at $x + dx$ is

$$\xi + \left(\frac{\partial \xi}{\partial x}\right) dx$$

The increase in volume of the slab due to the different displacements of its two faces is

$$A\left(\frac{\partial \xi}{\partial x}\right) dx$$

where A is the cross section of the slab. The original volume of the slab is $A\, dx$ so that the volume strain dV/V is

$$\frac{dV}{V} = \frac{\partial \xi}{\partial x}$$

Along with this change in volume there is a change in density and pressure of the gas. If ρ is the density of the gas at any position and time, or $\rho = \rho(x, t)$, and ρ_0 is the constant or standard density of the air, then the relative change in density $\delta(x, t)$ is given by

$$\delta = \frac{\rho - \rho_0}{\rho_0} \tag{9.14}$$

The change in volumes is related to the change in densities as given by

$$\frac{dV}{V} = \frac{V - V_0}{V} = \left(\frac{c}{\rho} - \frac{c}{\rho_0}\right)\bigg/\frac{c}{\rho} = \frac{\rho_0 - \rho}{\rho_0} = -\delta = \frac{d\xi}{dx} \qquad (9.15)$$

since the density $\rho = $ constant/volume $V = c/\rho$.

At a compression the value of $\partial \xi/\partial x$ is a negative maximum so that there is a positive increase in density or δ has its maximum value. Similarly, at a rarefaction, $\partial \xi/\partial x$ is a positive maximum and there is a maximum decrease in density of the gas.

We now need a relationship between the pressure and volume or density in the rarefactions and compressions. If the temperature is constant or the changes are isothermal, then from Boyle's law $PV = $ constant, whereas if the heat is constant or the changes are adiabatic, then $PV^\gamma = $ constant, where γ is the ratio of the specific heat at constant pressure to that at constant volume for the gas. Newton assumed isothermal changes and obtained a value for the speed of sound which did not agree with experiment. This discrepancy Newton adjusted in an ingenious but scientifically incorrect manner, as is given in *The Principia*.*

It was Laplace who, in 1816, pointed out that when sound waves pass through a gas the condensations and rarefactions respectively heat up and cool down the gas. The compressions and rarefactions take place so rapidly that there is no appreciable heat loss, and the condensations and rarefactions are adiabatic rather than isothermal. Thus the relation between pressure and volume of the gas through which a sound wave is passing is given by the adiabatic equation for a gas

$$PV^\gamma = \text{constant}$$

For air, γ is approximately 1.41. By differentiation of $PV^\gamma = $ constant, it follows that

$$V^\gamma dP + \gamma PV^{\gamma-1} dV = 0 \qquad (9.16)$$

Let $dP = p$ be the acoustic pressure or the change in pressure from the normal or standard pressure P_0. Using Eq. 9.15 with Eq. 9.16, it follows that

$$p = -\gamma P_0 \frac{dV}{V} = \gamma P_0 \delta \qquad (9.17)$$

where P_0 is used for the constant pressure P in Eq. 9.16. Thus the acoustic pressure is a maximum, where δ has its maximum value at a compression.

If ρ_0 is the average constant density of the gas in the thin section, then the motion of the gas is given by Eq. 9.11 as

$$\rho_0 \frac{\partial^2 \xi}{\partial t^2} = -\frac{\partial f}{\partial x} = -\frac{\partial p}{\partial x} \qquad (9.18)$$

* *The Principia.* A revision of Motte's translation by F. Cajori, University of California Press, 1934, p. 382.

where f is the force per unit area on the rod and is equivalent to the acoustic pressure p in the gas.

From Newton's second law it follows that

$$\rho_0 \frac{\partial^2 \xi}{\partial t^2} = \gamma P_0 \frac{\partial^2 \xi}{\partial x^2} \tag{9.19}$$

This is the wave equation, and the velocity of the progressive waves is given by

$$v = \sqrt{\frac{\gamma P_0}{\rho_0}} \tag{9.20}$$

This equation gives values that agree very well with the experimental value for the speed of sound in a gas.

9.7 Energy of a Plane Wave in a Gas

The energy in a plane wave is made up of the kinetic and potential energies of the displaced molecules. In an element of volume $dx\,dy\,dz$, in which the density of the gas molecules is ρ_0 and their displacement is ξ at time t, the kinetic energy is

$$W_{\text{K.E}} = \frac{\rho_0}{2} \left(\frac{\partial \xi}{\partial t} \right)^2 dx\,dy\,dz$$

The potential energy is measured by the amount of work done in compressing the element of volume. In Fig. 9.6 the work done in compressing the gas is given by the area VV_0ab, and this area is approximately equal to

$$\left(P_0 + \frac{p}{2} \right)(V_0 - V)$$

where P_0, V_0 are constants and p and V are variables.

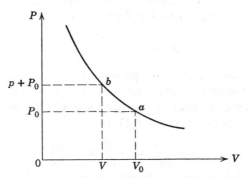

Fig. 9.6 Work done in compressing a gas.

If the wave is a simple harmonic one, then $P_0(V_0 - V)$ over any number of waves is zero. The remaining term is $[p(V_0 - V)]/2$, and this together with Eqs. 9.15 and 9.17 becomes

$$\frac{p}{2}(V_0 - V) = \frac{\gamma P_0 \delta^2}{2} V = \frac{\gamma P_0}{2}\left(\frac{\partial \xi}{\partial x}\right)^2 dx\, dy\, dz$$

where $V = dx\, dy\, dz$.

Thus the potential energy $W_{P.E}$ can be written, using Eq. 9.20, as

$$W_{P.E} = \frac{\rho_0}{2} v^2 \left(\frac{\partial \xi}{\partial x}\right)^2 dx\, dy\, dz$$

The total energy of a given volume of gas due to the passage of a plane wave through it is

$$W = \frac{\rho_0}{2} \iiint \left[\left(\frac{\partial \xi}{\partial t}\right)^2 + v^2 \left(\frac{\partial \xi}{\partial x}\right)^2\right] dx\, dy\, dz \qquad (9.21)$$

This expression is similar to that for the vibrating string given by Eq. 9.10.

The energy per unit volume or the energy density W_v is

$$\frac{W}{dx\, dy\, dz} = W_v \qquad (9.22)$$

The time average of this expression for the energy density at any point in the plane wave in the gas is

$$\overline{W}_v = \frac{1}{T}\int_0^T W_v\, dt \qquad (9.23)$$

Suppose a plane wave given by

$$\xi = A \sin k(x - vt)$$

travels through the gas in the positive X direction. The average energy density in the wave is

$$\overline{W}_v = \frac{\rho_0 v^2 A^2 k^2}{T}\int_0^T \cos^2 k(x - vt)\, dt$$

$$= \frac{\rho_0 v^2 A^2 k^2}{T}\int_0^T (\cos kx \cos kvt + \sin kx \sin kvt)^2\, dt$$

In evaluating this integral over one or more complete cycles, the $\cos^2 kvt$ and $\sin^2 kvt$ terms yield $T/2$, whereas the $\sin kvt$ or $\cos kvt$ terms yield zero. Thus

$$\overline{W}_v = \frac{\rho_0 v^2 A^2 k^2}{T}\frac{T}{2} = \frac{\rho_0 (vAk)^2}{2} \qquad (9.24)$$

Note that the average value of the kinetic energy over a cycle is equal to that of the potential energy.

The rate at which the wave energy passes perpendicularly through a unit area per second is called the intensity of the wave I; thus

$$I = \overline{W}_v v = \frac{\rho_0 v^3 A^2 k^2}{2} \qquad (9.25)$$

The intensity is also equal to the acoustic pressure p multiplied by the velocity $\partial \xi / \partial t$ of the gas molecules, or

$$I = p \frac{\partial \xi}{\partial t} = -\rho_0 v^2 \frac{\partial \xi}{\partial t} \frac{\partial \xi}{\partial x} \qquad (9.26)$$

As an exercise show that Eq. 9.26 gives Eq. 9.25 for the plane wave.

It is common to use a temperature of 20°C and standard atmospheric pressure in defining standards of acoustic intensity. For these conditions the density of air is 0.00121 gm/cm³, and the speed of sound in air is 34,300 cm/sec, giving $\rho_0 v$ equal to 41.5 gm/cm² sec. A unit of intensity of 10^{-16} watts/cm² which is barely audible is used in acoustic work, particularly as the I_0 in the intensity levels l of sounds, measured in decibels, where

$$l = 10 \log_{10} \frac{I}{I_0} \qquad (9.27)$$

For plane waves an equivalent expression for the intensity level can be given in terms of the acoustic pressure as

$$l = 20 \log_{10} \frac{p}{p_0} \qquad (9.28)$$

where p_0 is given as a root mean square (rms) pressure of 0.0002 dyne/cm².

9.8   Waves in Unlimited Solids

Up to the present we have discussed the velocity of transverse waves in a string, and of longitudinal waves in a rod and in a gas. Although the mathematical analysis of other types of wave motion is beyond the scope of this text, we should not leave this subject without giving some of the results. In an unlimited homogeneous isotropic medium the velocity of the longitudinal waves v_l is given by

$$v_l = \sqrt{\frac{k_m + 4n/3}{\rho}} \qquad (9.29)$$

where k_m is the bulk modulus, n the shear modulus, and ρ the density of the medium. The velocity of longitudinal waves in a rod is given by $v = \sqrt{E/\rho}$, where E is Young's modulus. In the definition of Young's modulus we are concerned only with the longitudinal strain and not with the lateral strain which, according to Poisson's ratio, is σ times as great. The distinction between a rod and an unlimited homogeneous medium is the fact that when there is a longitudinal strain in the rod there is a lateral strain, whereas in the unlimited homogeneous medium no lateral strain is present. The velocity of transverse waves in an unlimited homogeneous isotropic medium is given by $v_t = \sqrt{n/\rho}$. Thus we see that the velocity of longitudinal waves in an unlimited solid is larger than the velocity of the corresponding transverse waves. This is of importance in seismology where the waves created by earthquakes travel out in all directions. The first wave to reach a seismograph from an earthquake is the longitudinal one, and then follows the transverse one. If the time interval between the arrival of the longitudinal and that of the transverse waves is measured, then the distance of the earthquake from the seismograph can be determined. This, of course, involves a knowledge of the difference in velocity of the two types of waves. When the earthquake waves reach the surface, they create surface waves whose velocity we have not calculated. Since a perfect liquid does not possess any shear modulus, there is no transverse wave in the body of it but only a longitudinal wave whose velocity is $\sqrt{k/\rho}$. There are, of course, surface waves on a liquid as, for example, those on the surface of a body of water. These waves may appear to be transverse but are not really so. The individual particles of the water move in ellipses. Many causes contribute to the complicated waves on an ocean. Tides are primarily a result of the gravitational attraction of the moon, although at any place on the seashore they are greatly influenced by the local topography. Waves on an open ocean are most often caused by wind. The height of these depends on three factors; the speed of the wind, the length of time it has been blowing, and the area of open water over which it is blowing. The longest period of such waves is about 22 seconds, corresponding to a wavelength of half a mile and a speed of about 80 mph.

The most destructive waves on the ocean are produced by earth movements under the sea. These are commonly called tidal waves or tsunamis, a Japanese word meaning tidal waves, though these have no connection with tides. On the open ocean they have periods of more than 15 min, wavelengths of the order of 500 mi, and though their height or amplitude may only be a few feet, they carry enormous energy that may be released on the coastline. One of the most destructive tsunamis on the Bay of Bengal in 1876 caused about a quarter of a million casualties.

When a wave strikes a boundary, it is partially reflected and partially transmitted, the relative amplitudes of the two waves depending on the densities of the media on the two sides of the boundary. We shall now proceed to calculate the relative amplitudes of waves on reflection.

9.9   Reflection of Waves at a Boundary

Since it is easier to visualize transverse waves than longitudinal ones, we shall consider the reflection and the transmission of a transverse wave at the boundary of two ropes of different linear densities.

Suppose that two ropes, having masses per unit length of μ_1, μ_2 respectively, lie along the X axis in their equilibrium position and are joined at the origin, $x = 0$ (Fig. 9.7). The tension F in the two ropes is the same, and the velocity of the transverse wave in rope 1 is v_1 and in rope 2 is v_2. Let us investigate what happens to a wave traveling from the left toward the right when it arrives at the boundary. Any transmitted or reflected wave must have the same frequency f as the incident wave. For the incident wave in rope 1 the displacement at any time is given by

$$y_i = A_i \cos 2\pi f \left(\frac{x}{v_1} - t \right)$$

and for the reflected wave in rope 1 traveling to the left by

$$y_r = A_r \cos 2\pi f \left(\frac{x}{v_1} + t \right)$$

where A_i and A_r are the amplitudes of the incident and reflected waves respectively. The transmitted wave in rope 2 moving to the right is represented by

$$y_t = A_t \cos 2\pi f \left(\frac{x}{v_2} - t \right)$$

At the boundary, $x = 0$, the vertical displacement of the two ropes must be the same at every instant of time, or

$$y_i + y_r = y_t$$

From this it follows that

$$A_i + A_r = A_t$$

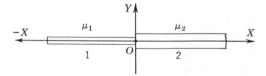

Fig. 9.7   Reflection of waves at a boundary.

Also at the boundary, $x = 0$, the vertical forces on the ropes must be the same. From Eq. 9.1 this gives

$$F\left(\frac{\partial y_i}{\partial x} + \frac{\partial y_r}{\partial x}\right)_{x=0} = F\left(\frac{\partial y_t}{\partial x}\right)_{x=0}$$

By substitution in the above equations, we have

$$-\frac{A_i}{v_1} + \frac{A_r}{v_1} = -\frac{A_t}{v_2}$$

or

$$A_i - A_r = \frac{v_1}{v_2} A_t$$

Substituting for A_t we have

$$\frac{A_i - A_r}{A_i + A_r} = \frac{v_1}{v_2}$$

If the coefficient of reflection R is defined as the ratio of the amplitude of the reflected wave to the incident wave, then

$$R = \frac{A_r}{A_i} = \frac{v_2 - v_1}{v_2 + v_1}$$

The velocities in the two ropes are

$$v_1 = \sqrt{\frac{F}{\mu_1}} \quad \text{and} \quad v_2 = \sqrt{\frac{F}{\mu_2}}$$

It follows that

$$R = \frac{1/\sqrt{\mu_2} - 1/\sqrt{\mu_1}}{1/\sqrt{\mu_2} + 1/\sqrt{\mu_1}} = \frac{\sqrt{\mu_1} - \sqrt{\mu_2}}{\sqrt{\mu_1} + \sqrt{\mu_2}} \qquad (9.30)$$

The reflecting power is defined as the ratio of the reflected to the incident energy. Since the waves used above are simple harmonic, their energy is proportional to the square of the amplitude. Thus the reflecting power is proportional to R^2.

If the mass per unit length μ_1 of rope 1 is smaller than μ_2 of rope 2, then the coefficient of reflection is negative and the amplitude of the reflected wave A_r is opposite to that of the incident wave. The reflected wave is 180° out of phase with the incident wave at the boundary. This is the case when a rope is tied to a wall (Fig. 9.8a) or when a sound wave is reflected at a closed end of an organ pipe. The displacements of the air molecules in the incident and reflected waves are in opposite directions at the closed end. A compression is reflected as a compression and a rarefaction as a rarefaction.

On the other hand, if μ_1 is greater than μ_2, then R is positive (Fig. 9.8b). There will be no change in phase at reflection, so A_r has the same sign

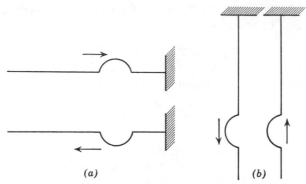

Fig. 9.8 Reflection at a fixed end; 180° change in phase, or R is negative ($\mu_1 < \mu_2$). (b) Reflection at a free end; no change in phase, or R is positive ($\mu_1 > \mu_2$).

as A_i. In acoustics this corresponds to reflection at the open end of an organ pipe, and for this a compression is reflected as a rarefaction, and vice versa.

The incident and reflected waves traveling in the same medium with the same velocity, frequency, and wavelength can interfere with each other in such a manner as to produce standing waves. We shall next investigate the conditions for setting up standing waves.

9.10 Standing Waves

Although progressive waves in various media have occupied much of our attention, it is perhaps well to remember that there are no such entities as infinite media. Sooner or later a wave meets a boundary and is reflected. The medium is then acted on by the two waves, the incident and the reflected. To investigate the action of these two waves, we shall consider a cord fastened at both ends. If the cord is plucked or excited as by a violin bow, standing waves are set up along the cord.

The incident and reflected waves whose wavelength is λ and whose velocity is v may be represented at any time by

$$y = A \sin \frac{2\pi}{\lambda} (x - vt - \alpha) + A \sin \frac{2\pi}{\lambda} (x + vt)$$

where there is a difference in phase of α, and the amplitude A of the two waves is equal. The equation above can be expanded to give

$$y = \left(A_1 \sin \frac{2\pi vt}{\lambda} + B_1 \cos \frac{2\pi vt}{\lambda} \right) \sin \frac{2\pi x}{\lambda} \qquad (9.31)$$

This is a solution of the wave equation, Eq. 9.3.

Fig. 9.9   The fundamental and first two harmonics in a standing wave.

Suppose that a cord of length L is rigidly clamped at $x = 0$ and $x = L$. At these two points the displacement y is zero at all times. The boundary condition, $y = 0$ at $x = 0$, is already satisfied by the equation. To satisfy the condition $y = 0$ at $x = L$ for all values of t, we must have

$$\sin 2\pi \frac{L}{\lambda} = 0$$

or

$$2\pi \frac{L}{\lambda} = n\pi$$

where $n = 1, 2, 3, \ldots$. If λ_1 is the value of λ corresponding to $n = 1$, then

$$\lambda_1 = 2L$$

Similarly

$$\lambda_2 = \frac{2L}{2}, \quad \lambda_3 = \frac{2L}{3}, \quad \lambda_4 = \frac{2L}{4}, \quad \cdots, \quad \lambda_n = \frac{2L}{n}$$

The corresponding frequencies are

$$f_1 = \frac{v}{\lambda_1} = \frac{v}{2L}, \quad f_2 = \frac{2v}{2L}, \quad f_3 = \frac{3v}{2L}, \quad \cdots, \quad f_n = \frac{nv}{2L}$$

The lowest frequency f_1 is called the *fundamental*, and the other possible frequencies $f_2, f_3, \ldots, f_n$ are called the *harmonics* or *overtones*. For the nth possible or allowable vibration

$$y_n = (A_n \sin 2\pi f_n t + B_n \cos 2\pi f_n t) \sin \frac{n\pi x}{L} \tag{9.32}$$

This is the equation of a standing wave. There are some points, called *nodes*, at which $\sin (n\pi x/L)$ or y is always zero and, therefore, the cord is always at rest at these points. There are other points called the *antinodes* or *loops* where $\sin (n\pi x/L)$ equals unity. Such points oscillate with maximum amplitude. The nodes and loops for the fundamental and the first two harmonics are shown in Fig. 9.9.

Figure 9.10 shows how two traveling waves moving in opposite directions can produce a standing wave. Analytically, the expression for the standing wave shown in Fig. 9.10 is given by Eq. 9.31.

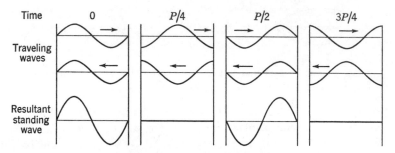

Fig. 9.10 Formation of standing waves by two similar waves traveling in opposite directions.

9.11   Vibration of a Cord with a Given Initial Displacement

As we have seen, a cord fixed at both ends is capable of vibrating with many different frequencies f_n related to each other in a simple manner.

We shall now examine this problem in a more general manner.

Consider a cord fixed at $x = 0$ and $x = L$. If the displacement y at position x and time t is written as $y = y(x, t)$ then $y = 0$ at the positions at $x = 0$ and $x = L$ may be written as $y(0, t) = y(L, t) = 0$. Let us assume that initially, $t = 0$, the cord is held at rest in a shape given by $y = f(x)$, or

$$y(x, 0) = f(x)$$

Also, at $t = 0$ the velocity of the cord is zero, which is expressed as

$$\partial y(x, 0)/\partial t = 0$$

Thus our problem is to solve Eq. 9.3, $\partial^2 y/\partial t^2 = (\mu/F)\partial^2 y/\partial x^2$, with the boundary conditions

$$y(0, t) = y(L, t) = 0; \qquad \partial y(x, 0)/\partial t = 0 \qquad \text{and} \qquad y(x, 0) = f(x)$$

This problem is typical of many problems in physics, including quantum mechanics, in which it is required to find the solution of a differential equation with some given boundary conditions. We assume, as in Section 9.3, that the variables x and t are separable or $y = X(x)T(t)$. Consider first the space part of the wave equation, Eq. 9.9, expressed as

$$\frac{d^2 X}{dx^2} = -c^2 X \qquad \text{with} \qquad X(0) = X(L) = 0 \qquad (9.33)$$

What we wish to find are the special values of c for which the differential equation is satisfied. This is a typical *eigenvalue* problem in which there is a

parameter c for which we are to find solutions when the boundary conditions are satisfied. The special values of the parameter c for which solutions exist are called the *eigenvalues* and the corresponding solutions the *eigenfunctions* or *eigensolutions*. (The word eigenvalue is a mixed German-English word having roughly the meaning of characteristic or proper value.)

The general solution of $d^2X/dx^2 = -c^2X$ is

$$X = A \sin cx + B \cos cx$$

That this is a solution may be readily verified by differentiation because it contains two arbitrary constants A and B, which are required in the solution of a second order differential equation. These arbitrary constants are determined from the boundary conditions of the problem. The boundary conditions at $x = 0$ and $x = L$ are

$$X(0) = 0 \text{ which gives } B = 0$$

and

$$X(L) = 0 \text{ which gives } A \sin cL = 0$$

The second equation is satisfied for $A = 0$, which is a trivial solution, and also for $\sin cL = 0$ which gives

$$cL = n\pi \qquad (n = 1, 2, 3 \ldots) \qquad (9.34)$$

For each value of n there is a corresponding value of c, the eigenvalue, so that the nth eigenvalue is

$$c_n = n\pi/L$$

and the corresponding eigenfunction is

$$X_n(x) = A_n \sin (n\pi x/L)$$

The values of the constants c_n are no longer arbitrary, but are given by Eq. 9.9, so that for the time portion we have

$$\frac{d^2T}{dt^2} = -\frac{F}{\mu} c_n{}^2 T \qquad (9.9)$$

with the boundary condition $dT/dt = 0$ at $t = 0$, which follows from $\partial y/\partial t = 0$ at $t = 0$ and $y = XT$.

The general solution of Eq. 9.9 is

$$T(t) = C_n \cos \left(\sqrt{\frac{F}{\mu}} c_n t \right) + D_n \sin \left(\sqrt{\frac{F}{\mu}} c_n t \right)$$

and the boundary condition $\partial T/\partial t = 0$ at $t = 0$ gives $D_n = 0$. Thus $T(t) = C_n \cos [(n\pi vt)/L]$ using $\sqrt{F/\mu} = v$ and Eq. 9.34.

Now the wave equation (Eq. 9.3) is linear and homogeneous and this has the important property that if y_1 and y_2 are solutions, then any linear combination of $c_1 y_1 + c_2 y_2$ is also a solution. Hence the general solution of Eq. 9.3 is a linear superposition of the solutions having the form

$$\sin\left(\frac{n\pi x}{L}\right) \cos\left(\frac{n\pi v t}{L}\right)$$

Thus the displacement y of the cord at any position x and time t, from $y = XT$, is given as

$$y(x, t) = \sum_{n=1}^{\infty} E_n \sin\left(\frac{n\pi x}{L}\right) \cos\left(\frac{n\pi v t}{L}\right) \tag{9.35}$$

where E_n is the single constant $A_n C_n$, and is found from the initial condition, that $y(x, 0) = f(x)$ at $t = 0$. Thus

$$f(x) = \sum_{n=1}^{\infty} E_n \sin\left(\frac{n\pi x}{L}\right) \tag{9.36}$$

The infinite series for $f(x)$ is called a Fourier sine series after Joseph Fourier, a distinguished French military engineer. In 1822 Fourier published an important book, *Théorie Analytique de la Chaleur*, in which he presented what has become known as the Fourier series. Fourier expressed the varying temperature in a substance at any point x and time t as a series of sines and cosines.

The question arises whether any arbitrary function $f(x)$ can be expressed as a series of sines and cosines. Mathematicians have shown that this is possible under certain conditions of continuity. For most physics problems, and all we shall be concerned with here, these conditions are met.

Our next problem is to find the coefficients E_n. This is done by multiplying $f(x)$ by $\sin(m\pi x/L)\, dx$ and integrating from 0 to L, giving

$$\int_0^L f(x) \sin\frac{m\pi x}{L}\, dx = \int_0^L \sum_n E_n \sin\frac{n\pi x}{L} \sin\frac{m\pi x}{L}\, dx$$

Now it can readily be shown that

$$\int_0^L \sin\left(\frac{n\pi x}{L}\right) \sin\left(\frac{m\pi x}{L}\right)\, dx$$

equals $L/2$ if $n = m$ and equals zero if $n \neq m$. For convenience this condition is often expressed in terms of the Kronecker δ (delta) symbol such that

$$\delta_{mn} = 1 \quad \text{for} \quad n = m$$
$$= 0 \quad \text{for} \quad n \neq m$$

Thus

$$\int_0^L \sin\left(\frac{n\pi x}{L}\right) \sin\left(\frac{m\pi x}{L}\right) dx = \frac{L}{2}\delta_{mn}$$

(Notice that if $n = m$ the integral has the integrand of $\sin^2 (n\pi x/L)\, dx$, which may be written as $[\cos (2\pi n x/L) - 1]\, dx/2$, of which the cosine integral is zero for the limits 0 and L, and the integral of $dx/2$ evaluated at the limits is $L/2$, whereas the integral for $n \neq m$ is zero.) Thus it follows that

$$\int_0^L f(x) \sin\left(\frac{n\pi x}{L}\right) dx = \frac{L}{2} E_n$$

or

$$E_n = \frac{2}{L}\int_0^L f(x) \sin\left(\frac{n\pi x}{L}\right) dx \qquad (n = 1, 2 \cdots) \tag{9.37}$$

Using this relationship, the solution of the equation for the vibrating cord may be written as

$$y(x, t) = \sum_n \left[\int_0^L \frac{2}{L} f(\beta) \sin\left(\frac{n\pi\beta}{L}\right) d\beta\right] \sin\left(\frac{n\pi x}{L}\right) \cos\left(\frac{n\pi vt}{L}\right) \tag{9.38}$$

where the expression in the square brackets is the E_n and has used the dummy symbol β so as not to confuse it with the x in the remainder of the equation.

As an example of the above consider a cord fixed at the ends $x = 0$ and $x = L$ with its center point pulled out a height h at $t = 0$ and then let go, as shown in Fig. 9.11. In order to evaluate E_n it is necessary to break the integral into the two parts from 0 to $L/2$ and $L/2$ to L. In the interval 0 to $L/2$ the equation of the initial configuration of the cord is

$$\frac{y}{x} = \frac{h}{L/2} \qquad \text{or} \qquad y = 2\frac{hx}{L}$$

and in the region from $L/2$ to L the equation is

$$\frac{y}{L - x} = \frac{h}{L/2} \qquad \text{or} \qquad y = \frac{2h(L - x)}{L}$$

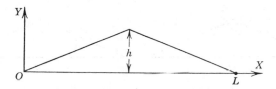

Fig. 9.11   The initial displacement of a cord.

Thus

$$E_n = \frac{2}{L}\left[\int_0^{L/2} \frac{2hx}{L} \sin\left(\frac{n\pi x}{L}\right) dx + \int_{L/2}^{L} \frac{2h}{L}(L-x)\sin\left(\frac{n\pi x}{L}\right) dx\right]$$

$$= \frac{8h}{\pi^2 n^2} \sin\left(\frac{n\pi}{2}\right) \tag{9.39}$$

as may be proved by integrating by parts. If n is an even integer then the corresponding E_n's are zero. If n is an odd integer, then recalling that $\sin \pi/2 = 1$ and $\sin 3\pi/2 = -1$ it follows that

$$E_n = \frac{8h}{\pi^2 n^2}(-1)^{(n-1)/2} \tag{9.40}$$

and the solution is given by Eq. 9.35 as

$$y(x, t) = \sum E_n \sin\frac{n\pi x}{L} \cos\frac{n\pi vt}{L} \tag{9.41}$$

Thus the displacement y at position x and time t for this problem is

$$y(x, t) = \frac{8h}{\pi^2}\left(\sin\frac{\pi x}{L}\cos\frac{\pi vt}{L} - \frac{1}{9}\sin\frac{3\pi x}{L}\cos\frac{3\pi vt}{L}\right.$$
$$\left. + \frac{1}{25}\sin\frac{5\pi x}{L}\cos\frac{5\pi vt}{L} - \frac{1}{49}\sin\frac{7\pi x}{L}\cos\frac{7\pi vt}{L} + \cdots\right) \tag{9.42}$$

Using $L = 5$ and $h = 1$, the first four terms are plotted in Fig. 9.12 for $t = 0$. We see from the above equation that even harmonics are absent for a cord started in the manner described. These even harmonics would imply a node at the center which does not exist since the cord is initially displaced at the center. Since the intensity of a wave is proportional to the square of its amplitude, then for the sound emitted by the string the fundamental would have an intensity 81 times the third harmonic and 625 times the fifth harmonic, etc.

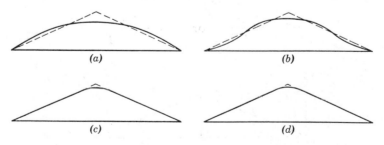

Fig. 9.12 The initial displacement of the cord given by the Fourier series, Eq. 9.42; (a) one term; (b) two terms; (c) three terms; (d) four terms. The dotted curve shows the initial displacement.

In an organ pipe, wind is blown past a lip and eddies are set up which produce the fundamental and the overtones. To compute the relative intensities of these is a difficult matter since the initial conditions are difficult to determine.

9.12   Velocity of a Group of Waves. Group Velocity

In every example discussed so far the velocity of the waves has not depended on their wavelength. All waves, whatever their wavelength, traveled with the same velocity, but this is not always so. In both surface water waves and in light waves traveling through some medium, the velocity of the waves depends on their wavelength, and there is then dispersion. The effect of this dispersion is frequently seen when a motorboat moves rapidly through a body of calm water. An inverted V-shaped wave front is produced with the point of the V at the moving boat. Individual small waves of various wave lengths are produced by the boat, and these move out with their respective velocities and interfere in such a manner as to produce the inverted V-shaped wave front. We shall now proceed to calculate the velocity of a group of waves produced by interference of waves of different wavelength traveling with different velocities through the medium.

Suppose that two harmonic waves of the same amplitude but having slightly different velocities, wavelengths, and frequencies are traveling along the positive X axis. If wave 1 of wavelength λ, velocity v, and amplitude A produces a displacement y_1 of the medium at a point x and time t, then y_1 may be expressed

$$y_1 = A \sin \frac{2\pi}{\lambda} (x - vt)$$

Similarly, if wave 2 of wavelength $\lambda + d\lambda$, velocity $v + dv$, and amplitude A produces a displacement y_2 of the medium at the same point x and time then

$$y_2 = A \sin \frac{2\pi}{\lambda + d\lambda} [x - (v + dv)t]$$

The resultant displacement y at the point x and time t is given by the principle of superposition of the waves as

$$y = y_1 + y_2 = A \sin \frac{2\pi}{\lambda} (x - vt) + A \sin \frac{2\pi}{\lambda + d\lambda} [x - (v + dv)t]$$

This expression can be transformed by using the trigonometric identity

$$\sin P + \sin Q = 2 \sin \frac{P + Q}{2} \cos \frac{P - Q}{2} = 2 \sin \frac{\text{sum}}{2} \cos \frac{\text{difference}}{2}$$

Hence

$$y = 2A \left\{ \sin \pi \left[\frac{x}{\lambda} + \frac{x}{\lambda + d\lambda} - \frac{vt}{\lambda} - \frac{(v + dv)t}{(\lambda + d\lambda)} \right] \cos \pi \right.$$

$$\left. \times \left[\frac{x}{\lambda} - \frac{x}{\lambda + d\lambda} - \frac{vt}{\lambda} + \frac{(v + dv)t}{(\lambda + d\lambda)} \right] \right\}$$

Assuming that $d\lambda$ and dv are small compared to λ and v respectively, we have

$$y = 2A \sin \frac{2\pi}{\lambda} (x - vt) \cos \pi \left[x \frac{d\lambda}{\lambda^2} - \frac{(v \, d\lambda - \lambda \, dv)t}{\lambda^2} \right] \qquad (9.43)$$

This complicated expression represents a wave traveling with a phase velocity v given by the sine term. The amplitude of the wave is changing as the wave travels along with the group velocity given by the cosine term. If Eq. 9.43 is plotted for some instant of time, then curves similar to those shown in Fig. 9.13 are obtained. The full curve is the sine term representing a wave of wavelength λ and velocity v whose amplitude changes in a manner shown by the dotted curve. This dotted curve is given by the cosine term. The points of maximum or minimum amplitude travel along the X axis with a velocity U called the group velocity, where

$$U = \frac{(v \, d\lambda - \lambda \, dv)/\lambda^2}{d\lambda/\lambda^2} = v - \lambda \frac{dv}{d\lambda} = \frac{dv}{d(1/\lambda)} \qquad (9.44)$$

where v is the frequency of the waves $= v/\lambda$ (see Problem 28). If, as is the case for water waves, the larger wavelengths have the larger velocity, i.e., $dv/d\lambda$ is positive, then the group velocity U is smaller than the phase velocity v of the waves. The individual waves move from the origin to the front, creating the larger crests and troughs which travel with the slower velocity U. The group waves are produced by the interference of the smaller waves advancing through the group from the origin to the front of the group where they die out and are replaced by new waves from the origin.

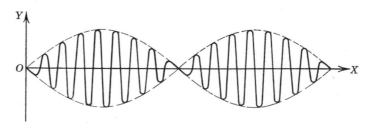

Fig. 9.13 Group velocity curves. The dotted line wave curve, the cosine curve, in Eq. 9.43, moves with the group velocity U, and the full curve, the sine term, moves with the phase velocity v.

In optics the concept of group velocity played an important role in reconciling the results obtained for the ratio of the velocity of light in air and in carbon disulfide by direct measurements with those obtained by refractive index methods. It has also been of importance in founding the modern theory of wave mechanics.

An interesting analogy to the bow wave produced by a fast-moving boat is that of the electromagnetic radiation given off by a charged particle moving in a medium with a speed greater than that of light in the medium. This radiation is called Čerenkov radiation.* It can be observed with electrons moving with almost the speed of 3×10^{10} cm/sec in lucite in which the speed of light is about 2×10^{10} cm/sec.

9.13   Dimensional Analysis of the Velocity of Sound in a Gas

To illustrate the method of dimensional analysis, we shall consider the problem of the velocity of sound in a gas. Let us assume that at any given temperature the velocity of sound v depends on the density ρ, the pressure p, and the viscosity η of the gas. Then the velocity v may be expressed

$$v = K\rho^a p^b \eta^c \tag{9.45}$$

where K is a dimensionless constant and a, b, c are dimensionless numbers which must be determined. Let us now put in the dimensions of the quantities in Eq. 9.45 in terms of mass, length, and time. We have:

$$[LT^{-1}] = [ML^{-3}]^a [ML^{-1}T^{-2}]^b [MT^{-1}L^{-1}]^c$$

Equating the exponents for the three-dimensional quantities, we have:

For M: $\qquad 0 = a + b + c$

For L: $\qquad 1 = -3a - b - c$

For T: $\qquad -1 = -2b - c$

Solving these three equations, it follows that

$$a = -\tfrac{1}{2} \qquad b = \tfrac{1}{2} \qquad c = 0$$

Hence from Eq. 9.45

$$v = K\sqrt{\frac{p}{\rho}}$$

*D. Halliday, *Introductory Nuclear Physics*. John Wiley and Sons, New York, 2nd ed., 1955.

Thus the velocity of sound does not depend on the viscosity of the gas through which the wave is passing. By measuring the velocity of sound in a gas of known pressure and density, it is possible to determine the constant K.

It can readily be appreciated that, if some physical quantity is a function of four or more other physical quantities, no unique solution can be obtained as was done in the above problem. It is usual then to form independent dimensional groups among the quantities. If a problem can be formulated in terms of p secondary quantities, such as density, viscosity, force, pressure, etc., and if these p quantities can be expressed in terms of q primary quantities which are frequently mass, length, and time, then $p - q$ independent dimensionless groups can be formed.

9.14   The Use of Models and Dimensional Analysis

Dimensional analysis has provided a powerful tool for investigating difficult problems dealing with airplanes, ships, etc. It is used to predict the forces on full-scale airplanes and ships from measurements of the corresponding forces on suitable models. Suppose that F is the resistance force exerted on a body having some characteristic length l when it is moved through a fluid of density ρ and viscosity η with a velocity v relative to the fluid. The length l might be the chord of a wing or some length associated with the beam of a ship. The force F is a function of the above variables and may be stated as follows:

$$F = f(l, \rho, \eta, v)$$

The force F may then be written

$$F = Kl^a\rho^b\eta^cv^d \tag{9.46}$$

where K is a dimensionless constant and the exponents are dimensionless constants which have to be determined. By inserting the dimensions of the various quantities, we have

$$[MLT^{-2}] = [L]^a[ML^{-3}]^b[ML^{-1}T^{-1}]^c[LT^{-1}]^d$$

Since the exponents of the three primary dimensional quantities must each be equal to zero, it follows that:

For M: $1 = b + c$

For L: $1 = a - 3b - c + d$

For T: $-2 = -c - d$

These equations may be solved to give

$$a = 2 - c \qquad b = 1 - c \qquad d = 2 - c$$

Equation 9.46 can then be written

$$F = l^{2-c} \rho^{1-c} \eta^c v^{2-c} = (l^2 \rho v^2) \left(\frac{\eta}{l \rho v}\right)^c$$

or

$$\frac{F}{l^2 \rho v^2} = f\left(\frac{\eta}{l \rho v}\right) \qquad (9.47)$$

Since in Eq. 8.41 we have introduced Reynolds' number R as $l \rho v / \eta$, we shall again use it here. Thus Eq. 9.47 may equally well be written

$$\frac{F}{l^2 v^2 \rho} = f_1\left(\frac{l \rho v}{\eta}\right) \qquad (9.48)$$

Each side of Eq. 9.48 represents a dimensionless group. Other equations having dimensionless groups on each side could have been obtained from the three equations between a, b, c, and d. From Eq. 9.48 it follows that the force exerted on a body in its motion through a viscous fluid can be represented by a curve showing the quantity $F/l^2 v^2 \rho$ plotted as ordinate against Reynolds' number $l \rho v / \eta$ as abscissa. Such a curve is called the characteristic curve and depends on the shape of the body but is independent of the size of the body. This curve defines the function f_1 in Eq. 9.48. Once the curve is constructed for the model, it is equally valid for the full-scale airplane or ship.

Suppose that conditions can be arranged so that $l \rho v / \eta$ is the same for an airplane wing as for its model. Then if F_1 is the measured force on the wing model having a chord length l_1, wind velocity v_1, and density ρ_1, the force F on the full-scale wing having a chord length l, a wind velocity v, and an air density ρ is given by Eq. 9.48 as

$$\frac{F}{l^2 v^2 \rho} = \frac{F_1}{l_1^2 v_1^2 \rho_1}$$

or

$$F = F_1 \frac{l^2 v^2 \rho}{l_1^2 v_1^2 \rho_1}$$

Now, if $l \rho v / \eta$ is the same for the wing and its model, then, from the above notation in which the subscript refers to the model,

$$\frac{l \rho v}{\eta} = \frac{l_1 \rho_1 v_1}{\eta_1}$$

If the model is placed in a wind tunnel where

$$\eta = \eta_1 \qquad \text{and} \qquad \rho = \rho_1$$

then in order for Reynolds' number to have the same value it is necessary that

$$v_1 = \frac{vl}{l_1} \qquad (9.49)$$

Now, since the ratio l/l_1 is usually very large, it follows from Eq. 9.49 that the velocity of the air in the wind tunnel must be much larger than the velocity of the plane in flight. These conditions cannot be carried out in practice. What is frequently done is to construct a wind tunnel in which the density of the air ρ_1 is larger than the density ρ at atmospheric pressure. By this means fairly large values of Reynolds' number can be obtained without excessive velocities. Thus results applicable to large planes can be obtained with relatively small models in pressure wind tunnels.

9.15   A Final Word

This book presents an elementary picture of classical mechanics. There are many more complicated methods of analysis in classical mechanics than have been given herein, but these must be left to more advanced treatises. What has been covered here should give you a good preparation for further study. Modern physics as it has been developed in the present century forms such an important part of the whole of physics that you may wonder why you should spend so much time on classical physics. The fact is that, although modern physics has made some breaks with classical physics, nevertheless, on the whole, there is a continuity of ideas running through all of physics. The laws of conservation of momentum and energy are just as valid today as they were before the turn of the century. Modern physics cannot really be understood without a sound knowledge of classical physics.

PROBLEMS

1. Show that the one-dimensional wave equation, Eq. 9.3, is satisfied by the following functions: (a) $y = A(x - vt)$; (b) $y = A(x - vt)^2$; (c) $y = A(\sqrt{x + vt})$; (d) $y = A \log_e (x + vt)$, where $v = \sqrt{F/\mu}$ and A is a constant. (e) Explain why the functions given in (a) to (d) are not useful in discussing wave motion.

2. Show that F/μ has the dimensions of the square of the velocity, using both the absolute and gravitational systems for the force F and mass per unit length μ.

3. Show that Eq. 9.35 for a standing wave is a solution of the wave equation, Eq. 9.3.

4. Show that the average energy per unit volume for a plane longitudinal wave of amplitude A and angular velocity ω in a homogeneous rod of density ρ is $(A^2\omega^2\rho)/2$.

5. Find (a) the velocity of sound in air at 1 atm of pressure and 20°C if $\gamma = 1.41$; (b) the velocity at 20°C and 2 atm of pressure; (c) the velocity at 100°C and 1 atm of pressure.

6. An explosion creates a sound wave which travels uniformly out in all directions from a point. Explain why the displacement ξ of the air at any distance r from the center may be written

$$\xi = \frac{A}{r}\sin k(r - vt)$$

To do this consider the energy per second crossing a spherical surface of radius r. Show that the intensity of the wave varies as the inverse square of the distance from the source.

7. A meter stick whose mass is 150 gm is clamped horizontally so that 80 cm projects over the edge of a table. When a 100 gm load is hung on one end, a further deflection of 2 cm occurs. The width of the stick is 3 cm, and its thickness is 0.5 cm. The stick is placed so that the width is originally horizontal. Find (a) Young's modulus for the rod; (b) the velocity of longitudinal waves in the rod.

8. A note whose frequency is 1000 vibrations per second has an intensity of 10 μw/m² (1 μw = 1 microwatt). Find the amplitude of the vibrations in the air created by this audible sound.

9. Corresponding to Eq. 9.19 show that:

$$\frac{\partial^2 \rho}{\partial t^2} = v^2 \frac{\partial^2 \rho}{\partial x^2} \qquad \frac{\partial^2 p}{\partial t^2} = v^2 \frac{\partial^2 p}{\partial x^2} \qquad \frac{\partial^2 \delta}{\partial t^2} = v^2 \frac{\partial^2 \delta}{\partial x^2}$$

10. A string of mass per unit length μ is fastened at $x = 0$ and $x = L$ such that the displacement y and any x and t is given by

$$y = \sum_1^\infty A_n \sin\left(\frac{\pi n x}{L}\right)\cos\left(\frac{\pi n v t}{L} - \alpha_n\right)$$

Show from Eq. 9.10 that the energy of the string is given by

$$W = \frac{\mu \pi^2 v^2}{4L}\sum_1^\infty n^2 A_n^2$$

11. Prove that the acoustic pressure or the pressure above normal atmospheric pressure in the standing wave given by $\xi = 2A \sin (2\pi x/\lambda) \cos (2\pi t/P)$ is

$$p = -\frac{4\pi v^2 \rho A}{\lambda} \cos 2\pi \frac{t}{P} \cos 2\pi \frac{x}{\lambda}$$

12. Prove that in a standing wave a pressure node occurs where the velocities of the particles or molecules of air are at a maximum and, conversely, that a velocity node occurs where the pressure is at a maximum.

13. Show that at the closed end of an organ pipe a compression is reflected as a compression, or that the reflected pressure wave is in phase with the incident pressure wave, and that the total excess pressure at the closed end is twice the excess pressure in each of the waves. (At a closed end the resultant velocity and displacement of the particles is zero.)

14. Show that at the open end of an organ pipe a compression is reflected as a rarefaction. (At the open end there is a pressure node and a velocity maximum or loop.)

15. The phase velocity of waves in deep water is proportional to the square root of their wave length. Show that the group velocity is equal to one-half the phase velocity.

16. About 1924 a French physicist, L. de Broglie, announced a new theory according to which a particle of mass m moving with a velocity V has associated with it waves whose wavelength is $\lambda = h/mV$. In this theory h is a constant, called Planck's constant. The phase velocity of these waves was given as $v = c^2/V$, where c is the velocity of light in a vacuum. This phase velocity is greater than c, the velocity of light, and this is contrary to the restricted theory of relativity. Prove that the group velocity of these waves is equal to the velocity V of the mass m. Use the equation of Chapter 1 showing that the mass m of a particle moving with a velocity V is given in terms of the rest mass of the particle as

$$m = \frac{m_0}{\sqrt{1 - (V^2/c^2)}}$$

17. About 1885 Michelson measured the velocity of light in air and in carbon disulphide and found their ratio to be 1.758. He also measured the refractive index n of carbon disulphide and found it to be 1.635. According to the wave theory of light the refractive index for a medium is the ratio of the phase velocity of a wave in air to that in the medium. Thus for carbon disulphide $n = c/v = 1.635$, where c is the velocity of light in air or a vacuum. Direct measurement of velocity of light gives the group velocity so that for carbon disulphide $c/u = 1.758$. Note that the group velocity is

$$u = v - \lambda_m \frac{dv}{d\lambda_m}$$

where λ_m is the wavelength of the light in the medium and $\lambda_m = \lambda/n$. Show that

$$\frac{c}{u} = n - \lambda \frac{dn}{d\lambda}$$

and that this theory is in agreement with Michelson's results for light of wave length $\lambda - 5.8 \times 10^{-5}$ cm at which the mean rate of change of refractive index with wavelength is -2.1×10^3/cm.

18. Prove the result given in Eq. 9.37

$$E_m = \frac{2}{L} \int_0^L f(x) \sin \frac{m\pi x}{L} \, dx$$

Also prove that in the Fourier coefficient

$$B_m = \frac{1}{\pi v m} \int_0^l \dot{y}_0 \sin \frac{\pi m x}{l} \, dx$$

where $m = 1, 2, 3, 4, \ldots$ and $\dot{y}_0$ is the initial velocity of the cord.

19. A cord of length l is rigidly attached at both ends and is plucked to a height h at a point $\frac{1}{3}$ from one end. If the velocity of transverse waves on the cord is v, show that in the subsequent vibration the third, sixth, and ninth harmonics are absent and that the displacement y at any distance x along the string at time t is given by

$$y = \frac{3^{5/2}h}{2\pi^2}\left(\sin\frac{\pi x}{l}\cos\frac{\pi vt}{l} + \frac{1}{4}\sin\frac{2\pi x}{l}\cos\frac{2\pi vt}{l} - \frac{1}{16}\sin\frac{4\pi x}{l}\cos\frac{4\pi vt}{l} \cdots\right)$$

20. The cord in problem 19 has a mass of 0.01 gm/cm and is subjected to a tension of 4×10^6 dynes. If its length is 50 cm and it is plucked to an initial height of $h = 2$ cm, find the ratio of the energy per unit length of the cord of the fundamental and the first two harmonics, assuming the energy to be proportional to the square of the amplitudes.

21. A cord of length l and rigidly attached at both ends is struck at its center. The velocity of the transverse waves in the string is v. If at time $t = 0$ the displacement y_0 of the cord is zero but the initial velocity of its center is V and this changes linearly to zero at both ends, show that the Fourier coefficients A_m are all zero and that the even harmonics, that is, those with a node at the center, are absent. Show that the displacement y at any point x at time t is given by

$$y = \frac{8Vl}{\pi^3 v}\left(\sin\frac{\pi x}{l}\sin\frac{\pi vt}{l} - \frac{1}{27}\sin\frac{3\pi x}{l}\sin\frac{3\pi vt}{l}\right.$$

$$\left. + \frac{1}{125}\sin\frac{5\pi x}{l}\sin\frac{5\pi vt}{l} - \frac{1}{343}\sin\frac{7\pi x}{l}\sin\frac{7\pi vt}{l} \cdots\right)$$

22. The cord in problem 21 is struck at a point $\frac{1}{4}$ of the length from one end. At time $t = 0$ the displacement of the cord is zero while the velocity changes linearly from V at $l/4$ to zero at both $x = 0$ and $x = l/2$ and is zero from $l/2$ to l. Show that the fourth and eight harmonics are absent, i.e., those having nodes at $x = l/4$. Show that the displacement y at any point x and time t is given by

$$y = \frac{8Vl}{\pi^3 v}\left[(\sqrt{2} - 1) \sin\frac{\pi x}{l} \sin\frac{\pi v t}{l} + \frac{1}{4}\sin\frac{2\pi x}{l}\right.$$

$$\left. \times \sin\frac{2\pi v t}{l} + \frac{1 + \sqrt{2}}{27}\sin\frac{3\pi x}{l}\sin\frac{3\pi t}{l}\cdots\right]$$

23. From dimensional analysis find the velocity of sound in a solid, assuming that the velocity v depends on the density ρ, Young's modulus E, and the acceleration of gravity g.

24. Waves on the surface of a liquid may be divided into two classes: those due to surface tension and those due to gravity. Surface tension waves are the short wavelength ripples, and their velocity is a function of the surface tension of the liquid, their wavelength λ, and the density ρ of the liquid. Gravity waves may be divided into two classes, depending on the relative size of the wavelength and the depth of the liquid. If the wavelength is short compared to the depth, the velocity of these deep sea waves depend on their wavelength λ and the acceleration due to gravity g. On the other hand, if the wavelength is large compared to the depth, the velocity depends on the depth h and the acceleration g due to gravity. By dimensional analysis find the velocity of the waves in each of the three cases.

25. By means of dimensional analysis and the data given below, find the complete equation for the velocity of sound in a solid rod. The velocity v depends on the weight per unit volume ρ_w, Young's modulus E, and the acceleration due to gravity g. The following data were found for a rod of steel: $v = 1.65 \times 10^4$ ft/sec $\rho_w = 490$ lbw/ft^3; $E = 4.16 \times 10^9$ lbw/ft^2. If the density of aluminum is 168 lbw/ft^3 and Young's modulus for this substance is 1.30×10^9 lbw/ft^2, find the velocity of sound in an aluminum rod.

26. Show that for a plane wave the intensity $I = p^2/v\rho_0$.

27. Show that for Eq. 9.27 and 9.28 the corresponding values for I_0 and p_0 are approximately equivalent.

28. Show that the group velocity of waves

$$U = v - \lambda\frac{dv}{d\lambda} = \frac{dv}{d\left(\dfrac{1}{\lambda}\right)}$$

where $v = v/\lambda$ is the frequency of the waves.

29. If the period of an oscillating drop of liquid depends on its radius r, its density ρ, and its surface tension σ, where σ has the dimensions of force per unit length, show the period is a constant $\times \sqrt{\rho r^3/\sigma}$ by the method of dimensions.

30. If $\mu_1 = 16$ gm/cm and $\mu_2 = 9$ gm/cm for two cords attached together at $x = 0$, as in Fig. 9.7, find the coefficient of reflection R and the reflecting power for a wave going from less to more dense cord and from the more to the less dense cord.

31. For the sinusoidal wave $y = A \sin 2\pi v(t - x/v)$ show the average energy per unit length of a cord is $2\pi^2 v^2 A^2 \mu$, where μ is the mass per unit length of the cord, and the average power transmitted by the wave is $2\pi^2 v^2 A^2 \mu v$.

MATHEMATICAL APPENDIX

Series Formula

$$e^x = 1 + x + \frac{x^2}{2!} + \frac{x^3}{3!} + \cdots = \sum_0^\infty \frac{x^n}{n!}$$

where

$$n! = n(n-1)(n-2) \cdots 3 \times 2 \times 1.$$

$$\sin \theta = \theta - \frac{\theta^3}{3!} + \frac{\theta^5}{5!} - \frac{\theta^7}{7!} + \cdots$$

$$\cos \theta = 1 - \frac{\theta^2}{2!} + \frac{\theta^4}{4!} - \frac{\theta^6}{6!} + \cdots$$

$$\sinh x = \frac{e^x - e^{-x}}{2} = x + \frac{x^3}{3!} + \frac{x^5}{5!} + \frac{x^7}{7!} + \cdots$$

$$\cosh x = \frac{e^x + e^{-x}}{2} = 1 + \frac{x^2}{2!} + \frac{x^4}{4!} + \frac{x^6}{6!} + \cdots$$

From above it is readily seen that

$$\frac{d(\sin \theta)}{d\theta} = \cos \theta \qquad \frac{d(\cos \theta)}{d\theta} = -\sin \theta \qquad \frac{d(e^x)}{dx} = e^x$$

$$\frac{d(\sinh x)}{dx} = \cosh x \qquad \frac{d(\cosh x)}{dx} = \sinh x$$

Complex Numbers

If $j = \sqrt{-1}$ and x and y are real numbers, then jy is said to be imaginary and $x + jy$ is a complex number. Such a complex number can be represented on a complex plane consisting of an axis of reals and an axis of imaginaries, Fig. A1. Consider the point P which has coordinates x, y and r, θ. From Fig. A1 it is seen that

$$x = r \cos \theta \qquad \text{and} \qquad y = r \sin \theta$$

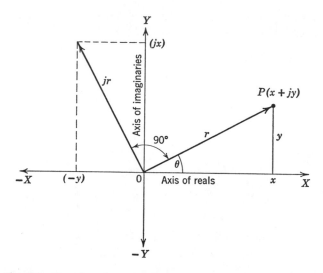

Fig. A1   Complex plane showing $r = x + jy$ and $jr = -y + jx$.

Thus the complex number $z = x + jy$ may be written as

$$z = x + jy = r(\cos \theta + j \sin \theta)$$

where $r = \sqrt{x^2 + y^2}$ and $\theta = \tan^{-1} y/x$.

Assuming that we may expand $e^{j\theta}$ from the above expansion for e^x, we have

$$e^{j\theta} = 1 + j\theta - \frac{\theta^2}{2!} - \frac{j\theta^3}{3!} + \frac{\theta^4}{4!} + \frac{j\theta^5}{5!} - \frac{\theta^6}{6!} - \frac{j\theta^7}{7!} \cdots$$

$$= \left(1 - \frac{\theta^2}{2!} + \frac{\theta^4}{4!} - \frac{\theta^6}{6!} + \cdots\right) + j\left(\theta - \frac{\theta^3}{3!} + \frac{\theta^5}{5!} - \frac{\theta^7}{7!} + \cdots\right)$$

$$= \cos \theta + j \sin \theta.$$

This is Euler's formula. It follows that

$$x + jy = r(\cos \theta + j \sin \theta) = re^{j\theta}$$
$$x - jy = r(\cos \theta - j \sin \theta) = re^{-j\theta}$$

If

$$z_1 = r_1 e^{j\theta_1} \qquad \text{and} \qquad z_2 = r_2 e^{j\theta_2}$$

Then

$$z_1 z_2 = r_1 r_2 e^{j(\theta_1 + \theta_2)} \qquad \text{and} \qquad \frac{z_1}{z_2} = \frac{r_1}{r_2} e^{j(\theta_1 - \theta_2)}$$

$$z^n = (re^{j\theta})^n = r^n e^{nj\theta} = r^n(\cos n\theta + j \sin n\theta)$$

If

$$z_1 = x_1 + jy_1 \qquad \text{and} \qquad z_2 = x_2 + jy_2$$

Then

$$\frac{z_1}{z_2} = \frac{x_1 + jy_1}{x_2 + jy_2} = \frac{x_1 + jy_1}{x_2 + jy_2} \cdot \frac{x_2 - jy_2}{x_2 - jy_2} = \frac{x_1 x_2 + y_1 y_2}{x_2^2 + y_2^2} + \frac{j(x_2 y_1 - y_2 x_1)}{x_2^2 + y_2^2}$$

The complex quantity $x_2 - jy_2$ is called the *complex conjugate* of $x_2 + jy_2$. Also

$$\frac{z_1}{z_2} = \frac{r_1}{r_2} e^{j(\theta_1 - \theta_2)} = \frac{r_1}{r_2} [\cos(\theta_1 - \theta_2) + j \sin(\theta_1 - \theta_2)]$$

so that

$$\frac{r_1}{r_2} \cos(\theta_1 - \theta_2) = \frac{x_1 x_2 + y_1 y_2}{x_2^2 + y_2^2}$$

and

$$\frac{r_1}{r_2} \sin(\theta_1 - \theta_2) = \frac{x_2 y_1 - y_2 x_1}{x_2^2 + y_2^2}$$

If θ_0 is the angle θ shown in Fig. A1 lying between 0 and 2π, then θ_0 is called the principal value of θ. Actually θ may be multiple-valued such that $\theta = \theta_0 \pm 2\pi k$, where $k = 0, 1, 2, 3, \ldots$, and the complex quantity z may be written as $z = re^{j(\theta + 2\pi k)}$

It may be readily proved from the above that

$$\cosh j\theta = \cos \theta \quad \text{and} \quad \sinh j\theta = j \sin \theta$$
$$\cos j\theta = \cosh \theta \quad \text{and} \quad \sin j\theta = \sinh \theta$$
$$\cosh^2 \theta - \sinh^2 \theta = 1$$
$$\cos \theta = \frac{e^{j\theta} + e^{-j\theta}}{2} \quad \text{and} \quad \sin \theta = \frac{e^{j\theta} - e^{-j\theta}}{2j}$$

If a real number n on the real axis of Fig. A1 is multiplied by j, the resulting imaginary number jn is on the axis of imaginaries, or the magnitude is unchanged but its position is rotated through $90°$. Similarly, multiplying n by j^2 gives $-n$ or produces a rotation of $180°$, multiplying by j^3 produces a rotation of $270°$, and by j^4 a rotation of $360°$ or $0°$. If the complex number $x + jy$ is multiplied by j, then on the complex plane the magnitude r is unchanged but the vector $\mathbf{r}$ is rotated through $90°$. This statement may be readily proved from the above.

EXERCISES. From Euler's equation show that

$$e^{2j\pi} = 1; \quad e^{j\pi} = -1; \quad e^{j\pi/2} = j;$$
$$\ln 1 = 2n\pi j \text{ where } n = 0, 1, 2, 3, \ldots$$
$$\ln(-1) = (2n + 1)\pi j; \quad \ln j = (4n + 1)\pi/2$$
$$\log_{10} j = 0.6822j; \quad \ln(-a) = j\pi + \ln(a)$$

find the three cube roots of -1. (Note $e^{j\pi/3} = -1$; $e^{3j\pi/3} = -1$; $e^{5j\pi/3} = -1$.)

ANSWERS TO ODD NUMBERED PROBLEMS

Chapter 1. 1. (a) 20 sec, 640 ft/sec; (b) speed becomes constant and acceleration zero; (c) 8 ft/sec; 3. -2.25 m/sec, 33.8 m; 7. $\sqrt{94}$, $3/\sqrt{94}$, $6/\sqrt{94}$, $-7/\sqrt{94}$; 9. $-4/9$; 21. (a) $\pi/2$ rad/sec, 2π rad/sec^2, (b) 0.9π rad/sec, (c) 2.14 m/sec^2 at 38.1° with radius, (d) 0.14π rad or 25.2°; 23. 47/13 m/sec, 90/13 m/sec, 61; 25. $(8\mathbf{i} - 6\mathbf{k})$ ft/sec, 10 ft/sec, $4\mathbf{i}$ ft/sec^2, 3.2 ft/sec^2, 2.4 ft/sec^2, 41.7 ft, $0.8\mathbf{i} - 0.6\mathbf{k}$, $0.6\mathbf{i} + 0.8\mathbf{k}$; 29. 0.943 m.; 31. 5.05 min.

Chapter 2. 1. 8 ft/sec^2, 7.5 lbw, 15 lbw; 3. (a) 3.58 m/sec^2, (b) 0.74 m/sec^2, (c) 3.2 m/sec^2 5. (a) 19.0 ft/sec^2; (b) 9.37 ft/sec^2; 7. 3.27×10^4 ft; 9. 24.4 ft; 11. 109 cm/sec^2, 108 gmw; 13. (a) 149 cm/sec; (b) 28.4 cm; (c) 2.25×10^8 dynes; 15. 4.57 m/sec^2, 1.47 m/sec^2; 21. (a) 17.9 sec; (b) 10,300 ft; (c) 42 sec; (d) 315 ft/sec; 23. (a) 6.93×10^4 ft; (b) 3,990 ft; (c) 33.8 sec; (d) 1.34×10^4 ft. 25. $T = (m_1 + m_2)g - a(m_2 - m_1)$; 27. 0.279, 0.995, 0.999,95, 0.999,999,5.

Chapter 3. 1. (a) 0.0285 m; (b) 1,120 J; (c) 3.74×10^4 Newtons, N. 3. 29·3 ft lbw 9.68ft /sec; 5. (b) 1.79 ft; (c) 1.68 ft; 7. (b) 8000 ergs; (c) spheres; (d) 500 ergs. 13. $- \mathbf{i}3z/y + \mathbf{j}3xz/y^2 - \mathbf{k}3x/y$; $\mathbf{i}(2ax - by) - \mathbf{j}(bx + 2cy)$; 15. (a) Curl $F \neq 0$; (b) Curl $\mathbf{F} = 0$, $V = -\int [F_1(x)\,dx + F_2(y)\,dy + F_3(z)\,dz]$; (c) Curl $\mathbf{F} = 0$; $V = -\int [F_x\,dx + F_y\,dy + F_z\,dz]$; 17. 64 lbw; 0.465 H.P.; 19. (a) 55 ft/sec; (b) 0.12 ft/sec^2; (c) 41.2 ft/sec. 21. 72/121; 23. (a) 100.5 gm; (b) 134.2 gm; (c) 1224 gm; (d) 4.33%.

Chapter 4. 1. $- GM/R$; 3. 7.35 N; 4.9 N; 9. 7.04×10^{22} kg; 11. (a) 1.64×10^4 ft/sec; (b) 690 sec; (c) 194 sec; 23. (a) 1.09; (b) 0.428; 25. 0; 0.101°; 27. (a) 1.22×10^{-12} cm.

Chapter 5.   1. (a) 300π cm/sec; $6 \times 10^5\pi^2$ cm/sec^2; (b) 166π cm/sec; $5.0\pi^2$ cm/sec^2;   7. $R(\theta - \sin \theta)$, $R(1 - \cos \theta)$; $R\omega(1 - \cos \omega t)$, $R\omega \sin \omega t$; $R\omega^2$;   9. (c) $x^2 + y^2 - \sqrt{3}xy = 25$;   19. 0.1 sec;   23. $y = -5.93e^{-2.68t} + 0.754e^{-37.3t} + 1.55 \cos (8t - 1.46)$;   25. (a) 1.76 amp; (b) 1.25 amp; (c) 31.4 μf, 2.14 amp.

Chapter 6.   1. 0.4 kg, 0.416 m;   3. 29.4 cm;   5. (a) 75 cm, 0; (b) 7.5 cm/sec, 15.5 cm/sec; (c) 150 cm, 155 cm; (d) 300 cm, 620 cm.   7. 125 r.p.m.;   15. 6.67×10^{-8} dynes cm^2/gm^2 ;   17. 7.31 ft/sec^2, 49.4 rad/sec^2;   19. 17.0 ft/sec, 0.380 lbw;   21. 0.75 ft, 95.6 ft lbw;   29. (d) 390 cm/sec, (e) 25·6 rad/sec, (f) $1\cdot52 \times 10^7$ ergs, $2\cdot61 \times 10^6$ ergs, (g) $1\cdot38 \times 10^6$ ergs, (h) 1.920×10^7 ergs.

Chapter 8.   1. 6.36 mm, 1.91×10^{-3} mm, 2.40×10^{-9} m^2;   7. 9.38×10^{-3} lb/ft^3;   17. 2.16×10^7 lbw/in^2, 0.291 ft lbw;   19. 2.55×10^4 lbw, 5.94 ft.   21. 1.34 sec, 1.24×10^4 dynes;   25. 400 cm/sec^2;   27. 0.00985 poises;   29. 1.87×10^{-4} poises;   31. 5.02 cm/sec, 5480 cm/sec.

Chapter 9.   5. (a) 340 m/sec; (b) 340 m/sec; (c) 384 m/sec;   7. (a) 2.68×10^{11} dynes/cm^2; (b) 5.18×10^5 cm/sec.   23. $v = K\sqrt{E/\rho}$;   25. 1.57×10^4 ft/sec.

INDEX